Paranormal
Borderlands of Science

Paranormal
Borderlands of Science

BEST OF
SKEPTICAL INQUIRER
VOLUME I

EDITED BY
KENDRICK FRAZIER

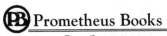 Prometheus Books

Essex, Connecticut

Ⓟ Prometheus Books

An imprint of Globe Pequot, the trade division of
The Rowman & Littlefield Publishing Group, Inc.
4501 Forbes Blvd., Ste. 200
Lanham, MD 20706
www.rowman.com

Distributed by NATIONAL BOOK NETWORK

British Library Cataloguing in Publication Information Available

Library of Congress Cataloging-in-Publication Data Available

ISBN 9781633889620 (pbk. : alk. paper) | ISBN 9781633889637 (epub)

♾ᵀᴹ The paper used in this publication meets the minimum requirements of American
National Standard for Information Sciences—Permanence of Paper for Printed Library
Materials, ANSI/NISO Z39.48-1992

Table of Contents

Book II: Inquiries Into Fringe-Science

Introduction

What an age of discovery this is! Our spacecraft have flown to Mars, Mercury, Venus, Jupiter, and Saturn. We've discovered a strange soil chemistry on Mars, well-ordered wind systems on Venus, superlightning bolts, giant auroras, and a thin system of rings at Jupiter, volcanoes in eruption on the Jovian moon Io, and more rings and a whole gaggle of new moons around Saturn. Our astronomers have recently discovered a gravitational "lens" in space, producing multiple images of a distant quasar. They've found a strange object 11,000 light-years away that is ejecting two high-speed jets of gas toward and away from the earth simultaneously. We've delved ever deeper into the inner nature of matter, finding new particles and mysteries every step of the way. We've dived to the ocean bottom where a new surface of the earth is being created. We've put together an elegant picture of how segments of our planet's surface scoot about the globe carrying oceans and continents on their backs. We've found new chemicals in the human brain responsible for relieving pain, enhancing pleasure, and improving concentration. We've unraveled the molecular nature of heredity and synthesized fully functional genes. And every step has revealed new puzzles and wonders. These are the frontiers of science.

There has also been a considerable flurry of activity in another area of endeavor. I call it the borderlands of science. The borderlands of science are not to be confused with the frontiers of science. In fact, in large part the activity that goes on in these borderlands has nothing to do with the true spirit of science. This is the land of pseudoscience, fringe-science, and the paranormal.

If the frontiers of science are a newly discovered ocean for exploration, the borderlands of science are a murky, backwatery swamp right here on land. If science is an advancing, forward progression of the frontiers of knowledge, the pseudosciences are a much-trodden farmed-out field at the rear flanks. If science is the cutting edge of discovery, pseudoscience is a dull blade left behind amid the sawdust and debris.

It's time to get to specifics. What are we talking about? One way to demonstrate is to list some of the subjects that seem most often—although not always—to be treated in a psuedoscientific way. Some of them have a strange hold on our imaginations and psyches. They include: claims of extrasensory perception, or ESP (alleged telepathy, clairvoyance, pre-cognition, and psychokinesis), claims of "psychics" generally ("psychic" metal-benders, "psychic" crime-solvers, "psychic" photographers, "psychic" surgeons, "psychic" archaeologists), fortunetelling, poltergeists, astral projection, remote-viewing, faith healing, aura reading, creationism, post-death experiences, reincarnation, spontaneous human combustion, birthdate-based biorhythms, astrology, UFOs, abductions by outer-space aliens, ancient astronauts, extraterrestrial ancestors, Bigfoot, Velikov-skianism, the Bermuda Triangle and pre-Viking Old World civilizations in America.

It is not the subject matter, however, that makes for pseudoscience. Each of these subjects can, to a degree at least, be treated in a responsible scientific fashion. All efforts to do so deserve support. When alleged ESP is tested in rigorously controlled experiments by scientifically trained investigators honestly intent on discovering all sources of perceptual and statistical bias and conscious and unconscious deception and when they meet all other standards of good science, then the practice deserves to be called parapsychology and distinguished from other personal claims of alleged psychics. When that description doesn't fit, it is pseudoscience. There is nothing inherently absurd about the idea that the earth may have been visited in the past by extraterrestrial beings, but to claim that it happened in the face of zero scientific evidence is pseudoscience. UFO reports can be examined in a responsible way, but to claim in the absence of reliable evidence that a momentary light in the sky is an alien spacecraft is pseudoscience.

More often that not the most publicized practitioners of the para-normal and pseudoscience are not real scientists at all. I'm not referring to credentials. Some have them. And people hardly need a scientific degree to apply common sense and reasoning to these matters. I refer instead to a number of characteristics that distinguish the way many of these people go about their supposed search for truth from the way a scientist acting in the spirit of science approaches his or her quest for knowledge about nature.

Not long ago I called a scientist acquaintance who had done some important and interesting work in a field of research I had been writing about. In the course of the conversation he volunteered to me that a study that cast doubt on the validity of some of his conclusions had recently been completed by a scientist at another institution. When I wasn't able to get a copy of the second scientist's research paper, I called back my scientist acquaintance and he gladly dug up his copy of the paper undermining part of his own work and sent it to me. He understood that science is a process of discovery subject to continual revision by newer and better information. He recognized that what was important was to get to the real truth of the matter rather than to defend at all costs one's own original conclusions if they don't seem supported. I can't imagine any of our modern pseudo-scientists or their chroniclers going to such an effort.

I don't want you to get the wrong impression. Science is a very human activity. Its practioners share all the flaws and shortcomings we all possess. It is a dynamic, highly competitive activity. Emotions often come into play. But there is also a strongly cooperative aspect of science. Most scientists have a sense of allegiance to the long-term honesty and integrity of the process. Science is, among other things, an organized means for testing the validity of ideas and claims. When a hypothesis fails to stand up to repeated tests, it is discarded—not always without last-ditch efforts to save it, but discarded nevertheless. Of course, it can always be rescued from the ash heap whenever new evidence warrants.

Pseudoscience operates by slightly different rules.

Science is a process of searching to understand nature and is always undergoing revision. There is a strong no-holds-barred, let-the-facts-fall-where-they-may attitude. Pseudoscience clings emotionally to comforting ideas long after they've been shown to be most likely wrong. The concepts of astrology were virtually set in concrete two thousand years ago and are clung to despite all our modern understanding of the stars and planets.

Scientists try to be careful to limit their claims to matters that they can support reasonably well with good evidence. Pseudoscientists make wild and exaggerated claims that go far beyond the evidence.

Scientists actively seek out comments and criticism from well-informed colleagues before publishing a claim. Pseudoscientists usually do all they can to avoid informed criticism before publication.

Scientists usually publish their research in scientific journals that take steps—peer review—to ensure that the work meets minimal standards of competence and accuracy before accepting it. Pseudoscientists usually go straight to the public, their claims first appearing in commercial books whose publishers make no independent efforts to verify accuracy, or in other publications that care only about a good story.

Scientists usually frame a hypothesis in such a way that if it is wrong it can be proved wrong. Pseudoscientists attempt to frame their claims in such a way that they cannot be proved wrong. They constantly shift the grounds for substantiation or purposely make vague or ambiguous statements that cannot be tested.

Scientists realize that the burden of proof is on the investigator making the claim. A hypothesis is not accepted as valid until it has stood up to many tests. Pseudoscientists place a burden of *dis*proof on the critics. They generally hold that their claim is true unless others "disprove" it.

Scientists accept that the stronger the claim made (that is, the more it contradicts previously demonstrated evidence), the greater the evidence must be for it before it can be accepted. Pseudoscientists thrive on the probability that the more sensational the claim, the more publicity it will get, and thus the more supporters it will gain among sympathetic segments of the public.

Scientists realize all information is imperfect. They try to avoid absolutes. They attempt to present an honest assessment of the amount of error attached to all measurements and of the degree of reliability associated with all claims. Pseudoscientists often present their claims as infallibly true. They make no effort to distinguish between the varying quality of evidence they use to support their claims.

Scientists build on other scientific work. They familiarize themselves with previous relevant research before attempting to extend or modify it. Pseudoscientists often ignore previous studies altogether, especially research results that conflict with their pet hypotheses.

When shown to be wrong, most scientists usually acknowledge that fact—not always immediately and not always gracefully, for they are human—and they usually either modify the work according to the criticisms or go on to other things. Pseudoscientists tend to be committed believers. They feel any criticism is only a sign of the closed minds and ignorance of scientists. They are quick to don the role of martyr. They appeal to public sympathies.

Science, all in all, is an error-correcting activity. Pseudoscience is an error-promulgating activity.

No wonder, then, that the borderlands of science, where pseudoscientists and their gazetteers so frequently trod, are littered with the wreckage of misbegotten ideas, misleading information, and misdirected hopes. What good scientific ideas do emerge in the borderlands—and some occasionally do—are quickly obscured and made unrecognizable by the accumulating layers of unevaluated, untested nonsense.

Not that any of this deters pseudoscientists and their ideas from reaching public prominence. To the contrary. They appeal straight to the public and they speak with confidence and certainty—no wishy-washy qualifications and none of that technical language that scientists are always putting in their writings to confuse us. They direct their appeal toward the personal interests and deepest psychological needs of the audience. The public eats it up, publishers find it profitable, television talk-show and entertainment programs know that audiences love it. Soon there is so much misleading information flying about that the consumer has no way of distinguishing tested ideas from half-truths and falsehoods, or responsible scientific speculation from outright fantasy. Many don't care. Those prone to accept exotic claims uncritically really aren't concerned about the truth. Others just see it all as harmless entertainment. Indeed, pseudoscientific ideas can lend a certain color and a kind of frivolous vibrancy to society. They can provide a temporary respite from life's serious matters, a sort of celebration of our cultural diversity. They can be fun.

This isn't the moment to argue that there is a darker side to it all. I think that is, or will be, clear. Cults everywhere make use of such gullibility and deception. And polls show that many of these claims have wide general acceptance in today's society. The implications for education alone are enormous.

I do think, however, that there is a large segment of the public that honestly does prefer accurate to inaccurate information and facts to nonfacts. Most people do want to base their opinions on some semblance of a realistic view of the world.

We hear so many claims so confidently asserted. Which ones deserve our attention and which don't? Is there perhaps something to astrology after all? What about all these people on televeision and in the newspapers claiming to be psychics? There are so many, isn't it likely some of them are the real thing? And some of them do get things right. How is that possible? And regardless of what you think of them, what about the research into psychic phenomena being carried out by academic scientists? They can't be fooled, can they? And besides, what about the time I had this dream about Uncle Ernie and the very next morning he called? Doesn't that mean something? And what about all those UFO reports? Obviously many are silly misrepresentations, but what about those not yet explained? For that matter, what about all these things that we are constantly hearing are "unexplainable by science." Isn't there possibly something really going on here?

This book is for people who are asking these questions in the true spirit of inquiry, who care about the truth, and who are willing to see their ideas and those of others subjected to the light of critical reasoning, scientific experimentation, and detailed analysis. And it's for scholars and teachers who find their students and friends frequently asking these questions and desire to give accurate and reliable information in response.

The chapters are written by responsible and respected scientists, philosophers, psychologists, educators, investigative journalists, and magicians (whose skill in understanding how we can be deceived is invaluable in this task). Among them are many of the internationally outstanding people involved in the critical investigation and analysis of fringe-science and paranormal claims. All, in my view, have something interesting and important to say.

They follow in a tradition that goes back at least to 1853, when the great English physicist Michael Faraday took time out from his regular research to design an ingenious apparatus for testing the claims of the widespread practice of "table turning"—a highly popular "paranormal" fad of the time. Faraday's elegant demonstration showed unequivocally that it was the force of the hands of the people touching the table (exerted subconsciously) that was responsible for its turning—not, as they thought, that their hands were merely following the motion of the table.

The chapters include reports of original research, critical essays, analytical articles, and investigative reports. They vary in style and approach. We haven't tried to establish a set mold, preferring instead to offer our authors a wide latitude in form of expression. What they all do have in common is the application of intelligence, reasoning, and critical inquiry to subjects too often spared from any sort of skeptical, informed attention.

These critiques are an outgrowth of an effort begun five years ago by a group of distinguished scientists, philosophers, educators, science writers, and other investigators. Concerned that the public was not very well informed about the actual facts behind the many kinds of paranormal and fringe-science claims prevalent in books and newspapers and on TV and radio, and even in parts of academia, they formed an organization called the Committee for the Scientific Investigation of Claims of the Paranormal. To my knowledge, it is the first such organized effort ever undertaken. This nonprofit organization attempts to encourage the critical investigation of such claims from a responsible, scientific point of view—in contrast to the sensationalized, credulous approach that marks so much of what we hear.

The Committee founded a quarterly journal, the *Skeptical Inquirer,* to disseminate accurate information about the results of these inquiries to the scientific community and the public. It is the only publication of its type in the world. The chapters in this book are taken from inquiries and investigations published in the *Skeptical Inquirer,* which it has been my pleasure to edit since 1977. For more information about the Committee or the *Skeptical Inquirer* write to Paul Kurtz, Chairman, Committee for the

Scientific Investigation of Claims of the Paranormal, Box 229, Central Park Station, Buffalo, N.Y. 14215.

We think you will find this book an entertaining and refreshingly well-informed and clear-headed examination of many of the most publiccized and exotic claims found today on the borderland of science. And I think you'll see why the borderlands—fascinating as they may be—aren't likely to be confused with the real frontiers.

Kendrick Frazier

BOOK I: INQUIRIES INTO PSI PHENOMENA

Psi Phenomena and Belief

Is Parapsychology a Science?

Paul Kurtz

I.

An observer of the current scene cannot help but be struck by the emergence of a bizarre new "paranormal world-view." How widely held this view is, whether it has penetrated science proper or is simply part of the popular passing fancy, is difficult to ascertain.

Many of those who are attracted to a paranormal universe express an antiscientific, even occult, approach. Others insist that their hypotheses have been "confirmed in the scientific laboratory." All seem to agree that existing scientific systems of thought do not allow for the paranormal and that these systems must be supplemented or overturned. The chief obstacle to the acceptance of paranormal truths is usually said to be skeptical scientists who dogmatically resist unconventional explanations. The "scientific establishment," we are told, is afraid to allow free inquiry because it would threaten its own position and bias. New Galileos are waiting in the wings, but again they are being suppressed by the establishment and labeled "pseudoscientific." Yet it is said that by rejecting the paranormal we are resisting a new paradigm of the universe (à la Thomas Kuhn) that will prevail in the future.

Unfortunately, the meaning of the term *paranormal* is often unclear. Literally, it refers to that which is "besides" or "beyond" the normal range of data or experience. Sometimes "the paranormal" is used as an equivalent of "the bizarre," "the mysterious," or "the unexpected." Some use it to refer to phenomena that have no known natural causes and that transcend normal experience and logic. The term here has been used synonymously with "the supernormal," "the supernatural," or "the miraculous." These definitions, of course, leave little room for science. They mark a limit to our knowing. Granted there are many areas at the present time that are unknown; yet one cannot on a priori grounds, antecedent to inquiry, seek to define the parameters of investigation by maintaining that something is irreducibly unknowable or inexplicable in any conceivable scientific terms.

Some use the term *paranormal* to refer to that which is "abnormal" or "anomalous," that is, that which happens infrequently or rarely. But there are many accidental or rare events that we wouldn't ordinarily call paranormal—a freak trainwreck, a lightning strike, or a meteor shower.

Some use the term *paranormal* simply to refer to the fact that some phenomena cannot be given a physical or materialistic explanation. In some scientific inquiries, physicalist or reductionist explanations are, indeed, not helpful or directly relevant—as, for example, in many social-science studies, where we are concerned with the function of institutions, or in historical studies, where we may analyze the influence of ideas or values on human affairs. But this surely does not mean that they are "nonnatural," "unnatural," or "paranormal"; for ideas and values have a place in the executive order of nature, as do flowers, stones, and electrons. Although human institutions and cultural systems of beliefs and values may be physical at root, they are not necessarily explainable in function as such. There seem to be levels of organization; at least it is convenient to treat various subject matters in terms of concepts and hypotheses relative to the data at hand. To say this in no way contravenes the physical laws of nature as uncovered in the natural sciences.

The term *paranormal*, however, has also been used in parapsychology, where something seems to contradict some of the most basic assumptions and principles of the physical, biological, or social sciences and a body of expectations based on ordinary life and common sense. C. D. Broad has pointed out a number of principles that parapsychologists would apparently wish to overthrow[1]: (*a*) that future events cannot affect the present *before* they happen (backward causation); (*b*) that a person's mind cannot effect a change in the material world without the intervention of some physical energy or force; (*c*) that a person cannot know the content of another person's mind except by the use of inferences based on experience and drawn from observations of his speech or behavior; (*d*) that we cannot directly know what happens at distant points in space without some sensory perception or energy of it transmitted to us; (*e*) that discarnate beings do not exist as persons separable from physical bodies. These general principles have been built up from a mass of observations and should not be abandoned unless and until there is an overabundant degree of evidence that would make their rejection less likely than their acceptance—if I may paraphrase David Hume.[2] Nevertheless, those who refer to the "paranormal" believe that they have uncovered a body of empirical facts that call into question precisely those principles. Whether or not they do remains to be seen by the course of future inquiry. These scientific principles are not sacred and may one day need to be modified—but only if the empirical evidence makes it necessary.

Some who use the term *paranormal* refer to a range of anomalous events that are inexplicable in terms of our existing scientific concepts and theories. Of course, there are many events not now understood. For example, we do not know fully the cause of cancer, yet we would hardly call it paranormal. There have been many reports recently of loud explosions off the Atlantic coast that remain unexplained and that some have hinted are "paranormal." (These may be due to methane gas, test flights, or distant sonic booms.) If we were to use the term *paranormal* to refer to that which is inexplicable in terms of current scientific theory, with the addition that it cannot be explained without major revisions of our scientific theory, this would mean that any major advance in science, prior to its acceptance, might be considered to be "paranormal." But then new developments in quantum theory or relativity theory, the DNA breakthrough, or the germ theory of disease would have been paranormally related. But this is absurd. There are many puzzles in science and there is a constant need to revise our theories; each new stage in science waiting to be verified surely cannot be called "paranormal."

In actuality, the term *paranormal* is without clear or precise meaning; its use continues to suggest to many the operation of "hidden," "mysterious," or "occult" forces in the universe. But this, in the last analysis, may only be a substitute for our ignorance of the causes at work. Although I have used the term because others have done so, I think that it ought to be dispensed with as a meaningless concept.

II.

It is clear that science is continually changing and growing. As new facts are discovered, existing concepts and theories must either be extended to account for them or be abandoned in favor of new and more comprehensive explanations.

In the current context, any number of new fields have recently appeared alongside the established sciences. These begin with a number of alleged anomalous events that proponents say cannot be readily explained in terms of the existing sciences. One may ask, Do these subjects qualify as sciences? One must always be open to the birth of new fields of inquiry. At first a new or proto science may be rejected by the existing body of scientific opinion; but in time, if it can make its case, it may become accepted as genuine. This has been a familiar phenomenon as new branches of inquiry emerge in the natural, social, and behavioral sciences. Unfortunately, not all of the claimants to scientific knowledge are able to withstand critical scrutiny, and many turn out to be pseudo or false sciences.

A classical illustration of this is phrenology, which swept Europe and America in the nineteenth century. It was formulated by F. J. Gall, and developed by his followers J. K. Spurzheim and G. Combe. According to the phrenologist: (1) the brain was the organ of the mind; (2) the mental powers of men could be distinguished and assigned to separate innate faculties; (3) these faculties had their seat in a definite region of the brain surface; (4) the size of each region is the measure to which the faculty forms a constituent element in the character of the individual; (5) the correspondence between the outer surface of the skull and the brain surface beneath it is sufficiently close to permit the scientific observer to ascertain the relative sizes of these organs by an examination of the head; and (6) such an examination provided a method by which the disposition and character of the subject could easily be ascertained. The theory was allegedly based on empirical observations from which generalizations were formulated. Gall and his associates examined the heads of their friends, men of genius, and inmates of jails and asylums in order to map the organs of intelligence, murder, sexual passions, theft, and so on. The theory seems quite mistaken to us today—not that behavioral functions may not be correlated in some sense with regions of the brain, but that they could be mapped by examining the exterior skull cap and that the permanent disposition of the persons could be so determined. Yet so great a degree of popularity did phrenology enjoy that in 1832 there were 29 phrenology societies in Great Britain alone, and several phrenology journals in America and Britain—all of which have virtually disappeared.[3] Indeed, I only know of one practicing phrenologist in North America. He tells me he is the leading phrenologist in the world and that he predicts a revival of the field!

The term *pseudoscience* has been used in many ways. One must be careful not to indiscriminately apply it to budding fields of inquiry that may have some merit. Perhaps it should be used for those subjects that clearly: (*a*) do not utilize rigorous experimental methods in their inquiries, (*b*) lack coherent testable conceptual framework, and/or (*c*) assert that they have achieved positive results, though their tests are highly questionable and their generalizations have not been corroborated by impartial observers.

There are a great number of candidates for "pseudoscience" today, many of them ancient specialties that still persist: numerology, palmistry, oneiromancy, moleosophy, aleuromancy, apantomancy, psychometry. And there are new ones constantly appearing. Perhaps some may in time develop testable and tested theories.

Astrology—which had all but died out by 1900 and is now very strong—is a good illustration of a pseudoscience. The principles of astrology remain largely unchanged from the days of Ptolemy (first century A.D.), who codified the ancient craft. And astrologers still cast their horoscopes and do their analyses very much as Ptolemy did, in spite of the fact that its original premises have been contradicted by modern post-Newtonian physics and astronomy. Most astrologers have considered astrology to be an occult field of paranormal study; others have attempted to develop it as a science. Yet astrology does not use rigorous experimental standards of inquiry by which it can reach conclusions, it lacks a coherent theory of what is happening and why, and it draws inferences and makes predictions that are highly dubious. Michel Gauquelin is a critic of traditional astrology on these grounds, though he has attempted to develop his own field of astrobiology. Based on careful statistical analysis, he has attempted to correlate personality characteristics with planetary configurations. Thus, for example, he maintains that there is a relationship between the position of Mars and the time and place of birth of sports champions. Thus far, the results of his study, in my judgment, are inconclusive, though his procedure is far different from the usual approach of astrologers.

Biorhythms appears to be another false science. It also claims to have its foundations in empirical data; yet when independent examination is made to see whether its predictions are accurate, the results appear to be negative.

III.

What are we to say about parapsychology? Is it a science or a pseudoscience?

Interest in psychic phenomena appears throughout human history, with reports abounding from ancient times to the present. There is a fund of anecdotal material—premonitions that seem to come true, apparent telepathic communication between friends or relatives, reports of encounter with discarnate persons, and so on—that leads many people to believe that there is some basis in fact for psi phenomena. It has been almost a century since the Society for Psychical Research was founded in 1882 in England by a distinguished group of psychologists and philosophers (including William James and Henry Sidgwick) who were hopeful of the chance of getting results from their careful inquiries. In October 1909, William James, a president of the Society, wrote "The Last Report: Final Impressions of a Psychical Researcher," summarizing his experiences.[4] The Society, he said, was founded with the expectation that if the material of "psychic" research were treated rigorously and experimentally then objective truths would be elicited. James reported:

> . . . Like all founders, Sidgwick hoped for a certain promptitude of results; and I heard him say, the year before his death, that if anyone had told him at the outset that after twenty years he would be in the same identical state of doubt and balance that he started with, he would have deemed the prophecy incredible.

Yet James relates that his experiences had been similar to Sidgwick's:

> For twenty-five years I have been in touch with the literature of psychical research, and have had acquaintance with numerous "researchers." I have also spent a good many hours (though far fewer than I ought to have spent) in witnessing (or trying to witness) phenomena. Yet I am theoretically no "further" than I was at the beginning; and I confess that at times I have been tempted to believe that the Creator had eternally intended this department of nature to remain *baffling*, to prompt our curiosities and hopes and suspicions all in equal measure, so that, although ghosts and clairvoyances, and raps and messages from spirits, are always seeming to exist and can never be fully explained away, they also can never be susceptible of full corroboration.

> The peculiarity of the case is just that there are so many sources of possible deception in most of the observations that the whole lot of them *may* be worthless . . . Science meanwhile needs something more than bare possibilities to build upon; so your genuinely scientific inquirer . . . has to remain unsatisfied. . . . So my deeper belief is that we psychical researchers

have been too precipitate with our hopes, and that we must expect to mark progress not by quarter-centuries, but by half-centuries or whole centuries.

Almost three-quarters of a century have elapsed since James's comments. Has any more progress been made? Since that time psychic research has given way to parapsychology, especially under the leadership of J. B. Rhine and the establishment of his experimental laboratory. Where there were before only a handful of researchers, now there are many more. We may ask, Where does parapsychology stand today? I must confess that for many researchers, both within and outside the field, not much further along than before.

One thing is clear: many researchers today at least attempt to apply experimental methods of investigation. This was not always the case; and the field today, as then, has been full of deception, conscious or unconscious—perhaps more than most fields of inquiry. There are a host of fraudulent psychics and researchers—including the Fox sisters (who were hailed as mediums, in whose presence raps were heard during seances, but who evidently admitted they had learned how to crack their toe knuckles), Blackburn and Smith (who deceived scientists into believing that telepathic communication occurred between them), Margery Crandon and Eustasia Palladino (both shown to be fraudulent mediums), the Soal-Goldney experiments on precognition (experiments now in disrepute), Walter J. Levy (who was exposed for faking the evidence on animal ESP at Durham n 1974), Uri Geller, Jean Girard, and Ted Serios (whose alleged abilities in psychokinesis and psychic photography are open to charges of trickery). Even some of the most sophisticated scientists have been taken in by illusionists posing as psychics. In spite of this there *are* many parapsychologists today who are committed to careful scientific inquiry—as Rhine's work illustrates—and the use of rigorous laboratory methods. Whether they ever achieve it is not always clear, and critics are constantly finding loopholes in their methodology.

What about the results? Are the hypotheses proposed by parapsychology testable? Have they been tested? Here there are also wide areas for dispute. Skeptics are especially unimpressed by the findings and believe that parapsychology has not adequately verified its claims—even though some parapsychologists believe that ESP, precognition, and PK have been demonstrated and need no further proof. I reiterate that, since the chief claims of parapsychology in these areas contravene the basic principles of both science and ordinary experience, it is not enough to point to a body of data that has been assembled over the years; the data must be *substantial*. This does not deny that there seems to be some evidence that certain individuals in some experiments are able to make correct guesses at above-chance expectations. The basic problem, however, is the *lack of replicability* by other experimenters. Apparently, some experimenters—a relative few—are able to get similar results, but most are unable to do so. The subject matter is elusive. It is rare for a skeptic to be able to replicate results, but it is even relatively rare for a *believer* in psi to get positive results. The problem of replicability has been dismissed by some parapsychologists who maintain that their findings *have* been replicated. But have they? For the point is that we cannot predict *when* or *under what conditions* above-chance calls will be made (with Zener cards, in precognitive dream labs, in remote-viewing testing situations); and one is much more likely to get negative results.

One explanation offered by parapsychologists for the difficulty in replication refers to the well-known "sheep/goat" distinction of Gertrude Schmeidler—that is, that those with a positive attitude toward psi (sheep) will get better results than those with a negative attitude (goats). Similar considerations are said to apply to the attitude of the experimenter. Is the explanation for this that when the experimenter is a believer he is often so committed to the reality of psi that he tends to weaken experimental controls? If so, perhaps we should distinguish between the donkey and the fox. The skeptic is accused of being so stringent that he dampens the enthusiasm of the subject. Yet parapsychologists Adrian Parker and John Beloff report on experiments at the University of Edinburgh by pro-psi experimenters that consistently score negative results. Most parapsychologists want positive results, but few receive them. Many or most people don't display ESP; or if they do, they do so infrequently. And those few that allegedly have the ability eventually seem to lose it.

According to John Beloff:

> There is still no repeatable experiment on the basis of which any competent investigator can verify a given phenomenon for himself.[5]

> The Rhine revolution . . . proved abortive. Rhine succeeded in giving parapsychology everything it needed to become an accredited science except the essential: the know-how to produce results where required.[6]

Adrian Parker writes:

The present crisis in parapsychology is that there appear to be few if any findings which are independent of the experimenter . . . It still remains to be explained why, if the experiment can be determined by experimenter psi, only a few experiments are blessed with success. Most experimenters want positive results, but few obtain them.[7]

Charles Tart says:

One of the major problems in attempting to study and understand paranormal (psi) phenomena is simply that the phenomena don't work strongly or reliably. The average subject seldom shows any individually significant evidence of psi in laboratory experiments, and even gifted subjects, while occasionally able to demonstrate important amounts of psi in the laboratory, are still very erratic and unpredictable in their performance.[8]

And Rhine himself says:

Psi is an incredibly elusive function! This is not merely to say that ESP and PK have been hard phenomena to demonstrate, the hardest perhaps that science has ever encountered . . . Psi has remained an unknown quantity so long . . . because of a definite characteristic of elusiveness inherent in its psychological nature . . . A number of those who have conducted ESP or PK experiments have reported that they found no evidence of psi capacity . . . Then, too, experimenters who were once successful may even then lose their gift. . . . All of the highscoring subjects who have kept on very long have declined . . .[9]

All of this means not only that parapsychology deals with anomalous events but that it may indeed be a uniquely anomalous science, for findings depend upon who the experimenter is. But even that is not reliable and cannot be depended upon. If any other science had the same contingent results, we would rule it out of court. For example, a chemist or biologist could not very well claim that he could get results in the laboratory because he believed in his findings, whereas his skeptical colleagues could not because they lacked this belief. We say in science that we search for conditional lawlike statements: namely, that if a, then b; whenever a is present, b will most likely occur. Yet in viewing the findings of parapsychology, the situation seems to be that we are not even certain that b occurs (there is a dispute about the reliability of the experiments). Moreover, we don't know what a is, or if it is present that b would occur; b may occur sometimes, but only infrequently. A high degree of replicability is essential to the further development of parapsychology. Some sciences may be exempt from the replicability criterion, but this is the case only if their findings do not contradict the general conceptual framework of scientific knowledge, which parapsychology seems to do. According to the parapsychologist, for example, ESP seems to be independent of space and does not weaken with distance; precognition presupposes backward causation; psychokinesis violates the conservation-of-energy law.

It is not enough for parapsychologists to tell the skeptic that *he*, the parapsychologist, on occasion has replicated the results. This would be like the American Tobacco Institute insisting that, based on its experiments, cigarette-smoking does not cause cancer. The neutral scientist needs to be able to replicate results in his own laboratory. Esoteric, private road-to-truth claims need to be rejected in science, and there needs to be an intersubjective basis for validation. Until any scientist under similar conditions can get the same results, then we must indeed be skeptical. Viewing what some parapsychologists have considered to be replication often raises all sorts of doubts. In the 1930s S. G. Soal attempted to replicate the findings of Dr. Rhine in Britain in regard to clairvoyance and telepathy. He tested 160 subjects, always with negative results, indeed with results far below mean chance expectations. After the tests were completed, he reviewed the data and thought he had found a displacement effect in two cases, which he considered evidence for precognition (that is, above-chance runs in regard to one or two cards before and after the target). Soal then went on to test these two subjects, Basil Shackleton and Mrs. Gloria Stewart, with what seemed to be amazing results. These results have often been cited in the parapsychological literature as providing strong proof for the existence of ESP. In 1941, in collaboration with the Society for Psychical Research, Soal designed an experiment with Shackleton that included 40 sittings over a two-year period. Among the people who participated were C. D. Broad, professor of philosophy at Cambridge, H. H. Price, of Oxford, C. A. Mace, C. E. M. Joad, and others. Broad described the experiment as follows:

> . . . Dr. Soal's results are outstanding. The precautions taken to prevent deliberate fraud or the unwitting conveyance of information by normal means . . . [are] seen to be absolutely water-tight.[10]

> . . . There can be no doubt that the events described happened and were correctly reported; that the odds against chance-coincidence piled up to billions to one . . .[11]

On the basis of his work in precognitive research, Soal was awarded a doctorate of science degree from the University of London. Even Rhine described the Soal-Goldney experiment as "one of the most outstanding researches yet made in the field . . . Soal's work was a milestone in ESP research."[12]

C. E. M. Hansel, in his work, found, on the contrary, that the Soal-Goldney experiments were full of holes, and he suggested the high results might be due to collusion between the experimenters and/or the participants, especially in the scoring procedures.[13] Broad responded to "Hansel and Gretel," denying the possibility of fraud. It now seems clear that Hansel was correct. And even parapsychologists now doubt the authenticity of these famous experiments. In a recent publication of the Society for Psychical Research, Betty Markwick reported that there is substantial evidence that extra digits were inserted into the "random number" sequences prepared by Soal to determine the targets in the Shackleton tests. These insertions coincided with Shackleton's guesses and apparently accounted for the high scores on the record sheets. Interestingly, Soal was present at every session in which the subject recorded high scores. The only exception was when he was absent, at which time the results were null.[14]

Thus the classical tests usually cited as "proof" of ESP often employed improper shuffling and scoring techniques or had other flaws in the protocol. More recent developments in parapsychology have been more hopeful in this regard. Parapsychologists have attempted to tighten up test conditions, to automate the selection of targets, to use random-number generators and ganzfeld procedures, and to design ingenious dream research and remote-viewing experiments.

One might consider the use of random generators in testing situations to be an advance over previous methods, except for the fact that it is still the experimenter who designs and interprets the experiment. Walter J. Levy, who fudged his results, it may be noted, used machines in his testing work. No wonder the critic is still skeptical of some recent claims made in this area. Great results have been heralded in ESP dream research. Yet here, too, there are many examples of failed replication. For example, David Foulkes, R. E. L. Masters, and Jean Houston attempted to repeat the results obtained at the Maimonides laboratory with Robert van Castle, a high-scoring subject, but they met with no success at all. Charles Honorton has reported what he considers to be impressive results using ganzfeld techniques (where subjects are deprived of sensory stimulation). To date there have been upward of 25 published studies.

Approximately a third have been significant, a third ambiguous, and a third nonsignificant. This may sound convincing. But given the sad experience in the past with other alleged breakthroughs, we should be cautious until we can replicate results ourselves. Moreover, we do not know how many negative results go unreported. (I should say that I have never had positive results in any testing of my students over the years.) Parker, Miller, and Beloff in 1976 used the ganzfeld method to test the relation of altered states of consciousness and ESP and reported nonsignificant results:

> A total of over 30 independent tests were conducted on the data without a single significance emerging. Whatever way we look at the results, they not only detract from the reliability of the ganzfeld, but also argue against the view that psychological conditions are the sole mediating variable of the experimenter effect.[15]

Similarly, Targ and Puthoff at the Stanford Research Institute, in widely reported remote-viewing experiments, have allegedly achieved results that have been replicated. But the critic has many unanswered questions about the method of target selection and the procedures for grading "hits." Given their shockingly sloppy work with Uri Geller, Ingo Swann, and other "super-psychics" in the laboratory, the skeptic cannot help but be unconvinced about their claimed results.

IV.

The accounts above have been introduced as a general comment on the field of parapsychological research: If parapsychology is to progress, then it will need to answer the concerns of its critics about the reliability of the evidence and the replicability of the results.

But difficulties become even more pronounced when we examine other kinds of inquiries that go on in this field; for the parapsychological literature contains the most incredibly naive research reports along with the most sophisticated. A perusal of the parapsychological literature reveals the following topics: clairvoyance, telepathy, precognition, psychokinesis, levitation, poltergeists, materialization, dematerialization, psychic healing, psychometry, psychic surgery, psychic photography, aura readings, out-of-body experiences, reincarnation, retrocognition, tape recordings of the voices of the dead, hauntings, apparitions, life after life, regression to an earlier age, and so on.

We now face a puzzling situation. There has been a marked proliferation of claims of the paranormal in recent years, many of them highly fanciful. Presumably, scientific researchers should not be held responsible for the dramatization of results by fiction writers. Yet in my view some parapsychologists have aided, whether consciously or unconsciously, the breakdown in critical judgment about the paranormal. I have not seen many parapsychologists attempt to discourage hasty generalizations based on their work. There are often extraordinary claims made about psychic phenomena, yet there are no easily determinable objective standards for testing them. Because parapsychologists are interested in a topic and do some research, it is said by some that, ipso facto, it is validated by science. (Lest one think that I am exaggerating, one should consult the *Handbook of Parapsychology*, the most recent comprehensive compilation in the field, which includes discussions of psychic photography, psychic healing, reincarnation, discarnate survival, and poltergeists, among other topics.) Professor Ian Stevenson, for example, of the University of Virginia, is well known for his work in reincarnation, which is of growing interest to many parapsychologists. After discussing the case of a young child who his parents think is a reincarnation of someone who had recently died, Stevenson says:

> Before 1960, few parapsychologists would have been willing to consider reincarnation as a serious interpretation of cases of this type [recall] . . . Today probably most parapsychologists would agree that reincarnation is at least entitled to inclusion in any list of possible interpretations of the cases, but [he added] not many would believe it the most probable interpretation.[16]

Rhine is himself much more cautious in his judgment and implies that only clairvoyance, precognition, and psychokinesis have been established and that adequate test designs have not been worked out for other areas. If one asks if parapsychology is a genuine science or a pseudo-science, it is important that we know if one is referring to the overall field or to particular areas. Surely the critic is disturbed at the ready willingness to leap to "occult" explanations in the name of science in some kinds of inquiry.

Although I have no doubts that Rhine is committed to an objective experimental methodology, I have substantive doubts about his views on clairvoyance, precognition, and PK. The problem here is that one may question not simply the reliability and significance of the data but the conceptual framework itself. Rhine and others have performed tests in which they maintain that they have achieved above-chance runs. What are we to conclude at this point in history? Simply *that* and no more. ESP is not a proven fact, only a theory used to explain above-chance runs encountered in the laboratory. Here I submit that the most we can do is simply fall back on an operational definition: ESP is itself an elusive entity; it has no identifiable meaning beyond an operational interpretation. Some researchers prefer the more neutral term *psi*, but this still suggests a psychic reality. Of special concern here is the concept that is often referred to in trying to explain the fact that some subjects have significant below-chance runs—"negative ESP," or "psi-missing"—as if in some way there is a mysterious entity or faculty responsible for both above-chance and below-chance guessing. All this seems to me to beg the question. If ESP is some special function of the mind, then we need *independent* verification that it exists, that is, replicable predictions.

One of the problems with ESP is that parapsychologists have noted a "decline" effect; namely, that even gifted subjects in time lose their alleged "ESP" ability. At this point, I must confess that I am unable to explain why there are significant above-chance or below-chance runs: to maintain that these are due to psi, present or absent, is precisely what is at issue. A problem for me is how many validated cases we actually have of significant below-chance runs in the laboratory. Rhine mentions some. But are they as numerous as above-chance runs? If so, perhaps the overall statistical frequencies begin to reduce, particularly if parapsychologists stop testing those who have shown psychic ability once they lose their alleged powers. We still need to come up with possible alternative explanations. Some that have been suggested are bias, poor experimental design, fraud, and chance. There may be others.

Rhine's reluctance to accept telepathy because of the difficulty in establishing test conditions is surprising to some. Of all the alleged psi abilities, this seems prima facie to be the most likely. Ordinary experience seems to suggest spontaneous telepathy, especially between persons who know each other very well or live together. If telepathy is ever established, I would want to find the mechanism for it—perhaps some form of energy transmission, though most parapsychologists reject this suggestion, possibly because they are already committed to a mentalistic interpretation of the phenomenon.

There are, as Rhine notes, very serious scientific objections to precognition—the notion that the future can be known beforehand (without reference to normal experience, inference, or imagination). The skeptical scientist believes that, where premonitions come true, coincidence is most likely the explanation. If one examines the number of times that premonitions do not come true, the statistics would flatten out. The conceptual difficulty with precognition is that, although we allegedly can know the future by precognition, we can also intervene so that it may not occur.

Louisa Rhine cites the following case to illustrate this:

> It concerns a mother who dreamed that two hours later a violent storm would loosen a heavy chandelier to fall directly on her baby's head lying in a crib below it; in the dream she saw her baby killed dead. She awoke her husband who said it was a silly dream and that she should go back to sleep as she then did. The weather was so calm the dream did appear ridiculous and she could have gone back to sleep. But she did not. She went and brought the baby back to her own bed. Two hours later just at the time she specified, a storm caused the heavy light fixture to fall right on where the baby's head had been—but the baby was not there to be killed by it.[17]

If the future is veridically precognized, how could one act to change it? There are profound logical difficulties with this concept. Some parapsychologists discuss a possible alternative explanation for the event: one parapsychologist suggests (without himself accepting it) that the dream itself might have contained enormous energy that forced the calm weather to change into a storm, which cracked the ceiling holding the light fixture. "This alternative, then, is not precognitive but of the mind-over-matter, or PK variety."[18]

This illustrates a basic problem endemic to parapsychology. The lack of a clearly worked out conceptual framework. Without such a causal theory, the parapsychologists can slip from one ad hoc explanation to another. In some cases we cannot say that telepathy is operating, it may be clairvoyance; and in others, if it is not precognition, then psychokinesis may be the culprit. (Even an ESP shuffle may be at work!) I fear that the central hypothesis of parapsychology, that mind is separable from body and that the "ghost in the machine" can act in uncanny ways, often makes it difficult to determine precisely what, if anything, is happening.

A number of familiar conceptual problems also concern psychokinesis. What would happen to the conservation-of-energy principle if PK were a fact? How can a mental entity cause a physical change in the state of matter? Comparing the alleged evidence for PK with the need to overthrow a basic, well-documented principle of physics is questionable. We read about Rhine's above-chance results in his die-rolling test: the results seem inconclusive. Recently a number of super-psychics, such as Uri Geller and Jean Girard, have made extraordinary claims for PK ability. Unfortunately, they have been uncritically welcomed by some parapsychologists and paraphysicists. Yet such super-psychics have been discredited, and what seems to be operating is probably magic and illusion, not psi.

Rhine at times expresses an underlying religious motive:

> What parapsychology has found out about man most directly affects religion. By supporting on the basis of experiment the psychocentric concept of personality which the religions have taken for granted, parapsychology has already demonstrated its importance for the field of religion . . . If there were no ESP and PK capacities in human beings it would be hard to conceive of the possibility of survival and certainly its discovery would be impossible . . . The only kind of perception that would be possible in a discarnate state would be extrasensory, and psychokinesis would be the only method of influencing any part of the physical universe . . . Telepathy would seem to be the only means of intercommunication discarnate personalities would have.[19]

Unfortunately, many parapsychologists appear to be committed to belief in psi on the basis of a metaphysical or spiritualist world-view that they wish to vindicate. Charles Tart, a former president of the American Parapsychological Association, admits this motive. Giving an autobiographical account of why he became interested in parapsychology, he says:

> I found it hard to believe that science could have *totally* ignored the spiritual dimensions of human existence . . . Parapsychology validated the existence of basic phenomena that could partially account for, and fit in with, some of the spiritual views of the universe.[20]

Of course, parapsychologists will accuse the skeptic of being biased in favor of a materialist or physicalist viewpoint and claim that this inhibits him from looking at the evidence for psi or accepting its revolutionary implications. Unfortunately, this has all too often been the case; for some skeptics have been unwilling to look at the evidence. This is indefensible. A priori negativism is as open to criticism as a priori wish-fulfillment. On the other hand, some constructive skepticism is essential in science. All that a constructive skeptic asks of the parapsychologist is genuine confirmation of his findings and theories, no more and no less.

I should make it clear that I am not denying the possible existence of psi phenomena, remote viewing, precognition, or PK. I am merely saying that, since these claims contravene a substantial body of existing scientific knowledge, in order for us to modify our basic principles—and we must be prepared to do so—the evidence must be *extremely strong*. But that it *is*, remains highly questionable.

In the last analysis, the only resolution of the impasse between parapsychologists and their critics will come from the *evidence* itself. I submit that parapsychologists urgently need at this juncture to bring their claims to the most hard-headed group of skeptics they can find. In a recent review, C. P. Snow forcefully argues for this strategy. He admits that there are a good many natural phenomena that we don't begin to understand and ought to investigate. Moreover, phenomena exist that are not explained by natural science but which do not contradict it. It is when such phenomena allegedly do so that we should take a hard look. Snow says:

> An abnormal number of all reported paranormal phenomena appear to have happened to holy idiots, fools, or crooks. I say this brutally, for a precise reason. We ought to consider how a sensible and intelligent man would actually behave if he believed that he possessed genuine paranormal powers. He would realize that the matter was one of transcendental significance. He would want to establish his powers before persons whose opinions would be trusted by the intellectual world. If he was certain, for example, that his mind could, without any physical agency, lift a heavy table several feet, or his own body even more feet, or could twist a bar of metal, then he would want to prove this beyond, as they say in court, any reasonable doubt.
>
> What he would not do is set up as a magician or illusionist, and do conjuring tricks. He would desire to prove his case before the most severe enquiry achievable. It might take a long time before he was believed. But men with great powers often take a long time for those powers to be believed. If this man had the powers which I am stipulating, it probably wouldn't take him any longer to be accepted than it did Henry Moore to make his name as sculptor.
>
> Any intelligent man would realize that it was worth all the serious effort in the world. The rewards would be enormous—money would accrue, if he was interested in money, but in fact he would realize that that was trivial besides having the chance to change the thinking of mankind.
>
> It would now be entirely possible for such a man to have his claims

considered with the utmost energy and rigor. For a number of eminent Americans of the highest reputation for integrity and intellectual achievement have set themselves to examine any part of the paranormal campaigns. The group includes first-class philosophers, astronomers, other kinds of scientists and professional illusionists. They are skeptical as they should be. This is too important a matter to leave to people who want to believe. So there they are, the challenge is down. It will be interesting to see if any sensible and intelligent man picks it up.[21]

This, then, is an invitation and a challenge to parapsychologists to bring their findings to the most thoroughgoing skeptics they can locate and have them examine their claims of the paranormal under the most stringent test conditions. If parapsychologists can convince the skeptics, then they will have satisfied an essential criterion of a genuine science: the ability to replicate hypotheses in any and all laboratories and under standard experimental conditions. Until they can do that, their claims will continue to be held suspect by a large body of scientists.

Notes

1. C. D. Broad, "The Relevance of Psychical Research to Philosophy," *Philosophy*, 24 (1949): 291-309.

2. David Hume, *Treatise on Human Nature*, 1739; *Essay Concerning Human Understanding*, 1748; *The Dialogues Concerning Natural Religion*, 1779.

3. Encyclopaedia Britannica, 11th ed., pp. 534 ff.

4. Gardner Murphy and Robert O. Ballou, eds., *William James on Psychical Research*, New York: Viking, 1960, p. 310.

5. John Beloff, "Parapsychology and Philosophy," *Handbook of Parapsychology*, ed. by B. Wolman, New York: Van Nostrand, 1977, p. 759.

6. ———, *Psychological Sciences: A Review of Modern Psychology*, New York: Barnes & Noble, 1973.

7. Adrian Parker, "A Holistic Methodology in Psi Research," *Parapsychology Review*, 9 (March-April 1978): 4-5.

8. Charles Tart, "Drug-Induced States of Consciousness," *Handbook of Parapsychology*, op. cit., p. 500.

9. J. B. Rhine, *The Reach of Mind*, New York: Wm. Sloane, 1947, pp. 187-189.

10. C. D. Broad, "The Experimental Establishment of Telepathic Precognition," *Philosophy*, 19 (1944): 261.

11. ———, "The Relevance of Psychical Research to Philosophy," reprinted in *Philosophy and Parapsychology*, ed. by Jan Ludwig, Buffalo, N.Y.: Prometheus, 1978, p. 44.

12. J. B. Rhine, op. cit., p. 168.

13. C. E. M. Hansel, *ESP: A Scientific Evaluation*, New York: Scribner, 1966.

14. Betty Markwick, "The Soal-Goldney Experiments with Basil Shackleton: New Evidence of Data Manipulation," *Proceedings of the Society for Psychical Research*, 56 (1978): 250-278; D. J. West, "Checks on ESP Experimenters," *Journal of the Society for Psychical Research*, 49 (Sept. 1978): 897-899.

15. Adrian Parker, op. cit., p. 4.

16. Ian Stevenson, "Reincarnation: Field Studies and Theoretical Issues," *Handbook of Parapsychology*, op. cit., p. 657.

17. L. E. Rhine, "Frequency of Types of Experience in Spontaneous Precognition," *Journal of Parapsychology*, 18 (2) (1954): 199.

18. Douglas Dean, "Precognition and Retrocognition," in *Edgar D. Mitchell, Psychic Explorations: A Challenge for Science*, ed. by John White, New York: Putnam, 1974, p. 155.

19. J. B. Rhine, op. cit., pp. 209, 214.

20. Charles Tart, *Psi: Scientific Studies of the Psychic Realm*, New York: E. P. Dutton, 1977, vii-viii.

21. C. P. Snow, "Passing Beyond Belief" (a review of *Natural and Supernatural: A History of the Paranormal*, by Brian Inglis), *Financial Times*, London (Jan. 28, 1978).

Science, Intuition, and ESP

Gary Bauslaugh

The power of rationality, as evidenced by the scientific revolution, has led us to a confused and dichotomous view of life. On one hand, we are reluctant to believe that anything is real unless its objective nature can be specified; on the other hand, we recognize important aspects of life that cannot be precisely defined and measured—for example, truth, beauty, and wisdom. Surely these things are real: civilized life would be untenable if they were not. But what is their place, what is their reality, in a world in which objectivity so dominates?

The omnipotence of reason in our society has long bothered many artists and intellectuals. They argue that judgments of an intuitive or aesthetic nature must be different from, and not merely an extension of, logical analysis. And there is some physiological evidence that the intellect does work in two fundamentally different ways—that there are two separate and distinguishable modes of human thought, a "rational" mode and an "intuitive" mode.[1] The rational mode is considered to be that which functions in a linear, objective, scientific sense; the intuitive, in a holistic, aesthetic, visionary sense. The rational mode is needed to deal sensibly with the objective world; the intuitive is required to function intelligently in the abstract world.

The idea of two modes of thought, whether or not it is physiologically supportable, is indeed attractive; for it allows for some kind of resolution, or at least mutual accommodation, of the often antagonistic notions of rationality and intuition. If they are both regarded as being legitimate, but different, ways of thinking, then they can be used in a mutually supportive manner. The scientist ought not to scorn intuitive judgment; he uses it himself in virtually every response he makes to his environment. The artist ought not to denigrate rationality, for rationality must be a significant component of his approach to life. Mutual understanding, however, is not prevalent. Scientists, many of them at least, seem intent on proving that, to use Einstein's phrase, "God does not play dice with the universe." Artists, and others, are convinced that there is a ghost in the machine, but their approach is often unnecessarily and unfortunately hostile to rationality. Many, however, sense that each approach has validity, and they search for ways of accommodating both.

Proponents of belief in extrasensory perception (ESP) claim a unique affinity to both ways of thinking. They supposedly incorporate rationality into their view of reality (in the experiments of the parapsychologists), yet they also make considerable claims to a world beyond rationality, a world of mystical, unexplainable forces, a world in which normal, mundane concerns, such as mechanism and cause and effect, are superfluous. They believe that they have discovered a physical reality that extends beyond current scientific understanding, one which perhaps is inherently not amenable to standard scientific analysis. This view has considerable appeal to those who are disturbed by their conflicting feelings in regard to intuition and rationality, because it seems to offer a plausible reconciliation of the two. Science, through parapsychological experiments, is seen to be in the service of revealing a higher truth—a world beyond the strictures of traditional science. Both the perceived need for compatibility with rationality and the intuitive need for something beyond can be seen to be satisfied. The fact that many scientists denounce the psychic movement is regarded as further confirmation of the reluctance of the scientific establishment to accept any reality beyond the narrow confines of objective analysis.

There is a human need to find an alternative to the mechanistic vision of life that is presented by science. But in presenting a spurious synthesis of science and a paranormal world beyond science, psychic proponents offer a superficially attractive but counterfeit fulfillment of this need.

* * * * *

Believers in paranormality frequently draw an analogy between ESP skeptics and Galileo's inquisitors. In the classic story of rigid and single-minded belief, the inquisitors refused to look through Galileo's telescope, where they could have seen objective evidence for some of his astronomical claims; in the most famous incident, he wanted to show them the moons of Jupiter. Many modern scientists, it is claimed, are like the inquisitors in their refusal to consider openly and willingly the evidence for ESP. But psychic proponents are, in fact, much more like the inquisitors, and the skeptics like Galileo. I think it is predilection to belief (or predetermined belief) that characterizes psychic proponents, just as it did the inquisitors; and it is skepticism, an unwillingness to accept rigid and unsubstantiated belief, that characterizes ESP skeptics, just as it did Galileo. Intelligent skepticism has always been a prerequisite for the overthrow of entrenched belief.

And one more word about the inquisitors. I do not wish to do them too great an injustice by comparing them to psychic proponents. The inquisitors' position was not a foolish one and it was not trivial. They were not simple people or narrow-minded bigots. They had a legitimate concern, a concern that Plato had when he brought reason and rationality and philosophic thinking to Western thought; a concern that we, on the whole, seem to be unaware of. Plato had warned about capitulation to empiricism, which he understood would destroy the mythic, moral vision of the world that was attainable only through pure thought. He had warned against replacing the mythic world with simple facts about nature, obtained merely by experiment. Empiricism is soulless and amoral and therefore limited. Our complete acceptance of rational objectivity as a mode of thought makes it difficult for us to understand Plato's and the inquisitors' legitimate concerns about empiricism. They were trying to protect a beautiful and morally superior and long-standing mythic vision of the world; the same cannot be said, I think, about psychics and parapsychologists.

What I am saying in regard to ESP is this: ESP is defended on scientific grounds, but, as I shall try to show, the attempt is misguided. Alternatively, ESP is defended on mythic grounds; the attempt there is similarly misguided. One can agree, of course, that, as Hamlet said, "there are more things in Heaven and Earth than are dreamt of in your philosophy," and no one can deny that there are vast areas of metaphysics that no one knows anything about. But the vision of the universe presented by parapsychology is limited, sterile, amoral, and has none of the beauty and ethical value of a true mythic-poetic vision. In other words, parapsychology makes bad science and it makes bad poetry, and by claiming an affinity to both science and poetry it does damage to both.

* * * * *

In regard to the question of scientific validation of ESP, I cannot, of course, review such a topic in any detail in this paper. Excellent reviews have been presented and are constantly being written.[2] I should like, instead, to concentrate briefly on one point of relevance: that of the tenuous nature of scientific validation in general. Scientists are reluctant to make the claim of absolute validation, because, contrary to popular understanding, the nature of their enterprise does not often lead to the revelation of absolute truth. Scientists are fallible people, and nature often acts as though it were capricious. It is difficult to establish truth or reality through even the best scientific research. Science does not constantly dispense truths; rather, it grinds slowly toward better and better approximations of truth. And it does not go directly toward the truth; it discovers many false and misleading pathways. A flurry of work in parapsychology, for example, and numerous scientific reports of its existence, do not prove, as it is frequently claimed, the reality of its existence.[3] Such reports exist independently of the reality of psychic phenomena; they exist and would exist in any case. Proof requires far more substantial and conclusive evidence.

Those unfamiliar with science often ascribe unwarranted validity to the results of scientific research. An excellent example of the elusiveness of truth in science was recently provided by those engaged in research on a substance called polywater.[4] This substance, described first by a Russian chemist in 1961, was thought to be a new and potentially important form of water. The implications of this discovery were enormous; and hundreds of scientists worked intensively on it, using the most sophisticated equipment available, until it was discovered in 1973 that polywater was really only ordinary water containing silicon impurities from the glass containers that held it. It took 12 years to establish that this substance, which existed in measurable quantities right before the scientists' eyes, was not a new form of one of the simplest substances on earth.

One must be particularly skeptical about conclusions drawn from research in the behavioral sciences. If the complexity of nature is such that chemical analyses of even the most basic and simple substances on earth can be fraught with misdirection and error, how much more difficult to establish the truth when dealing with the most complex organisms on earth. Furthermore, it is difficult to avoid having predetermined belief when dealing with human problems, and evidence has a curious way of supporting belief. Because of the very real problem of self-deception (as well as intentional deception of others), one should always seriously question the reliability of those who busy themselves finding evidence that confirms strongly held beliefs.

Perhaps the most striking example of the improper use of scientific research in the behavioral sciences was provided by Sir Cyril Burt, the eminent British psychologist, who died in 1971.[5] Burt won many honors and awards for his work and continued to be held in high regard after his death, until the disclosure that his most important work—I.Q. tests of identical twins that supposedly established the objective reality of the hereditary basis for intelligence—was completely invalid and possibly fraudulent. It was discovered that Burt's collaborators in the work were fictitious, and there is reason to believe that some of the twins reported on did not even exist. The results, upon close examination, show unmistakable evidence that they were altered to give the desired results. Even Burt's strongest supporters now admit that this is so.[6]

Whether or not Burt was intentionally fraudulent will never be known. It is felt by many that he simply committed one of the fundamental mistakes of scientific research: he let his private beliefs pervade the realm of objectivity. Burt believed in inherited intelligence; he believed in it long before he reported on the experiments with twins. When one knows the truth of what one is trying to prove, then proof becomes an annoying inconvenience. When things do not work out quite as expected, one begins to influence the results, ever so slightly at first, or to ignore certain results, just to get it all over with. Then, of course, the objective value of such work is completely lost.[7]

The problem with much parapsychological research is similar to that which probably affected Burt's work: many parapsychologists strongly believe in the reality of what they are trying to prove. In some instances, adequate controls are simply abandoned; in other cases, results are influenced in very subtle ways. Because simple experiments seem not to work, for example, very complex experiments are devised. The complex ones merely allow more scope for spurious nonpsychic influences upon the results (and for fraud).

The debate will probably go on indefinitely, because no one can ever prove that ESP does not exist. New experiments will always be devised— complex and sophisticated experiments—and the eager enthusiasts of the psychic world will proclaim that the ultimate and final proof is at hand. What is one to do, then, when faced with this barrage of new and supposedly conclusive experiments? I suppose one must look through the telescope. But is there really a Galileo among all of these parapsychologists? Personally, I am inclined to adopt the view of Hume (from his essay *Of Miracles*):

> Fools are industrious in propagating imposture; while the wise and learned are in general content to deride its absurdity without informing themselves of the particular facts by which it may be distinctly refuted.

* * * * *

If it is true that there are two fundamental modes of human thought— the rational-objective-scientific mode and the intuitive-mythic-poetic mode—then our tendency to consider rationality as the only legitimate mode is indeed limiting. Of the two forms of human thought, the intuitive must be the fuller, richer form: it offers the only true potential for human wisdom. Wisdom cannot consist simply of logical virtuosity. Intuitive thought encompasses all experience; it can include rational knowledge, but it also incorporates the nonrational aspects of human nature, such as human feeling and aesthetic sensibility. If there is, in fact, such a thing as human wisdom, then our intuitive sense must constitute the essence of it. And it is the intuitive-aesthetic-visionary sense of the world that becomes dissatisfied, I think, by an excessive dependence upon rationality. It is this intuitive sense that suggests there must be more to existence than the enormous storehouse of factual information that we have accumulated by rational, objective inquiry. It is our intuition that says there must be more; that, as Plato and Galileo's inquisitors feared, rational empiricism allows no higher vision of the reason for existence, no unifying, mythic vision of experience. It is our intuition that demands a mythic vision of life.

The scientific revolution has left us without mythic vision. It has left us with a metaphysical void that, I think, science and rationality can never fill. We see in our society many ways in which an attempt is made to fill the void, or to come to terms with it. Some of these involve simply denying rationality and ultimately become indistinguishable from madness.[8] Others attempt to do what I think must be done, to develop a mythic vision of life that somehow accommodates rationality or allows a concomitant rational approach to the objective aspect of life. In most instances, however, I think that a critical mistake is made. The mistake is to compromise mythic vision by imposing rationality upon it. It does not work, because myth is fundamentally different from rationality and it is hostile to the imposition of rationality. Some churches, for example, in their abandonment of traditional ritual and myth, are removing the essence of religious spiritual experience. They are agreeing with the rationalists that their vision must be more objective, relevant, and rational. But they are left with soulless social agencies.

One reason—at least what I take to be a reason—for widespread belief in the paranormal and the accompanying growth of pseudoscience is that psychic investigation represents another attempt to arrive at a mythic vision that somehow incorporates rationality. The metaphysical void can be found more acceptable if there are forces beyond our comprehension that act upon our lives, and psychic phenomena suggest such forces. But if they are not merely irrational forces, but are empirically verified forces, then they can coexist in harmony with rationality. That is why people wish to believe and why they so desperately seek objective validation. But they have limited what they can see, and feel, and sense, by the limitations imposed by rational inquiry. Parapsychology ultimately presents a sterile vision of merely unexplained paranormal forces, which become evident only through statistical manipulations or in uncontrolled circumstances. The psychic movement has compromised its vision by seeking objective validation and has employed corrupt objectivity in its eagerness to prove the vision. It has neither objectivity nor vision of mythic dimension.

* * * * *

Those who believe in the paranormal do harm (I have suggested) to the concept of rationality, and they provide a confining intuitive vision of existence. A humanly satisfying intuitive vision does not arise from attempts to imitate rationalism, as the liberalized religions do, or to distort and corrupt rationality as pseudoscience does. None of these ultimately satisfies the human need for an intuitive-mythic vision of life that can coexist with rationality.

It is only, I think, through a deeper understanding of the central role of intuitive vision in human existence, and of its proper relationship to rationality, that we can begin to come to terms with the metaphysical void, if we can ever do that at all. Pseudoscience, particularly in the guise of parapsychology, detracts and distracts from the search for richer visions of experience. It substitutes titillating and shallow speculations, and spuriously rational ones, for intuitive vision. It is in poetry and aesthetics and in the complexities of human feelings and intellect that we ought to search for more satisfactory reasons for existence. We shall find little solace in our frantic preoccupation with rational inquiry, and less still in the frantic preoccupation that many have with the spurious forms of rational inquiry that characterize the world of psychics and parapsychology.

Notes

1. See, for example, R. Ornstein, "The Split and the Whole Brain," *Human Nature* 1 (5) (May 1978):76-83.

2. For example, *ESP: A Scientific Evaluation,* by C. E. M. Hansel. Also see Martin Gardner's many essays on the topic in the *New York Review* and the SKEPTICAL INQUIRER.

3. Arthur Koestler, for example, in *The Roots of Coincidence* (New York, Random House, 1972), states that parapsychology had received the "final seal of respectability" when the Parapsychological Association was accepted into affiliate status with the American Association for the Advancement of Science in 1969. He suggested that the field had progressed so far that, rather than being of questionable status, now it was at the point of being "laid open to the charge of scientific pedantry." Koestler's aggressive writing on this topic reminds me of Cyril Burt, whom I discuss later in this essay. Both Koestler and Burt are eloquent in their denunciation of those they perceive to be narrow-minded skeptics. Both write with the conviction of committed believers. The AAAS and the scientific world in general are, of course, far from accepting the validity of ESP. The "Science and the Citizen"

column in the April 1979 *Scientific American* and the Spring 1979 SKEPTICAL INQUIRER report on a current movement within the AAAS directed at critically reviewing the affiliation with the Parapsychological Association.

4. For a brief review of the polywater phenomenon, see Leland Allen in the *New Scientist* 59 (Aug. 16, 1973):376-380.

5. Burt provides a particularly interesting example for this essay, since he was a powerful defender of the psychic movement. He wrote many articles denouncing skeptics such as Hansel. In *Science and ESP* (J. R. Smythies, ed., Atlantic Highlands, N.J., Humanities Press, 1967), he states: "Professor Hansel's book is certainly the fullest statement to date of the case against parapsychology and, if this is the best the critics can do, parapsychology would seem to be in a fairly strong position." One can understand Burt's annoyance with Hansel. Hansel had quoted (p. 26) Burt's statement that experiments on the telepathic abilities of two Welsh schoolboys provided the final evidence for telepathy. In a detailed review of the conditions under which the experiments were conducted, Hansel shows how he had two Welsh schoolgirls duplicate the feats of the boys, using an Acme "silent" dog whistle to signal each other. Such whistles are commonly used to summon sheep dogs in the Welsh countryside. Children can hear the whistles clearly but adults cannot.

6. For a brief review of the flaws in Burt's work, see Nicholas Leland Allen in the *New Scientist* 59 (Aug. 16, 1973):376-380.

7. One of the strongest statements I have seen on the hazards of deception in scientific research was written by Cyril Burt himself, in *Science and ESP* (op. cit.). The irony speaks for itself. In discussing mediums, Burt said: "Far more frequent, however, and far more subtle, are the effects of unconscious self-deception—a proclivity which even trained investigators seem at times to underestimate. The tendency to heighten one's statements so as to make them more interesting or enhance one's own importance as the subject of some memorable experience, the desire to avoid qualification or reservations as indicative of an irresolute judgement, and above all perhaps the insistent need to adjust our observations and our recollections to fit our dominant hopes and wishes—these are all ingrained and natural tendencies of the human mind, as unconscious as they are automatic. It needs a long and arduous discipline to turn a man into an exact, objective and truly scientific reporter."

8. I am referring to the ways in which a nonrational vision of life is imposed upon the objective world. Nonrational visions constitute the greater part of poetic-aesthetic sensibilities; they are the essence of imagination; they contribute enormously to the richness of one's existence; they can lead to many complex insights into human behavior. But when these visions of experience are confused with rationality, when one's world becomes directed and dominated by nonrational vision, then madness, or something indistinguishable from madness, ensues.   •

Believing in ESP:
Effects of Dehoaxing

Scot Morris

I believed in ESP. I was a teenager and had read one of those "Incredible Tales" paperbacks, and I believed. A patient teacher pointed out the fallacies and flimsy evidence on which I was basing this belief, and I began to wonder. Then there was a newspaper story about a poltergeist—very exciting and mysterious. I believed, and tried to convince others, until two weeks later when I read in a follow-up story that the boy confessed to fooling his parents and the investigators "for a little excitement."

The experience was embarrassing, but it taught me a valuable lesson—that I could be fooled. I was determined not to be fooled again. I am convinced that the best way to develop a healthy skepticism toward the many incredible tales one hears in life is not to go about disbelieving everything blindly, but first to *believe*, with all one's heart, and then suddenly and dramatically be disabused of the idea. The lesson, like a pie in the face, is never forgotten.

In a lab session of Introductory Psychology that I taught at Southern Illinois University, part of the course curriculum was to teach students some principles of scientific method so that they could learn to evaluate any kind of evidence—"Incredible Tales" books, technical experiments, TV commercials, and personal experiences—and be able to see whether someone's conclusions are warranted or not.

In most classes there was a keen interest in ESP. The majority of students did not believe in it but were curious; a few had come to believe after studying about it or after an intense personal experience. But a surprising number believed in ESP because of an article in the Sunday supplement or an impressive stage telepathy act. This was discouraging. It seemed that students could be convinced so easily that they apparently made no effort to come up with counterexplanations. Many apparently found it easier to believe in ESP than to admit that they couldn't explain something or that they could be fooled. With the help of some fellow psychology instructors, I worked out an exercise and lecture that I hoped would teach the students to be more critical and selective in deciding on all their beliefs—from the existence of ESP, to the validity of an experiment, to the efficiency of the latest commercial "miracle ingredient."

ESP in Front of Their Eyes

My colleague Steve Werk came to class as a visiting lecturer for the first hour of a lab session. I introduced him as an expert in ESP who had just returned from Duke University. He talked about the different forms of ESP, then went through a pack of Zener cards to see if anyone could score better than chance in guessing the symbols—star, circle, square, cross, and wavy lines. Steve then began an "ESP demonstration."

His first feat was to receive a number between 1 and 20 that the class had decided upon while he was out of the room. He then spoke of a friend with whom he had developed an especially strong ESP bond. His friend was at home, Steve said, but was expecting a long-distance telepathic communication sometime during the hour.

Steve had a student volunteer pick a card from a standard deck—the two of clubs, let us say. He then selected three volunteers—to assure there were no accomplices—and told them to go call his friend from the public phone. "Call 755-8472 and ask for Mr. Black," he said. "Tell him you are in the ESP experiment with me and ask what card we have selected. He'll know what you mean." Steve wrote the name and number on a slip of paper, gave them a dime, and they left the room. He asked for quiet, closed his eyes, and held his fingertips to his brow.

A few moments later the excited students returned: "*He got it! He got it! The two of clubs!*"

In the final demonstration Steve distributed a half-dozen sheets of paper, with envelopes, to six students in different parts of the room. "Write a question on the paper," he said, "a question about the future. Show your question to the people sitting near you so there can be no dishonesty." Each student then folded his paper, placed it in the envelope, and sealed it. I collected the six envelopes and gave them to Steve. He asked for silence, picked up the top envelope and pressed it against his forehead. He closed his eyes. "I see this is a question about sports. Yes . . . it's a question about basketball. Someone wants to know whether we will win the regional championship."

The student who had written that question let out a gasp. I asked him whether Steve's wording was correct; it was. The students looked at each other in amazement. Steve opened the envelope to verify the question and answered it: "Yes, we will win," he said, or "No, we won't," or whatever came to his mind. Since the questions were about the future he was free to improvise.

After extemporizing about the basketball tournament, he held the second envelope to his forehead. "I see . . . it's a question about scholarship and . . . numbers," he said. "Someone wants to know: 'Will I keep my four-point average?'" Another student gasped.

At the end of the envelope demonstration the first lab hour was about over, so I thanked Steve and he left.

This was our ESP demonstration. In the next couple of years we became so fascinated with its effects that we repeated it to over a half-dozen different classes, and once to a Student Activity Club audience of about 150 persons.

Of course it was all a hoax. The demonstrations were well-known parlor tricks. In the first trick, I was the accomplice. My helpful "Ready now, Steve?" or "OK, are you ready?" was a code phrase that told him the number. In the long-distance telepathy feat the accomplice was another psychology instructor, Paul Fox. When he got a call for *Mr. Black*, he knew the card was the two of clubs. The first letter meant *two* (since *B* is the second letter of the alphabet), and the last letter, *k*, meant *clubs* (because *clubs* starts with a *k*-sound). Had the card been the four of diamonds, the students would have been told to ask for *Mr. David*. Steve and Paul, of course, had practiced the code in advance so that Steve could think up a plausible name within a few seconds of seeing any card in the deck.

In the last demonstration, one of the students in class was the accomplice. He wrote a prearranged question on his paper—for example, "Will we win the basketball tournament?" When I collected the envelopes, I made sure the accomplice's was on the bottom of the pile. Steve had no idea what was in the top envelope—until he read it while pretending to verify the wording of the basketball question. He noted the question, "Will I keep my four-point average?" and pretended to receive it from the second envelope.

The three parlor tricks were frighteningly impressive. During all our classroom demonstrations, no student ever doubted that he had seen real ESP—at least not out loud.

"I have something to tell you . . ."

I always dehoaxed the students during the next lab hour or the next class meeting. I would start by asking for a show of hands—how many now believed in ESP, or had their beliefs strengthened by Steve's demonstration? It was almost embarrassing. Usually about 80 percent raised their hands. I asked them to keep their hands up while I wrote some figures on the board: A = ace, B = 2, C = 3, etc. It was the code for the telephone trick. I explained how the name *Black* had carried all the information our accomplice had needed to guess the two of clubs.

The students looked disappointed and embarrassed. Slowly, sheepishly, they began to lower their hands.

"Everything you saw in the demonstration was a hoax," I said. "You were fooled. Steve and the man who answered the phone are friends of mine who teach in other classes. Now, why did we—your friendly psychology instructors—do this to you? Because we wanted you to see how easy it is to believe something that's not true; how readily you will jump to a false conclusion when you want to believe it and can't think of any other explanation. The envelope demonstration was a trick too; but I'm not going to say how it was done, because I want you to experience the feeling that, even though you can't explain something, that doesn't make it supernatural. Why not just say, 'I don't know how it was done,' and leave it at that? We tricked you because we want you to think about the way you reach conclusions and the type of evidence you accept as proof of something."

I followed the dehoaxing with a skeptical lecture about ESP. It was admittedly one-sided, because I felt the students had already seen many pro-ESP accounts in popular magazines and paperbacks. I made the lecture as powerful and persuasive as possible, because it reflected my own beliefs at the time and because we later wanted to measure the effects of the lecture (and, separately, the dehoaxing experience and the ESP demonstration) on students' belief in ESP and other controversial issues.

The Lecture

There is room here only to sketch the outlines of the anti-ESP lecture. It lasted about 50 minutes and consisted of four major sections, the first three corresponding to the major types of "proof" people cite when explaining why they believe in ESP.

Stage ESP. I pointed out that ESP research has generally shown that if the phenomenon exists it is an elusive, long-run, slightly-above-chance sort of thing. No serious believer would dare to announce, "Ladies and gentlemen, I will now guess the next card correctly." In short, if it's done on stage or by appointment, there's a trick to it. Several secrets of professional mind-readers were revealed, including "cold reading," the artificially inflated "hit rate," use of accomplices and codes, safe predictions, researched predictions, self-fulfilling predictions, and the *post hoc* selection of events to "fit" previous predictions.

Personal experience. "My Aunt Gertie had a dream and it came true."

Perhaps. But is she remembering selectively, or embellishing the tale ever so slightly for the sake of a good story? And how many dreams didn't come true? Personal experiences, first- or second-hand, can be very persuasive, but they always beg the hidden question, "What is the probability of this coincidence happening by chance?" The question is often unanswerable in principle because of the vagaries of human perception, memory, and subconscious "editing," especially during stress. It would be wonderful if someone kept a record of all the failures—but imagine calling the features editor at the newspaper to report, "The dog next door started howling at 1:20 last night. Not much else happened—its master didn't die or anything—but I thought you ought to know."

Laboratory studies. This section examined pitfalls in some scientific research, including the tales of Clever Hans and Lady Wonder (a telepathic horse investigated by J. B. Rhine); the claims that ESP doesn't work well when experimental controls are tightened or in the presence of skeptics; publishers' and researchers' selectivity in reporting the most encouraging results; statistical problems with "the decline effect," "psi-missing," and the multiple reanalysis of random data until significant "above chance" patterns are found. Examples from C. E. M. Hansel's *ESP: A Scientific Approach* showed how classic "definitive" ESP studies could have been fudged, and, though this doesn't prove fraud was used, it does mean that such experiments should not be considered conclusive.

If ESP exists... Finally, I asked the class to assume that ESP does exist, and then to explain why, after so many years of research, it has not shown itself when it had the chance. One might expect the power to make more appearances in situations of strong motivation and emotion than in simple digit-guessing tasks. But, then, why do Las Vegas casinos continue to operate, year after year, always showing the expected amount of profit? When TV shows and magazines have offered cars and cash prizes to anyone who could receive a secret message, why has no one won? Does materialism inhibit the natural gift of ESP? Then, when Charles Lindbergh's baby was kidnapped and its whereabouts was a national obsession in 1932, why did none of the 1,300 dream reports solicited by the Harvard Psychology Clinic even come close to identifying the baby's true whereabouts?

In conclusion, I said, it is difficult to say definitely that ESP does or does not exist, but the "evidence," when examined very closely, does not turn out to be very compelling.

Measuring the Effects

The lecture above, along with the ESP demonstration and the dehoaxing (telling the students they had been hoaxed and explaining why it was done) took about two hours of class time. Some warned that deceiving the students would only teach them to distrust the instructors and the rest of the introductory course. This didn't seem to happen. Several students began to ask for advice and opinions about other controversial issues; several previously quiet students showed new interest in the class and began to enter into discussions for the first time; and many volunteered that it was the most enjoyable, instructive lesson of the year.

But were our theatrics having any effect on students' beliefs? Before we took a second class through the experience, we constructed a questionnaire. To assess students' beliefs in ESP, we included statements about each of five types of ESP (telepathy, clairvoyance, precognition, psychokinesis, and predictive-telepathic dreams), and a sixth about ESP in general:

ESP Items

ESP exists. (ESP)

Some people, by telepathy, can tell what another person is thinking. (telepathy)

Some people's dreams enable them to know about unseen events or the future. (dreams)

Some people have mental powers such that they can be aware of events taking place at a distance from them. (clairvoyance)

Some people have mental powers enabling them to tell the future. (precognition)

Some people can influence the roll of dice by concentration. (psychokinesis)

(Note: Key words in parentheses are for reference in this paper and were not included in the questionnaire.)

The students were to rank the strength of their belief in each statement on an eight-point scale from *absolutely believe* (8) to *absolutely disbelieve* (1).

We wondered whether the exercise would teach the students skepticism for ESP statements only, or a more general attitude of skepticism, as we had hoped. For example, would their experience also make them more skeptical of astrology, Ouija boards, and ghosts?

We selected 15 more beliefs—supernatural, or at least "borderline-science"—and included statements about these in the questionnaire.

Supernatural Items

The star constellation under which a person is born can tell about his future and/or personality. (astrology)

Some UFOs (unidentified flying objects) are really flying saucers. (flying saucers)

Some mediums can get in contact with the spirits of dead people. (spiritualism)

Under hypnosis, people behave differently from the way they do in the normal state. (hypnotic state)

Human beings have souls. (souls)

There is a God. (God)

Some religious persons can cure people of illnesses to a degree not attributable to mere "power of suggestion." (faith healing)

The Ouija board has the power to give some people answers to questions they could not get otherwise. (Ouija boards)

Some people are able to tell colors by feeling them. (D.O.P.—dermo-optical perception)

Some people, by the use of a forked twig, can locate underground water. (dowsing)

It is possible to tell about a person's history or personality by the lines on his hands. (palmistry)

Horoscope books can tell about your future. (horoscopes)

Some people can tell about your personality by looking at your handwriting. (graphology)

Ghosts exist. (ghosts)

The menstrual cycle of females is related to the phases of the moon. (moon)

Finally, we wondered whether our trickery would have any effect on more neutral, natural beliefs, e.g., that man is descended from apes, or that smoking causes cancer. We wanted to include some "natural" items so that students wouldn't immediately interpret the questionnaire as a "survey of foolish beliefs." We included 11 statements about issues that are controversial but not supernatural:

Natural Items

Machines exist that measure electrical brain waves. (EEG)

Smoking causes lung cancer. (smoke—cancer)

Moths are attracted to light. (moths)

Scientists will find life on Mars. (Mars)

The only addicting effect of marijuana is psychological. (marijuana)

Some people are born unable to distinguish colors. (color-blindness)

Kennedy was shot by a lone assassin: Lee H. Oswald. (Kennedy)

Intelligence is at least partially determined by heredity. (intelligence)

Some animals can find their way in the dark by listening to echoes of their own voices. (animal sonar)

The pill is the most effective birth-control device. (the pill)

Man is descended from apes and lower animals. (evolution)

We called the questionnaire *A Survey of Controversial Issues*, and mixed the ESP, supernatural, and natural items throughout. We administered the survey to many different groups of students—sometimes once, usually twice or three times, with the testings counterbalanced to occur at various possible points in the sequence: sometimes before the ESP demonstration, sometimes after it, sometimes after the dehoaxing and/or the anti-ESP lecture. In this way, we could compare various groups or sets of groups to assess the separate effects of our manipulations. The data were analyzed by analyses of variance. The 6 ESP items were grouped for a single, average ESP belief for each subject; similarly, the 15 supernatural items yielded a supernatural-belief score, and the 11 natural items a natural-belief score for each subject.

The students took the questionnaire anonymously, but there was a space for listing sex and birthdate, and we used these data to match up the forms when questionnaires were given twice to the same class.

To determine the base rate of college students' beliefs independent of our meddling, we gave the survey to 200 Introductory Psychology students who did not take part in the other experiments (Fig. 1).

Students' belief in the natural items was highest of all, averaging 5.84 on the 8-point scale. There was significantly less belief in ESP (mean = 4.64), and still less belief in the supernatural items (mean = 3.89). This basic ordering of scores was obtained under almost every condition (usually significant with p k 0.01), though in some testings, especially just after the ESP lecture, ESP belief was elevated as high as natural belief. There were no significant sex differences.

We did not assess changes within subjects from before to after the demonstration, since we felt this would encourage spurious changes.

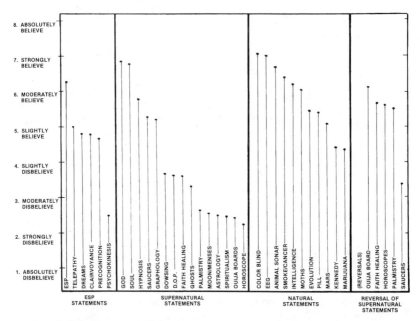

FIGURE 1: Beliefs. The mean beliefs of 200 college students on items on the Survey of Controversial Issues. Items are arranged within four categories in descending order of degree of belief.

Instead, we compared 55 students tested for the first time after seeing the ESP demonstration and found that they had a significantly higher belief in ESP than did a control group of 70 students who had not seen the demonstration ($p < 0.05$). The ESP demonstration had no significant effect on natural or supernatural beliefs. Further proof that the demonstration was effective came from a control group of 50 students who took the survey twice in two weeks, with no intervening manipulations. None of the belief-systems changed significantly during this period.

Generalization from one supernatural belief to others is not surprising; a person who believes in one paranormal phenomenon, such as ESP, tends to believe in others. In a group of 70 students participating in the survey for the first time, the correlation between ESP belief and supernatural belief was 0.58. Belief in ESP was also related to natural beliefs, though less strongly ($r = 0.25$). There was no relationship between supernatural and natural beliefs ($r = -0.05$).

Generalization. When we gave the survey to 39 students who had just seen Steve's impressive ESP demonstration, belief in ESP was elevated to an average of 5.72. After this testing I dehoaxed the students, presented the anti-ESP lecture, and then passed out the questionnaire again. Belief in

ESP had dropped to 2.6. Even though the lecture and dehoaxing dealt only with ESP, they were followed by a significant drop in supernatural belief as well, though this was not as dramatic as the drop in ESP belief. Natural beliefs did not change significantly.

These findings suggest a *generalization of skepticism*: teaching someone to be skeptical of one belief makes him somewhat more skeptical of similar beliefs, and perhaps slightly more skeptical of even dissimilar beliefs.

Memory. With one group we appended a few questions to the second version of the questionnaire (administered after the dehoaxing and the anti-ESP lecture), which asked students to recall how much they had believed in ESP on the *first* testing, taken just after the ESP demonstration. In other words, they were to recall the answers they had given to the ESP items just an hour before. On every question the students minimized the subjective change—they "pulled" their old ESP belief down toward their new level of disbelief, and assumed that their old level of belief in ESP was less than it actually was (Fig. 2). Later, we repeated this test twice on larger groups, and in both cases this *tempered recall effect* was significant ($p <$

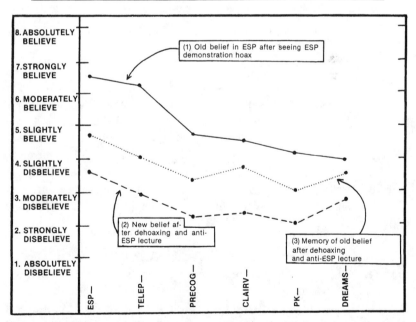

FIGURE 2: **Tempered recall effect.** Belief in six ESP statements (1) dropped considerably an hour later, after students were dehoaxed and heard a lecture critical of ESP. (2) On this second testing, students were asked to recall their earlier levels of belief. They consistently "pulled down" their old belief toward their new one, minimizing the subjective change. Data are presented in the order of descending degree of belief on the first testing.

0.01). In one study the old and new beliefs were 5.83 and 3.23, while memory of the old belief was 4.18; in another, the figures were, respectively, 5.39, 2.88, and 4.49. *The memory of one's old belief tends to split the difference between one's present belief and one's actual old belief.*

Reversals. The dehoaxing and anti-ESP lecture seemed to have a clear effect in making the students more skeptical of the propositions in the *Survey of Controversial Issues.* But were students becoming indiscriminately skeptical, disbelieving everything, no longer willing to take a positive stand, no matter what the issue? Would a student, for example, say that he didn't believe in flying saucers and also that he didn't believe in an opposite statement, that "flying saucers are all from earth"? To find out, we included toward the end of the questionnaire five statements that reversed earlier supernatural items:

Reversal Items

Belief in horoscopes is just a superstition. (anti-horoscopes)

It is impossible to tell about a person's history or personality by the patterns of lines on his hands. (anti-palmistry)

Belief in the Ouija board is a superstition. (anti-Ouija board)

Faith healing is just "power of suggestion." (anti-faith healing)

"Flying saucers" are all from earth. (anti-flying saucers)

We found that after the dehoaxing and the anti-ESP lecture, belief in these five reversal items rose instead of dropping. Apparently the students were not just checking the questionnaire blindly, disbelieving everything.

Separating the Effects. We had established that students became more skeptical of ESP and related beliefs after being dehoaxed and hearing the anti-ESP lecture; but which of these experiences was most important? Was it the hour-long attack on ESP by a psychology instructor or the simple realization that one had been fooled? We tried to separate these influences in our next experiment.

At an hour when two introductory labs, A and B, were in session, we arranged to have all students come into one room for Steve's ESP demonstration. Afterwards, all students answered the questionnaire.

The following week I went to Class A as a visiting lecturer and dehoaxed the students in the usual way—I told the students that they had been duped the previous week, explained how the telephone trick was done, and discussed the reasons for the hoax. I told them that Instructor B was at that moment explaining the hoax to his class, after which I would talk about ESP to the combined classes. In fact, however, the instructor in Class B was going over quiz scores, stalling for a few moments while Class A was dehoaxed.

We then combined the classes, quickly and quietly, and I delivered the anti-ESP lecture. Everyone in the audience had seen the ESP demonstration, but only half of them had learned it was a hoax.

After the lecture, the students answered the questionnaire for the second time. I then explained about the previous week's hoax for the benefit of Class B, so that everyone left knowing the truth of the matter.

For Class B students, who did not know that the previous week's demonstration was a hoax, the anti-ESP lecture alone produced a significant drop in ESP beliefs of 1.66 points, with a slight generalization of

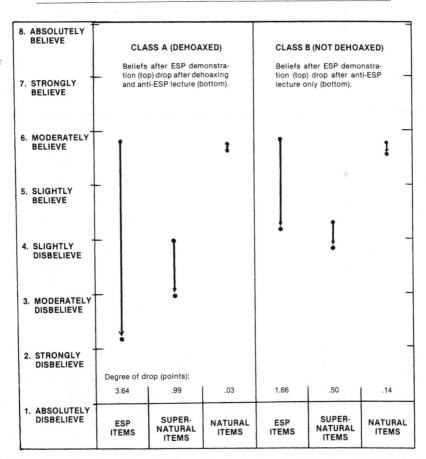

FIGURE 3: **Effect of dehoaxing.** Being told that one has been fooled is critical. In dehoaxed students, belief in ESP dropped 3.64 points. In students who were not dehoaxed, ESP belief dropped significantly less, only 1.66 points.

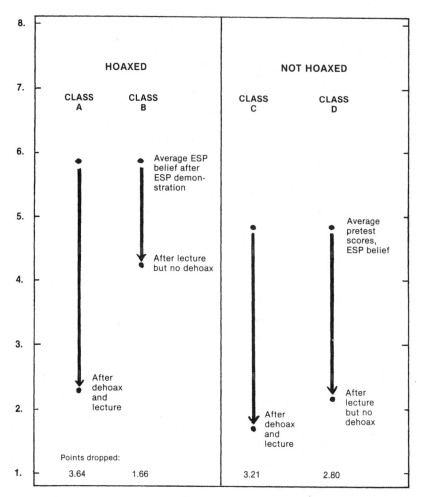

FIGURE 4: Effect of hoaxing. The greatest drop in ESP beliefs occurs after all three experiences—being hoaxed, then dehoaxed, and hearing an anti-ESP lecture. But final belief in ESP is slightly (but not significantly) lower in groups that have never experienced an ESP hoax.

skepticism to supernatural beliefs. Class A students, however, had a much more dramatic drop in ESP belief (3.64 points), as well as a greater generalization of skepticism (Fig. 3).

The dehoaxing experience was apparently crucial: a three-minute revelation that they had been fooled was more powerful than an hour-long denunciation of ESP in producing skepticism toward ESP.

Effect of Hoaxing. To measure the separate effects of the ESP demonstration alone, we repeated this experiment on two more classes, C and D, which also met simultaneously. But these classes never saw Steve's demonstration. The only time they took the survey was after the anti-ESP lecture. The difference was that Class D received the anti-ESP lecture alone, but Class C was "dehoaxed"—I told them how easily another class had been fooled—before the two classes were combined for the anti-ESP lecture.

In Class C students, there was a drop in ESP belief of 3.21 points, almost as great as in students who had been hoaxed.

In Class D students, the anti-ESP lecture alone produced a drop of 2.80 points in ESP belief (Fig. 4). This drop was significantly greater than the 1.66-point drop in Class B students who saw the ESP demonstration. This is understandable. Students in Class B saw the anti-ESP lecture as "equal time for the other side," but it was just an academic argument to them because they still thought they had seen real ESP the previous week.

The amount of "drop" in ESP belief was greater for the hoaxed students than for the nonhoaxed students. This was not terribly surprising, since in hoaxed students (Class A) the "drop" was measured from the inflated ESP belief taken just after the demonstration, whereas in non-hoaxed students (Class C) the "drop" was measured from the everyday pretest level. We did not find, as we had hoped, that the experience of knowing one has been duped produces more skepticism than one would have without the experience. The absolute levels of belief in ESP after it was all over were, in fact, slightly lower for the students who had not been duped, though the difference was not statistically significant. Our belief scale may not discriminate well at the low end. Also, there is the unanswered question of how long the effects last. We were unable to do follow-up studies. It may be that hoaxed students remember the lesson longer, but we don't know.

Anecdotal evidence suggests this may be the case, that nothing can quite undo the psychological effect of believing that one has seen ESP before his very eyes. E. J. Dingwall, a member of the Society for Psychical Research, in Britain, thinks everyone should go the route of first believing in something paranormal and then being disabused of it. He says that, as a young man, for three days he actually believed that one spirit medium had powers to communicate with the departed, and then he bothered to look under the table and found out how she was doing it. The revelation, he said, was so strong it lasted him for the rest of his life.

Our exercise that started as a lesson in scientific method aroused so many theoretical questions about the nature of belief that we indulged our curiosities for almost two years on several hundred Introductory Psychology students. Did they learn anything more than the fact that their instructor was a very curious fellow? Did they, the next time they saw a newspaper account of the paranormal, examine it with caution, reason, and determination not to be fooled again? I hope so. I sincerely hope so.   •

Superstitions: Old and New

William Sims Bainbridge and Rodney Stark

Ours is an age of science, yet it also is an age of superstition. Continuing vehement opposition to the theory of evolution by fundamentalists has aroused consternation within the scientific community. New cults promote modern forms of magic, and belief in paranormal phenomena is widespread. Rationality seems assaulted on the one side by the ancient myths of traditional religion and on the other by pseudoscience and novel occultisms. In this paper, we show the connection between the old superstitions and the new. We will find that our society is faced not with a choice between rationality and mysticism but, at best, with a choice of which style of supernaturalism science must endure.

The analyses that follow will use several sets of sociological data to evaluate the effect of traditional religion on attitudes toward supernatural and paranormal notions. We will demonstrate three specific points:

1. Among the general public, there is a strong tendency for members of conservative and fundamentalist religious groups to reject Darwin's theory of evolution.

2. There is a strong tendency for "born again" Christians to reject several occult and pseudoscientific notions.

3. Cults and individual occult activity tend to be most common where traditional churches are the weakest.

The first of these three points is no surprise, but the relationship between religion and rejection of Darwin deserves to be documented with precision. The second and third points show that the influence of traditional religion is complex, in some ways opposing science but in other ways defending rationality against several modern superstitions. Of all traditional religious beliefs, those antagonistic to the theory of evolution are most clearly antiscientific, so we will begin by examining them.

Surveys of Attitudes Toward Darwin

Many writers have described the fundamentalist campaign against the teaching of biological evolution in the schools, focusing on hard-core groups of creationists who constitute a small, though active, minority. We will show that there is vast opposition to Darwin by masses of ordinary citizens. While antagonism to other parts of science may also be felt by fundamentalists, it is the theory of evolution that receives the most concerted attack. Undoubtedly, this reflects the head-on competition between Darwin and Genesis. We must *not* assume, however, that fundamentalists reject science as a whole any more vehemently than do other groups.

One of the first really good surveys of religious beliefs was a study of Northern California church members carried out by Charles Y. Glock and Rodney Stark in 1963. A very long questionnaire was completed by 3,000 persons who belonged to a variety of denominations and religious sects. Recently, we have reanalyzed this massive data set (Bainbridge and Stark, forthcoming d) and found that even after 16 years it continues to reveal new insights. One of the questions asked was: "Do you tend to agree or disagree with Darwin's theory of evolution—which maintains that human beings evolved from lower forms of animal life over many millions of years?" Table 1 shows the percentages in each of a number of groups that completely rejected Darwin.

Table 1 is based on responses from 2,871 white church-members, and it shows a consistent pattern. Only 11 percent of those in liberal Protestant denominations rejected Darwin, and 29 percent in moderate denominations. Two denominations have many qualities of the more radical sects, the Missouri Synod Lutherans and the Southern Baptists. About two-thirds of their members say the theory could not possibly be true. Five other Protestant sects show high levels of rejection, all the way up to 94 percent of the Seventh Day Adventists.

As we scan Table 1 from the most liberal denominations to the highly conservative fundamentalist sects, we see steeply rising opposition to Darwin. Finally, the table compares the entire Protestant group with Catholics. The Protestant total includes 20 members of two sects that were too tiny for reliable separate statistics—the Four-square Gospel Church and the Gospel Lighthouse. Overall, the Protestants and Catholics show almost exactly the same levels of rejection, which suggests that a significant fundamentalist minority lies hidden within Catholicism.

TABLE I

Rejection of the Theory of Evolution in the 1963
Survey of 2,871 Northern California Church Members

Church Membership	Percent who say that the theory could not possibly be true
Protestant Denominations	
1,032 liberal Protestants (Congregationalists, Methodists, Episcopalians, Disciples of Christ	11
844 moderate Protestants (Presbyterians, American Lutherans, American Baptists)	29
Sectlike Protestant Denominations	
116 Missouri Synod Lutherans	64
79 Southern Baptists	72
Protestant Sects	
44 Church of God	57
37 Church of Christ	78
75 Nazarenes	80
44 Assemblies of God	91
35 Seventh Day Adventists	94
Totals	
2,326 Protestants	30
545 Catholics	28

Several religious groups were not included in the Northern California survey. Sociologist Armand Mauss has provided us with data from his study of 958 Mormon residents of Salt Lake City and 296 Mormons living in San Francisco. His questionnaire included the same question about Darwin's theory used by Glock and Stark that was reported in Table 1. In Salt Lake City, where the Mormon church is the dominant religious group, 56 percent feel that Darwin's theory could not possibly be true. In San Francisco, where the Mormon church is a small minority, one finds a greater variety of people who are or were members. Mauss surveyed 106 people who had left the Mormon church after having been raised in it, and only 13 percent of these defectors completely rejected Darwin's theory. There were also 67 persons who recently had been converted to the church, and just 28 percent of these new Mormons rejected Darwin. But of 122 lifelong Mormons, 59 percent felt the theory could not possibly be true. Thus, of people who have been committed Mormons for some time, who have long received Mormon teaching and continue to accept it, 56 percent or more reject Darwin absolutely.

TABLE 2

Agreement with the Theory of Evolution in the 1973
Survey of 1,000 San Francisco Area Residents

Religious Affiliation	Percent who agree that "Man evolved from lower animals"
Conservative Protestants (Missouri Lutherans, Southern Baptists, members of sects)	10
Moderate Protestants (Presbyterians, American Lutherans, American Baptists)	43
Liberal Protestants (Congregationalists, Methodists, Episcopalians, Disciples of Christ)	53
All Protestants	39
Catholics	38
Those with *no* religious preference	74
Agnostics and atheists	81

Note: The original study used a quota sample technique, so these figures are based on data adjusted to reflect the general population of the area.

Since the Glock and Stark study was a survey of church members, it did not include atheists, agnostics, or persons who were indifferent to religion. Recent research, also carried out in Northern California, updates the 1963 findings and extends them to citizens who are not church members. In 1973, a team of sociologists surveyed 1,000 residents of the San Francisco area (Wuthnow 1976). As we shall see later, the West Coast is the least religious region of the country, containing many who have only the slightest contact with traditional Christian influences. One question in the 1973 survey asked whether respondents agreed that "man evolved from lower animals." Table 2 shows the percentage accepting Darwin in each of several groups.

Only 10 percent of conservative Protestants agree with this simple statement of the theory of evolution. Fully 94 percent of this group accept the alternative view that "God created the first man and woman." Presumably, the 4 percent of conservatives who accepted *both* theories are especially agreeable people. Acceptance of Darwin rises sharply to 43 percent among the moderate Protestants and to 53 percent among liberal Protestants. Again, total figures for Protestants are almost exactly the same as for Catholics. Darwin is accepted by 74 percent of those with no religious preference and by 81 percent of agnostics and atheists.

The results in Table 1 and Table 2 should not be overinterpreted. It would be a mistake to conclude that fundamentalists oppose all science. Here, they have registered opposition to but a single theory, one that directly contradicts the Bible. But it would be an equally great mistake to conclude that religious liberals and the irreligious possess superior minds of great rationality, to see them as modern personalities who have no need of the supernatural or any propensity to believe unscientific superstitions. On the contrary, we shall see that *they are much more likely to accept the new superstitions.* It is the fundamentalists who appear most virtuous according to scientific standards when we examine the cults and pseudosciences proliferating in our society today.

The University of Washington Study

In 1979 we administered a long questionnaire to 1,439 students at the University of Washington in Seattle (Bainbridge and Stark, forthcoming b, c). Response to one question taken from a Gallup poll was particularly revealing: "Would you say that you have been 'born again' or have had a 'born again' experience—that is, a turning point in your life when you committed yourself to Christ?" In 1977, Gallup found that 34 percent of American adults claimed to be "born again," an indication that conservative Christianity remains strong in our society. One might think that undergraduate students in our university would be modern, secular persons. But 26 percent said they were "born again." Of course, some of these students may have been lapsed Christians who had abandoned conservative religion after having been "born again" some time in the past. Therefore, in our analysis we focus on the 245 students who said they had had the "born again" experience and also said their religious beliefs were currently "very important" to them.*

After removing all the "born agains," we were left with 251 ordinary Catholics and 319 ordinary Protestants. Another 241 respondents claimed to have no religious affiliation. We will not consider the Jewish students here, nor the members of numerous "other" groups, because they represent very small minorities in this sample.

Before analyzing occult and pseudoscientific opinions, we need to understand something more about the ordinary Catholics and Protestants. These young people are not necessarily members of churches. For many of them, religion is an unimportant detail of their family backgrounds. One indicator of this is the fact that many Protestants could not spell the names of their denominations! Several claimed to be "Lutherns," while others were "Piscables" or "Presbitarians." Perhaps the best quantitative measure of real religious involvement is church attendance. Not surprisingly, zero percent of those with no religion attend church every week. Among ordinary Protestants, only 5 percent attend weekly, as do 25 percent of ordinary Catholics. Among the 245 "born agains," 68 percent attend church at least once a week. Thus, while the "born agains" do appear to be a profoundly religious group, the ordinary Protestants and Catholics include many whose religion is only nominal and may have no influence on their lives.

We included an evolution question in our student survey, the one used in the 1973 San Francisco area study. Only 20 percent of the "born again" university students agreed that "man evolved from lower animals," showing that conservative religious opposition to Darwin is strong even among the student body of a leading secular university. Almost equal percentages of ordinary Catholics and ordinary Protestants agree with the

---

*Several earlier research studies, including some by each author of this article, failed to find a significant impact of religion on patterns of deviant behavior. We now realize that those studies simply did not include enough highly committed members of cohesive fundamentalist groups to show the results reported here. As this article demonstrates, data covering the entire nation do reveal a strong religious influence against occultism. Research performed in the Pacific region, where religion is very weak, might miss this effect unless it includes individuals, like "born agains," who are involved in an intense religious social movement that counteracts the regional irreligiousness. In the parts of the country where religion is strong, even members of moderate denominations may be protected from occultism by their religion.

statement, 74 percent and 76 percent, respectively. The highest level of acceptance, 93 percent, is found, as expected, among those with no religion. Before we leap to the false conclusion that the students with no religion are rational, scientific secularists, we should look at Table 3, which reports the distribution of responses toward some recently popular superstitions.

TABLE 3

Attitudes Toward the Occult in the 1979 Survey of
University of Washington Undergraduates

Percent Giving the Indicated Response

Questionnaire Item	241 with no religion	Ordinary Christians 251 Catholics	319 Protestants	245 Born Again Christians
Agrees: "Some Eastern practices, such as Yoga, Zen, and Transcendental Meditation, are probably of great value"	73	66	60	28
Agrees: "UFOs are probably real spaceships from other worlds"	67	66	60	43
Agrees: "Some occult practices, such as Tarot reading, seances, and psychic healing, are probably of great value"	16	22	12	6
Agrees: "I myself have had an experience that I thought might be an example of extra sensory perception"	59	57	55	44
Respondent thinks that ESP "definitely exists"	26	29	17	17
Respondent very strongly dislikes "occult literature"	38	34	36	65
Respondent very strongly dislikes "Your Horoscope"	29	24	23	53

The first four items in Table 3 come from a list of statements to which each of the students was supposed to respond by checking one of four boxes: strongly agree, agree, disagree, strongly disagree. We did not give students the opportunity to say they were "not sure" or "undecided" or had "no opinion." Among those who checked the "agree" box are undoubtedly many who were just barely willing to give the particular statement the benefit of the doubt. Therefore, it may not be valid to take the figures as exact measures of *actual belief* in each statement, but it is valid to compare *levels of support* for each idea.

The first statement refers to Yoga, Zen, and Transcendental Meditation, practices alien to Western traditions but deeply rooted in Eastern religion. Seventy-three percent of those with no religious affiliation are willing to say that these oriental cults are probably of great value; 66 percent of the ordinary Catholics and 60 percent of ordinary Protestants agree. Thus, those with no religion or only nominal religion are especially likely to accept deviant, exotic alternatives to Christianity, just as they are likely to accept Darwin. Apparently, they are open to new ideas of many kinds, rather than accepting Darwin because they are well-informed secular rationalists. Among the "born again" Christians, only 28 percent express positive evaluation of the oriental cults.

The statement that UFOs are probably real spaceships also receives much stronger support from those with no religion or only nominal religious affiliation. The percentage of "born agains" accepting the statement is higher this time, perhaps reflecting the fact that belief in UFOs does not necessarily have religious implications. But only a minority of "born agains" think UFOs might be spaceships, compared with clear majorities in the three other groups.

Deviant magical practices, such as Tarot reading, séances, and psychic healing, are very unpopular. But even at this extreme, "born agains" show by far the lowest level of acceptance.

One might predict that religious people are more likely than the irreligious to believe in extrasensory perception. For one thing, people who believe humans have souls that transcend the material world already share some of the assumptions behind belief in ESP. Furthermore, guidance from God, and even prayer to God, would seem to be communications by the spirit rather than by the senses. Two questionnaire items about ESP show only slight differences among the four groups of respondents, but in both cases those with no religion are significantly more likely than "born agains" to favor the idea of extrasensory perception.

Part of the questionnaire consisted of a long list of things (ranging from foods to movies to other consumables and mass media) and asked students to indicate how much they liked each one on a seven-point scale from "0" (do not like) to "6" (like very much). Table 3 concludes by comparing the percentages who gave zero ratings to "occult literature" and to "Your Horoscope." In both cases, "born agains" were nearly twice as likely as other students to reject these forms of occultism.

As a whole, Table 3 shows two very interesting things. First, it reveals that "born agains" are much less likely than other students to accept radical cults and pseudoscientific beliefs. Second, it reveals that the group with no religious affiliation is receptive to these unscientific notions. On three of the seven items, in fact, those with no religion are the most favorable toward occultism. Those who want to blame the fundamentalists for their opposition to Darwin ought to praise them for their responses reported in Table 3. Those who hope that a decline in traditional religion would inaugurate a new Age of Reason ought to think again. Having seen results from three large questionnaire surveys, we will now examine the relationship between religion and the occult through geographic analysis.

Geography of the Supernatural

Working from an incomplete census of religious organizations carried out in 1971, Stark (forthcoming) has been able to calculate good estimates of the total number of church members in every state and many major cities. For our present purposes, the raw figures are not very revealing, so we have calculated *rates* of church membership per thousand residents in each geographic area. Figure 1 gives these rates for the nine regions of the United States, defined by the Bureau of the Census. Nationally, there were about 115 million church members in a total population of 205 million, for a rate of 560 church members per thousand. As Figure 1 shows, the nine regions vary considerably in the proportions of citizens who formally belong to a church.

Seven of the nine regions have church-member rates at or above the national average. In the West South Central region, nearly two-thirds of the residents are members of a church. The remaining third are not atheists, of course, but mostly unchurched Christians who profess religious belief but have no continuing participation in a religious group. The vast, but thinly settled, Mountain region has a rate of 530 per thousand, slightly below the national rate of 560. But the main feature of this map is the Pacific region, which includes Alaska and Hawaii, as well as California, Oregon, and Washington. Here, the church-member rate is 360 per thousand, 200 less than the national rate. Again, this does not mean that the majority in the Pacific region are atheists but that the majority are unchurched Christians. But whatever people's private religious faith, the rates shown in Figure 1 do measure the *strength of organized religion*.

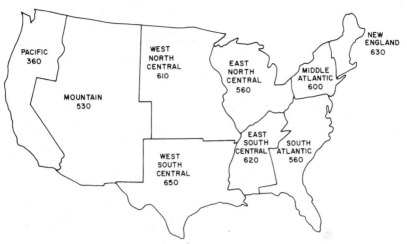

FIGURE 1. Church membership across the nation in members per thousand residents

This is not the place to explain all the religious trends of recent years, but we should point out that the Pacific rates do not necessarily reflect any greater rationality in the West. We think the weakness of conventional religion in the Pacific region may be due to nothing more subtle than the migration patterns in this area. Preliminary research by our associate Kevin Welch suggests that geographic mobility (moving from one home to another) often tears people away from one church community without placing them in contact with a united group of new neighbors who will invite them to join a specific conventional church near their new home. Regions with high rates of migration will therefore show low rates of church membership. We are still engaged in research on the causes of defection from the conventional churches; but certainly much of it flows from sources other than personal convictions opposed to religion, and many among the unchurched want contact with the supernatural.

Our questionnaire research suggests that strong religion prevents occultism. Therefore, we would expect to find that interest in deviant cults and in the paranormal was greatest where the churches are weakest—in the Pacific region. In fact, this is the case. Perhaps everyone interested in cults already knows that Southern California is a hotbed of them. But our research has shown that cults and occultism are strong all the way up the coast, through Oregon, Washington, British Columbia, to Alaska (Stark, Bainbridge, and Doyle 1979; Bainbridge and Stark, forthcoming a). Let us look at some of the evidence.

Recently, we have developed several measures of cult activity that can be analyzed geographically. One such is a list of 501 headquarters of cults drawn from J. Gordon Melton's monumental *Encyclopedia of American Religions* (1978). The cults are of many kinds, including New Thought, Theosophist, Spiritualist, psychedelic, psychic, pagan, witchcraft, satanic, flying saucer, occult, and communal groups. California is the headquarters for 167 cults, a third of the total, proving its reputation as cult capital of the nation. But less populous states in the Pacific and Mountain regions also have more than their share of cults. The 167 in California yield a rate of 7.9 per million population. Sparsely inhabited Nevada, with only 6 cults, achieves a rate of 10 per million. New Mexico has a rate of 9.1 per million. Colorado has 6.0, Arizona 5.9, and Oregon 4.8. There are 192 cult headquarters in the Pacific region, for a rate of 6.6 per million, nearly three times the national rate of 2.3. Figure 2 shows cult rates for the nine geographic regions.

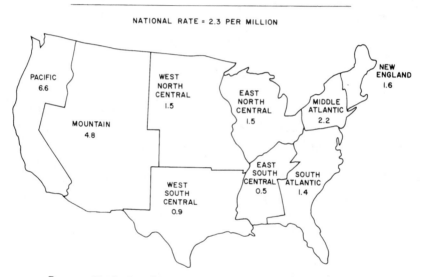

NATIONAL RATE = 2.3 PER MILLION

FIGURE 2. Distribution of 501 cult headquarters in cults per million residents

Figure 2 demonstrates that cult headquarters do indeed cluster where the conventional churches are the weakest—on the Pacific shore. Apparently, when Christianity loses its grip on large numbers of people, deviant religious alternatives arise and get hold of some of the unchurched. There are two ways in which this process of transformation takes place. First, strong traditional churches can use their influence in the community to punish involvement in deviant supernatural activities; but when churches lose their power, people are liberated to follow their religious enthusiasms. Second, religion serves important social and psychological functions for many people and, when the conventional churches fail in this task, people will seek satisfaction from other sources. This means not only

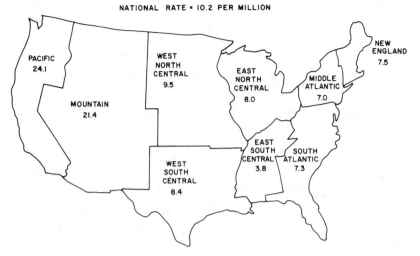

FIGURE 3. Distribution of 2086 Fate Magazine readers' stories given in stories per million residents

that many will join formal cult movements but also that others will seek personal contact with the supernatural realm. An excellent indicator of this latter course is found in the pages of *Fate* magazine.

Fate is a central medium of communication for the American occult, reaching far more than 100,000 readers with each monthly issue. Among the standard features are columns of readers' stories, submitted for token rewards, supposedly proving the existence of supernatural events and beings. Founded at the beginning of 1948, *Fate* quickly began the first of these, called "True Mystic Experiences," a potpourri of paranormal reports. In 1954, *Fate* added "My Proof of Survival," stories of apparitions, near-death experiences, and messages from the beyond. From January 1960 through September 1979, *Fate* published a total of 2,086 of both types by authors with American addresses. Figure 3 shows the distribution of the stories per million population for each of the nine regions.

The rates shown in Figure 3 are higher than those in Figure 2, simply because it tabulates more than four times as many items. But the shape of the distribution is almost identical. The Pacific region is highest. The Mountain region comes second. Other regions fall near the national average, except for the very low East South Central region. There is no direct connection between these two measures of cult activity. We have no reason to believe that any significant number of the 2,086 stories were written by members of the 501 cults. Rather, we think they are individual testimonials in favor of occultism, both permitted and stimulated by weakness in conventional religion.

Conclusion

Traditional religion is not simply the enemy of rationality and science, but it plays an ambivalent role. True, fundamentalists show high levels of rejection of the theory of evolution. But they also reject a wide range of occult or pseudoscientific ideas that may threaten the progress of human culture. Persons with no religious affiliation are often among the first to toy with novel or exotic supernatural notions and are not the secular rationalists we might want to think them. Cults flourish precisely where the conventional churches are weakest, in the western parts of the country. Here, too, numbers of unchurched people seek private contact with the supernatural, as shown in the distribution of *Fate* magazine "mystic experiences" and "proofs of survival." Therefore, a further decline in the influence of conventional religion may not inaugurate a scientific Age of Reason but might instead open the floodgates for a bizarre new Age of Superstition.

References

Bainbridge, William Sims
    1978   *Satan's Power: Ethnography of a Deviant Psychotherpy Cult.* Berkeley: University of California Press.
    1979   "In Search of Delusion." SKEPTICAL INQUIRER 4 (1): 33-39.
Bainbridge, William Sims, and Rodney Stark
    1979   "Cult Formation: Three Compatible Models." *Sociological Analysis,* in press.
    1980   "Scientology: To Be Perfectly Clear." *Sociological Analysis,* in press.
    Forthcoming a   "Client and Audience Cults in America."
    Forthcoming b   "The 'Consciousness Reformation' Reconsidered."
    Forthcoming c   "Friendship, Religion, and the Occult."
    Forthcoming d   "Sectarian Tension." *Review of Religious Research,* in press.
Glock, Charles Y., and Rodney Stark
    1966   *Christian Beliefs and Anti-Semitism.* New York: Harper.
Godfrey, Laurie R.
    1979   "Science and Evolution in the Public Eye." SKEPTICAL INQUIRER 4 (1): 21-32.
Stark, Rodney
    Forthcoming   "Estimating Church Membership Rates for Ecological Areas."
Stark, Rodney, and William Sims Bainbridge
    1979   "Of Churches, Sects, and Cults: Preliminary Concepts for a Theory of Religious Movements." *Journal for the Scientific Study of Religion,* in press.

1980   "Networks and Faith: Interpersonal Bonds and Recruitment to Cults and Sects." *American Journal of Sociology,* in press.

Forthcoming a   "Sect Transformation and Upward Mobility: The Missing Mechanisms."

Forthcoming b   "Towards a Theory of Religion: Religious Commitment." *Journal for the Scientific Study of Religion,* in press.

Stark, Rodney, William Sims Bainbridge, and Daniel P. Doyle

1979   "Cults of America: A Reconnaissance in Space and Time." *Sociological Analysis,* in press.

Wuthnow, Robert

1976   *The Consciousness Reformation.* Berkeley: University of California Press.

Einstein and ESP

Martin Gardner

Einstein is frequently mentioned in the literature of parapsychology as a great scientist who, in contrast to so many of his colleagues, believed that psi phenomena had been demonstrated by the work of J. B. Rhine and his successors. In 1930 when Upton Sinclair published his book *Mental Radio,* Einstein contributed a brief preface to the German edition. In the American edition, the preface reads as follows:

> I have read the book of Upton Sinclair with great interest and am convinced that the same deserves the most earnest consideration, not only of the laity, but also of the psychologists by profession. The results of the telepathic experiments carefully and plainly set forth in this book stand surely far beyond those which a nature investigator holds to be thinkable. On the other hand, it is out of the question in the case of so conscientious an observer and writer as Upton Sinclair that he is carrying on a conscious deception of the reading world; his good faith and dependability are not to be doubted. So if somehow the facts here set forth rest not upon telepathy, but upon some unconscious hypnotic influence from person to person, this also would be of high psychological interest. In no case should the psychologically interested circles pass over this book heedlessly.

Parapsychologists, and journalists who write about the paranormal, often refer to this preface, sometimes quoting from it, as evidence of Einstein's belief in ESP. R. A. McConnell, for example, in his influential article, "Parapsychology and Physicists" (*Journal of Parapsychology,* Vol. 40, September 1976), lists Einstein, along with William Crookes, Oliver Lodge, and other physicists, as one of the "Titans" who were sympathetic toward psi research. A portion of Einstein's preface is quoted.

An even longer quotation appears in chapter 7 of *Mind-Reach*, the recently published book by Russell Targ and Harold Puthoff. The chapter is about their work with Uri Geller—tests which they are convinced demonstrated beyond any doubt the clairvoyant powers of the Israeli psychic. To put their Geller experiments in perspective, and to argue that Geller's ability is not unique, they bring up Sinclair's book, quote from Einstein's preface, and ask, "Why then has this treasure trove of a book been neglected for the past forty-five years?"

It is not my purpose to explain here why I do not believe Sinclair's book should be taken seriously, because I have summarized my reasons in chapter 25 of my 1952 book, *Fads and Fallacies in the Name of Science.* If the reader will consult this book's index for page references to Sinclair, they will understand why I regard him as sincere and honest, but incredibly gullible. He had only the dimmest grasp of scientific method and (in my opinion) was an unreliable observer and reporter of the uncontrolled, informal ESP-tests he conducted with his wife.

My purpose now is merely to reproduce, with the permission of the Einstein estate, a letter that Einstein wrote in 1946 to Jan Ehrenwald, and which came into my hands by way of physicists John Stachel and E. T. Newman. Dr. Ehrenwald is a British psychoanalyst now living in New York City where he is a consulting psychiatrist to Roosevelt Hospital. For thirty years he has been studying psi phenomena and seeking a neurological basis for it. He is the most distinguished of a trio of living psychoanalysts (the other two are Jule Eisenbud and Montague Ullman) who are firm believers in psi. Next year Basic Books will publish Ehrenwald's latest book, *The ESP Experience: A Psychiatric Validation.*

A translation of Einstein's letter (the original is in German) follows:

Dear Dr. Ehrenwald: 13 May 1946

I have read with great interest the introduction to your book,[1] as well as the story of all the unpleasant experiences you have suffered, as many others among us have. I am happy that you succeeded in emigrating to this country, and I hope that you will find here the possibilities for fruitful work.

Several years ago I read the book by Dr. Rhine. I have been unable to find an explanation for the facts which he enumerated. I regard it as very strange that the spatial distance between (telepathic) subjects has no relevance to the success of the statistical experiments. This suggests to me a very strong indication that a nonrecognized source of systematic errors may have been involved.

1. Dr. Ehrenwald had sent Einstein a copy of his book, *Telepathy and Medical Psychology* (London: Allen and Unwin, 1946).

I prepared the introduction to Upton Sinclair's book because of my personal friendship with the author, and I did it without revealing my lack of conviction, but also without being dishonest. I admit frankly my skepticism in respect to all such beliefs and theories, a skepticism that is not the result of adequate acquaintance with the relevant experimental facts, but rather a lifelong work in physics. Moreover, I should like to admit, that, in my own life, I have not had any experiences which would throw light on the possibility of communication between human beings that was not based on normal mental processes. I should like to add that, since the public tends to give more weight to any statement from me than is justified, because of my ignorance in so many areas of knowledge, I feel the necessity of exercising utmost caution and restraint in the field under discussion. I should, however, be happy to receive a copy of your publication.

With many regards,
Albert Einstein

It is worthy of note that Einstein's main reason for skepticism is the fact, so often emphasized by Rhine, that reported psi forces do not decline with distance. All of the four known forces of nature—gravity, electromagnetism, the strong force, and the weak force—diminish in strength as they radiate from a source. Rhine has always considered this proof that psi forces lie entirely outside the bounds of known physical laws. In recent years, attempts to explain psi's independence of distance (as well as time!) have been varied—currently fashionable attempts draw on quantum mechanics—but none has been satisfactory or amenable to confirmation. Einstein found it easier to apply Occam's razor and adopt the simpler explanation: namely that some sort of bias, of which experimenters were unaware, entered into the experimental designs of psi experiments and accounted for the statistical results. If so, the failure of psi to decline with distance and time would be easily accounted for.

Einstein mentioned in his preface one possible source of bias in Upton Sinclair's tests. To spell it out: perhaps Mrs. Sinclair unconsciously suggested to her husband what he should draw, or he unconsciously suggested to her what she should draw. To give another instance, consider the possible role of hand-recorded errors in Rhine's early and poorly controlled PK experiments with dice. If bias were introduced by recording errors on the part of assistants who knew the target number (tests by psychologists have shown how common such recording errors are), it obviously would not matter in the least whether the subject was ten feet from the tumbling cubes, or ten miles, or in a submarine ten fathoms

down, or on a spaceship ten thousand miles out. It would not even matter
if the dice were shaken and tossed ten hours after the subject had concen-
trated his PK energy on the target.

This independence of psi from time and space continues to be a
dramatic and troubling aspect of psi research. A splendid example is
provided by the latest remote-viewing tests of Puthoff and Targ. In their
project "deep Quest," psychic superstars Hella Hammid and Ingo Swann
were in a minisub, submerged off the coast off Catalina Island. They
managed to describe the target sites, 500 miles away on land, as accurately
as they had done in previous tests on land when the targets were nearby. In
Mind-Reach the authors report that Ms. Hammid also did just as well in
her remote viewing when the targets were randomly selected *after* she had
made her report.

In his characteristically simple, humble, commonsense fashion, Einstein went directly to the hub of the matter. After a century of reporting of
results by parapsychologists, the indifference of psi to all the rules that
govern known forces continues to be (along with replication failures by
unbelievers) a major reason why the majority of psychologists remain, like
Einstein, highly skeptical of the reported extraordinary results.   •

A Second Einstein ESP Letter

The previous issue of this magazine (Fall/Winter 1977) published my
note on Einstein's attitude toward the brief introduction he wrote for
Upton Sinclair's book *Mental Radio*. The note included a letter that
Einstein had written to the psychoanalyst and parapsychologist Dr. Jan
Ehrenwald.

Dr. Ehrenwald has kindly allowed me to have a copy of a second letter he received from Einstein, which contains further comments on parapsychology. I have obtained permission from the Einstein Estate to
publish the following translation:

8 July 1946

Dear Mr. Ehrenwald:

I have read your book with great interest. It doubtlessly represents a
good way of placing your topic in a contemporary context, and I have no
doubt that it will reach a wide circle of readers. I can judge it merely as a
layman, and cannot say that I have arrived at either an affirmative or negative conclusion. It seems to me, at any rate, that we have no right, from a

physical standpoint, to deny a priori the possibility of telepathy. For that sort of denial the foundations of our science are too unsure and too incomplete.

My impressions concerning the quantitative approach to experiments with cards, and so on, is the following. On the one hand, I have no objection to the method's reliability. But I find it suspicious that "clairvoyance" [tests] yield the same probabilities as "telepathy," and that the distance of the subject from the cards or from the "sender" has no influence on the result. This is, a priori, improbable to the highest degree, consequently the result is doubtful.

Most interesting, and actually of greater interest to me, are the experiments with the mentally retarded nine-year-old girl and the tests by Gilbert Murray. The drawing results seem to me to have more weight than the large scale statistical experiments where the discovery of a small methodological error may upset everything.

I find important your observations that a patient's productivity in psychoanalytic treatment is clearly influenced by the analyst's "school." This portion of your book alone is worth careful attention. I cannot fail to note that some of the experiences you mention arouse the reader's suspicion that unconscious influences along sensory channels, rather than telepathic influences, may be at work.

At any rate, your book has been very stimulating for me, and it has somewhat "softened" my originally quite negative attitude toward the whole of this complex of questions. One should not walk through the world wearing blinders.

I cannot write an introduction, as I am quite incompetent to do so. It should be provided by an experienced psychologist. You may show this letter privately to others.

Respectfully yours,
(signed) A. Einstein.

The book, which Dr. Ehrenwald had sent to Einstein in the form of page proofs and for which Einstein declined to write an introduction, was *Telepathy and Medical Psychology*. It was published in England by Allen and Unwin in 1947 and in the United States the next year by W. W. Norton. The introduction was written by Gardner Murphy. (Dr. Ehrenwald's latest book, *The ESP Experience*, was published earlier this year by Basic Books.)

Let me add that I find Einstein's remarks entirely admirable. He is less dogmatic in his negative attitude toward parapsychology than he had been when he wrote his previous letter. He believes one should keep an open mind, but he is still strongly put off by the reported evidence that ESP does not decline with distance. With great tact and politeness he informs Dr. Ehrenwald that unconscious but quite normal sensory channels, rather that ESP, may be causing the effects that Dr. Ehrenwald attributes in his book to telepathic contact between analyst and patient. Finally, with characteristic humility, he points out that the entire field is one in which he has no competence.

I wish to thank Martin Ebon for providing a translation of Einstein's letter. It has been approved and slightly edited by Dr. Ehrenwald.   ●

Belief in ESP Among Psychologists

Vernon R. Padgett, Victor A. Benassi, and Barry F. Singer

In a recent letter to *Science* magazine, Charles Tart (1980) pointed out that 65 percent of American college professors think ESP is "a likely possibility" or "an established fact," according to a survey by Wagner and Monnet (1979). But *which* professors believe this about ESP?

Psychologists ranked lowest among academic disciplines on an extraordinary-belief scale, according to Alcock (1975). Evans (1973) reported that psychologists tend to be "frightfully goatish." Finally, the Wagner and Monnet survey gives the following breakdown by academic discipline of those giving a positive response to the question "Do you consider ESP a likely possibility?" or "Do you consider ESP an established fact?":

Humanities, arts and education	73-79%
Social science, excluding psychologists	66%
Natural science	55%
Psychologists	34%

Why do psychologists rank lowest? It may be, first, because psychologists are most familiar with the experimental literature, which in over a half century has failed to demonstrate the existence of ESP, as even a noted parapsychologist admits (Beloff 1978); second, because psychologists tend to know about factors influencing belief formation; and, third, because many psychologists, including these authors, have themselves caused college students to believe in the existence of several "psychic phenomena" simply by performing sleight-of-hand magic tricks such as those described in Randi's *Magic of Uri Geller* (1975).

Following Tart's argument that those who are most educated are those who set the norms in society, it is interesting to note that those who should know most about ESP are those who believe in it least.

Why do the majority of people, as Tart points out, believe in ESP? One reason might be that their sources of information are poor. This appears to be borne out by the Wagner and Monnet survey: newspapers are given as the primary source of information on ESP. Research done by Benassi and Singer provides a second example. College students were asked to list their sources of information on ESP. "Scientific media" was frequently cited. When asked what these were, the *Reader's Digest, National Enquirer,* and von Daniken's *Chariots of the Gods?* were the sources most frequently named.

Another reason for belief in ESP is the heuristic biases that limit our ability to process information accurately (Tversky and Kahneman 1973; Nisbett and Ross 1980). These may account for a substantial amount of our occult belief.

The existence of ESP may someday be demonstrated—but it will be with independently replicable laboratory experiments, not by polls of college professors. It may be "normal" to believe in ESP, but what is "normal" is not necessarily true.

References

Alcock, J. 1975. "Some Correlates of Extraordinary Belief." Paper read at the 36th Annual Meeting of the Canadian Psychological Association, Quebec:

Beloff, J. 1978. "Why Parapsychology Is Still on Trial." *Human Nature* 1 (12):68-74.

Evans, Christopher 1973. "Parapsychology: What the Questionnaires Revealed." *New Scientist* 57:209.

Nisbett, Richard, and Lee Ross 1980. *Human Inference: Strategies and Short-comings of Social Judgments.* Englewood Cliffs: Prentice-Hall.

Randi, James 1975. *The Magic of Uri Geller.* New York: Ballantine.

Tart, Charles 1980. "Is the Paranormal 'Normal'?" *Science* 207:712.

Tversky, Amos, and Daniel Kahneman 1974. "Judgment under Uncertainty: Heuristics and Biases." *Science* 185:1124-1131.

Wagner, M. W., and M. Monnet 1979. "Attitudes in College Professors toward Extrasensory Perception." *Zetetic Scholar* 5:7-16. ●

Do Fairies Exist?

Robert Sheaffer

Do fairies exist? At first the question seems too absurd to ask. Of course not! So deeply-ingrained is our cultural prejudice against fairies that we use the phrase "fairy tale" to refer to a story we deem to be totally unworthy of belief.

But should this be so? After all, in recent years many persons carrying the highest academic and professional credentials have publicly stated their belief in the reality of such mysterious alleged phenomena as UFOs, ESP, and metaphysical spoon-bending. Are fairies any more absurd than these?

If you say "Yes!" it is probably only because you, like so many others, are unaware of the large body of reliable eyewitness testimony, *from our own Twentieth Century,* of persons who report seeing creatures, often complete with wings, Pan's pipes, and ruffled Elizabethan collars cavorting among the flowers. We even possess *actual photographs* of tiny winged-creatures, dressed in flowing robes, dancing in the forest, and gathering hare-bells. These photos have been subjected to the most careful expert scrutiny and have been pronounced to be authentic. *The evidence for fairies is of the very same kind* as that which has persuaded so many sober and learned persons of the reality of UFOs and similar phenomena. Hence we scientific UFO researchers must take reports of fairy sightings very seriously, indeed.

Like sightings of UFOs, fairy sightings appear to occur in waves. Many waves have peaked in bygone centuries, when no accurate records of sightings were kept, and hence little further investigation is possible. But many people seem unaware that a major "flap" of fairy sightings occurred as recently as 1917-1921. Fortunately for present-day researchers, these sightings were painstakingly researched and documented by a learned man of the highest caliber: none other than the celebrated Sir Arthur Conan Doyle.

Everyone is familiar with Doyle as the author of the perennially popular stories about Sherlock Holmes, the master detective. Not as well known today is Doyle's real-life role as a scientific investigator of the paranormal, especially in Spiritualism and related fields. Present-day scientific researchers of the paranormal, who follow in Doyle's footsteps (especially in his uncanny ability to sniff out hoaxes and deceptions), owe him an unacknowledged debt of gratitude.

Doyle set up, in essence, a "clearing house" for fairy sightings, where individuals could report their experiences without fear of ridicule. Names were never released without the witness's permission. Over a period of several years he compiled an impressive list of firsthand close-encounter fairy sightings, from entirely reliable witnesses. Here are some examples.

The Reverend Arnold J. Holmes of the Isle of Man reports the following close-encounter sighting, which occurred on a deserted road one night: "My horse—a spirited one—suddenly stopped dead, and looking ahead I saw . . . what appeared to be a small army of indistinct figures—very small, clad in gossamer garments. . . . I watched spellbound, my horse half mad with fear . . . one "little man" of larger stature than the rest, about fourteen inches high, stood at attention until all had passed him dancing, singing with happy abandon, across the Valley fields towards St. John's Mount" (Doyle, 1972). In this instance *the horse* reacted to the fairies' presence, proving beyond all doubt that these creatures were not merely illusory. UFO investigators consider reports of animal reactions to UFOs to be a strong indication that a sighting is authentic. We must find it equally convincing when examining the fairy evidence.

Mrs. Rose of Southend-on-Sea reported: "I see them constantly here in the shrubbery by the sea . . . the gnomes are like little old men, with little green caps, and their clothes are generally neutral green. The fairies themselves are in light draperies . . . the fairies appear to be perpetually playing, excepting when they go to rest on the turf or in a tree. . . . I have seen the gnomes arranging a sort of moss bed for the fairies. . . . " Some would dismiss the report of Mrs Rose because she is a "repeater," claiming to have witnessed the phenomenon on more than one occasion. But some of our best UFO evidence was likewise gathered from repeaters. The McMinnville photos of 1950 are a notable example, and the prolific Danish UFO photographer Jorma Viita has sighted UFOs no less than sixteen times in one year. Both of these series of photographs are taken very seriously by scientific UFO investigators. Hence there is no reason to doubt the observations of Mrs. Rose.

Mrs. Ethel Enid Wilson of Worthington said, "I have often seen them on fine sunny days playing in the sea, and riding on the waves, but no one I have ever been with at the time has been able to see them, excepting once my little nephews and nieces." This aspect of subjective visibility is frequently encountered in UFO reports. "They were like little dolls, quite small, with beautiful bright hair, and they were constantly moving and dancing about."

Fairy sightings, like UFO sightings, constitute a truly worldwide phenomenon. Sightings poured in not only from all over England, but from Canada, the United States, and other countries as well. From the Maori districts of New Zealand, Mrs. Hardy reports: "One evening when it was getting dusk I went into the yard to hang the tea-towels on the clothes-line. As I stepped off the verandah, I heard a soft galloping coming from the direction of the orchard . . . suddenly a little figure, riding a tiny pony, rode right under my uplifted arms. I looked round, to see that I was surrounded by eight or ten tiny figures on tiny ponies . . . at the sound of my voice they all rode through the rose trellis across the drive . . . I heard the soft galloping dying away into the distance."

Sir Arthur expresses a certain reservation about the inclusion of little fairy horses in this fairy sighting, although he notes that such horses are also mentioned in other accounts. Always of a skeptical and cautious mind, Doyle carefully observes, "I have convinced myself that there is overwhelming evidence for the fairies, but I have by no means been able to assure myself of these adjuncts."

Are all fairy sightings single-witness cases? Not at all! Mr. Lonsdale, in the company of Mr. Turvey, was sitting perfectly still in the garden of the latter's estate in Bournemouth, England, one warm summer day, when "Suddenly I was conscious of a movement on the edge of the lawn. . . . In a few seconds a dozen or more small people, about two feet in height, in bright clothes and with radiant faces, ran on to the lawn, dancing hither and thither." He whispered to Mr. Turvey, "Do you see them?"; his companion nodded. For four or five minutes these fairies danced about "in sheer joy." One of them grabbed a croquet hoop like a horizontal bar and gaily tumbled round and round. They remained until they were frightened off by a servant bringing tea.

Here we have the firsthand testimony of two entirely credible persons, who report watching fairy revels at close range in full daylight. Clearly this is no misidentification of some ordinary phenomenon! Doyle also chronicles an even more remarkable three-witness sighting, which we will examine later. Testimony such as this would be accepted in any court of law. How long can the skeptics continue to scoff at such overwhelming evidence? We cannot deny the existence of fairies unless we also are willing to throw out all the equally credible reports of phenomena such as UFOs and the Loch Ness Monster. In arguing for the fairies' existence, Doyle observes that "Victorian science would have left the world hard and clean and bare, like a landscape in the moon." How apt a description of those narrow souls we refuse to consider the evidence for UFOs and fairies!

In Doyle's book *The Coming of the Fairies* we find many more such reports. Even if the above accounts were the only ones on record, open-minded scientists would have to take the fairy phenomenon very seriously, indeed. But the most convincing fairy evidence in Doyle's files has not yet been presented: *a series of five photographs was obtained, each one showing one or more winged fairies frolicking in the forest.* The negatives of these photographs have been carefully examined by expert photographers, who agree that they show no signs of trickery whatsoever!

The story of these remarkable photographs begins in the village of Cottingley in Yorkshire, in the summer of 1917. Elsie Wright, sixteen, and her cousin Frances Griffiths, ten, both claimed that they often saw fairies in the Cottingley beck and glen while they were outside playing. Elsie implored her father to let them borrow his newly acquired camera (a type which holds only a single-exposure glass plate). When he agreed to her request, the girls returned later that afternoon with a plate that, when developed, contained the image of Frances standing behind a grouping of five winged, dancing fairies. The fairies, which appear to be about six inches high, are dressed in flowing robes and seem to be dancing in sheer joy, to the music played on a Pan's pipe by the fairy in the center.

A second photograph was obtained two months later, which shows Elsie beckoning onto her knee "the quaintest goblin imaginable." The gnome (its more accurate designation), like the fairies, appears to be less than a foot in height, and it, too, has wings. It is wearing a pointed hat, a ruffled Elizabethan collar, and is carrying a Pan's pipe.

Three years later, in the summer of 1920, the girls obtained three more clear and distinct photographs of fairies. (The elves, goblins, brownies, and nymphs, which the girls also report seeing, were apparently unwilling to permit themselves to be photographed.) The second series of photographs depicts Frances and the leaping fairy; a fairy offering posy of hare-bells to Elsie; and most remarkable of all, the fairies and their sun-bath. This last photograph clearly shows the fairies relaxing in the tall grass, standing in and around a cocoon-like object which is said to be a "magnetic bath," of which the fairies partake in inclement weather. The respected Conan Doyle found the photo of the fairy sun-bath to be the most impressive of all: "Any doubts which had remained in my mind as to honesty were completely overcome, for it was clear that these pictures, specially the one of the fairies in the bush, were altogether beyond the possibility of fake."

Photographic experts echoed Sir Arthur's endorsement. Mr. H. Snelling of Middlesex, a photographer of many years' experience who was himself a specialist in trick photography, examined the negatives and found no evidence of trickery whatsoever. In fact, he declared himself willing to stake his reputation that the photo had not been faked! The photos were also taken to an office of the Kodak corporation in England. The Kodak experts stated that while a skilled photographer *might* be able to fabricate photos such as these, they found no evidence whatsoever to indicate that the fairy photographs were not authentic.

As if the photographic evidence alone presented by Frances and Elsie were not totally overwhelming, they have given verbal testimony, *in conjunction with a third independent witness*, which makes the Cottingley sightings among the most solidly attested observations of *any* paranormal phenomenon. Frances and Elsie, accompanied by a "clairvoyant" whom Doyle does not name (identified elsewhere as Geoffery L. Hodson, a leading Theosophist writer), ventured into the fairy glen on several different evenings. On these occasions *all three of them* reported seeing a wide variety of manifestations of fairy phenomena. Surely three people cannot all share the same hallucination!

Many famous UFO researchers place great emphasis on the psychic aspect of UFOs; for example, see Dr. Jacques Vallee's recent book, *The Invisible College*, in which "The Psychic Component" of the UFO phenomenon is explored. Hence the observations of as gifted a clairvoyant as Mr. Hodson are especially significant.

Here is a summary of their field observations during the month of August 1921:

—*Water Nymph:* "In the beck itself . . . an entirely nude female figure with long fair hair . . . I was not sure whether it had any feet or not. . . . It showed no consciousness of my presence, and, though I waited with the camera in hand in the hope of taking it, it did not detach itself from the surroundings in which it was in some way merged."

—*Wood Elves:* "Two tiny wood elves came racing over the ground past us as we sat on a fallen tree trunk. . . . As Frances came up and sat within a foot of them they withdrew, as if in alarm, a distance of eight feet or so. . . . These two live in the roots of a huge beach tree."

—*A Brownie:* "He is rather taller than the normal, say eight inches . . . bag-shaped cap, almost conical, knee breeches, stockings, thin ankles, and large pointed feet—like gnomes' feet."

—*Fairies:* "There is a tinkling music accompanying all this. . . . Frances sees a small Punch-like figure, with a kind of Welsh hat, doing a kind of dancing by striking its heel on the ground and at the same time raising his hat and bowing. Elsie sees a flower fairy, like a carnation in shape . . . I see what may be described as a fairy fountain about twenty feet ahead. . . . This was also seen by Frances."

—*Goblins:* "A group of goblins came running towards us from the wood to within fifteen feet of us. They differ somewhat from the wood elves. . . ."

"All three of us keep seeing weird creatures as of elemental essence."

It is interesting to note that as of the time of this writing, more than fifty-five years after those remarkable nocturnal sightings in the fairy glen, all three witnesses—Frances, Elsie, and Mr. Hodson—are still living. (Perhaps there is something in the fairy aura that is conducive to longevity? Mr. Hodson is today more than ninety years old and is living in New Zealand. Elsie and Frances are about seventy-five and seventy, respectively.) And all three of them continue to adhere to their story: they insist that the fairy photographs are authentic! As the noted UFO writer Jerome Clark observed in a recent issue of FATE magazine (February 1977): "It now remains for the skeptics to explain (1) how Elsie and Frances could have faked the pictures in the first place and (2) why they continue to insist they did not, long after they have no reason to lie." The skeptics have totally ignored Mr. Clark's ringing challenge. They know when they are beaten.

Multiple independent witnesses. A series of photographs. Worldwide sightings. Close encounter cases. Certainly Conan Doyle did not exaggerate when he described the evidence for fairies as "overwhelming." As he observes, with logic worthy of his detective Sherlock Holmes: "these numerous testimonies come from people who are very solid and successful in the affairs of life. One is a distinguished writer, another an opthalmic authority, a third a successful professional man, a fourth a lady engaged in public service, and so on. To wave aside the evidence of such people on the ground that it does not correspond with our own experience is an act of mental arrogance which no wise man will commit." In short: credible persons reporting incredible things.

Fairies *do* exist, as the evidence most convincingly demonstrates. The truly scientific mind cannot permit itself to reject such solid evidence out of hand, merely because it does not fit in with popularly-accepted patterns of thought. Tiny gossamer-draped fairies and gnomes, often with wings and Pan's pipes, unquestionably *do* exist, as the evidence so unambiguously shows. To deny such evidence is tantamount to denying the existence of UFOs, the Loch Ness Monster, Bigfoot, and Spontaneous Human Combustion, the evidence for which has been so painstakingly gathered by so many forward-thinking researchers over so many years.

The time has come to end more than a half-century of buffoonery. Do we have to give a day in court to the man who says winged fairies are all nonsense? Hell! One could expend all one's energy confronting the skeptics. Why waste time on people who have not bothered to learn the basic facts? It's their problem!

To remedy this situation, I hereby announce the formation of an organization which is long overdue: APRON, the Anomalous Phantasmagorical Research and Observation Network. This is a loose association of skilled field investigators and Tarot card readers, in which dedicated professionals will pool their talents to give reliably-witnessed fairy encounters the sensible scientific scrutiny they have so long been denied.

Among the principal activities of APRON will be public relations, compiling a computer catalog of close encounters of the third kind, public relations, publication of a newsletter to disseminate the latest scientific fairy research findings, public relations, computerized edge enhancement and profile analysis of fairy photographs, as well as the usual fund-raising and promotional activities, and public relations.

As president and founder of APRON, as well as the chief gossamer guru, I will be kept quite busy making radio and TV appearances, granting newspaper interviews, attending scientific fairy conferences, and presenting my research findings to foreign governments as well as to the United Nations. Please do not bother me by sending in any more reports of fairy sightings. I won't have time to read them.

References

For further fairy research, see:

Doyle, Sir Arthur Conan. 1921. *The Coming of the Fairies.* New York: Samuel Weiser Inc., reprinted 1972.

Gardner, Edward L. 1945. *A Book of Real Fairies: The Cottingley Photographs and their Sequel.* London: Theosophical Publishing House, Ltd., reprinted 1966.

(Both of these books are available from Samuel Weiser, Inc., 734 Broadway, New York, N.Y. 10003.)

The original materials concerning the Cottingley incident, including the photographs, correspondence, and related items, have been donated to the Institute of Dialect and Folk Life Studies, Brotherton Collection, The Brotherton Library, University of Leeds, Leeds LS2 9JT, England. •

Tricks of the Psychic Trade

Cold Reading:
How to Convince Strangers That You Know All About Them

Ray Hyman

Over twenty years ago I taught a course at Harvard University called "Applications of Social Psychology." The sort of applications that I covered were the various ways in which people were manipulated. I invited various manipulators to demonstrate their techniques—pitchmen, encyclopedia salesmen, hypnotists, advertising experts, evangelists, confidence men, and a variety of individuals who dealt with personal problems. The techniques which we discussed, especially those concerned with helping people with their personal problems, seem to involve the client's tendency to find more meaning in any situation than is actually there. Students readily accepted this explanation when it was pointed out to them. But I did not feel that they fully realized just how pervasive and powerful this human tendency to make sense out of nonsense really is.

Consequently, in 1955 I wrote a paper entitled "The Psychological Reading: An Infallible Technique For Winning Admiration and Popularity." Over the years I have distributed copies of this paper to my students. The paper begins as follows:

> So you want to be admired? You want people to seek your company, to talk about you, to praise your talents? This manuscript tells you how to satisfy that want. Herein you will find a "sure-fire" gimmick for the achievement of fame and popularity. Just follow the advice that I give you, and, even if you are the most incompetent social bungler, you cannot fail to become the life of the party. What is the secret that underlies this infallible system? The secret, my friend, is a simple and obvious one. It has been tried and proven by practitioners since the beginnings of mankind. Here is the gist of the secret: To be popular with your fellow man, tell him what he wants to hear. He wants to hear about himself. So tell him about himself. But not what you know to be true about him. Oh, no! Never tell him the truth. Rather, tell him *what he would like to be true about himself!* And there you have it. Simple

and obvious, but yet so powerful. This manuscript details the way in which you can exploit this golden rule by assuming the role of a character reader.

I will include essentially the same recipe for character reading in this paper that I give to my students. In addition I will bring the material up to date, describe some relevant research, and indicate some theoretical reasons why the technique "works." My purpose is not to enable you to enhance your personal magnetism, nor is it to increase the number of character readers. I give you these rules for reading character because I want you to experience how the method works. I want you to see what a powerful technique the psychological reading is, how convincing it is to the psychologist and layman alike.

When you see how easy it is to convince a person that you can read his character on sight, you will better appreciate why fortune tellers and psychologists are frequently lulled into placing credence in techniques which have not been validated by acceptable scientific methods. The recent controversy in *The Humanist* magazine and *THE ZETETIC* over the scientific status of astrology probably is irrelevant to the reasons that individuals believe in astrology. Almost without exception the defenders of astrology with whom I have contact do not refer to the evidence relating to the underlying theory. They are convinced of astrology's value because it "works." By this they mean that it supplies them with feedback that "feels right"—that convinces them that the horoscope provides a basis for understanding themselves and ordering their lives. It has personal meaning for them.

Some philosophers distinguish between "persuasion" and "conviction." The distinction is subtle. But for our purposes we can think of subjective experiences that persuade us that something is so and of logical and scientific procedures that convince, or ought to convince, us that something is or is not so. Quite frequently a scientist commits time and resources toward generating scientific evidence for a proposition because he has already been persuaded, on nonscientific grounds, that the proposition is true. Such intuitive persuasion plays an important motivational role in science as well as in the arts. Pathological science and false beliefs come about when such intuitive persuasion overrides or colors the evidence from objective procedures for establishing conviction.

The field of personality assessment has always been plagued by this confusion between persuasion and conviction. In contrast to intelligence and aptitude tests, the scientific validation of personality tests, even under ideal conditions, rarely results in unequivocal or satisfactory results. In fact some of the most widely used personality inventories have repeatedly failed to pass validity checks. One of the reasons for this messy state of affairs is the lack of reliable and objective criteria against which to check the results of an assessment.

But the lack of adequate validation has not prevented the use of, and reliance on, such instruments. Assessment psychologists have always placed more reliance on their instruments than is warranted by the scientific evidence. Both psychologist and client are invariably persuaded by the results that the assessment "works."

This state of affairs, of course, is even more true when we consider divination systems beyond those of the academic and professional psychologist. Every system—be it based on the position of the stars, the pattern of lines in the hand, the shape of the face or skull, the fall of the cards or the dice, the accidents of nature, or the intuitions of a "psychic"— claims its quota of satisfied customers. The client invariably feels satisfied with the results. He is convinced that the reader and the system have penetrated to the core of his "true" self. Such satisfaction on the part of the client also feeds back upon the reader. Even if the reader began his career with little belief in his method, the inevitable reinforcement of persuaded clients increases his confidence in himself and his system. In this way a "vicious circle" is established. The reader and his clients become more and more persuaded that they have hold of a direct pipeline to the "truth."

The state of affairs in which the evaluation of an assessment instrument depends upon the satisfaction of the client is known as "personal validation." Personal validation is, for all practical purposes, the major reason for the persistence of divinatory and assessment procedures. If the client is not persuaded, then the system will not survive. Personal validation, of course, is the basis for the acceptance of more than just assessment instruments. The widespread acceptance of myths about Bigfoot, the Bermuda Triangle, ancient astronauts, ghosts, the validity of meditation and consciousness-raising schemes, and a host of other beliefs is based on persuasion through personal validation rather than scientific conviction.

Cold reading

"Cold reading" is a procedure by which a "reader" is able to persuade a client whom he has never before met that he knows all about the client's personality and problems. At one extreme this can be accomplished by delivering a stock spiel, or "psychological reading," that consists of highly general statements that can fit any individual. A reader who relies on psychological readings will usually have memorized a set of stock spiels. He then can select a reading to deliver that is relatively more appropriate to the general category that the client fits—a young unmarried girl, a senior citizen, and so on. Such an attempt to fit the reading to the client makes the psychological reading a closer approximation to the true cold reading.

The cold reading, at its best, provides the client with a character assessment that is uniquely tailored to fit him or her. The reader begins with the same assumptions that guide the psychological reader who relies on the stock spiel. These assumptions are (1) that we all are basically more alike than different; (2) that our problems are generated by the same major transitions of birth, puberty, work, marriage, children, old age, and death; (3) that, with the exception of curiosity seekers and troublemakers, people come to a character reader because they need someone to listen to their conflicts involving love, money, and health. The cold reader goes beyond these common denominators by gathering as much additional information about the client as possible. Sometimes such information is obtained in advance of the reading. If the reading is through appointment, the reader can use directories and other sources to gather information. When the client enters the consulting room, an assistant can examine the coat left behind (and often the purse as well) for papers, notes, labels, and other such cues about socioeconomic status, and so on. Most cold readers, however, do not need such advance information.

The cold reader basically relies on a good memory and acute observation. The client is carefully studied. The clothing—for example, style, neatness, cost, age—provides a host of cues for helping the reader make shrewd guesses about socioeconomic level, conservatism or extroversion, and other characteristics. The client's physical features—weight, posture, looks, eyes, and hands provide further cues. The hands are especially revealing to the good reader. The manner of speech, use of grammar, gestures, and eye contact are also good sources. To the good reader the huge amount of information coming from an initial sizing-up of the client greatly narrows the possible categories into which he classifies clients. His knowledge of actuarial and statistical data about various subcultures in the population already provides him the basis for making an uncanny and strikingly accurate assessment of the client.

But the skilled reader can go much further in particularizing his reading. He wants to zero in as quickly as possible on the precise problem that is bothering the client. On the basis of his initial assessment he makes some tentative hypotheses. He tests these out by beginning his assessment in general terms, touching upon general categories of problems and watching the reaction of the client. If he is on the wrong track the client's reactions—eye movements, pupillary dilation, other bodily mannerisms—will warn him. When he is on the right track other reactions will tell him so. By watching the client's reactions as he tests out different hypotheses during his spiel, the good reader quickly hits upon what is bothering the customer and begins to adjust the reading to the situation. By this time, the client has usually been persuaded that the reader, by some uncanny means, has gained insights into the client's innermost thoughts. His guard is now down. Often he opens up and actually tells the reader, who is also a good listener, the details of his situation. The reader, after a suitable interval, will usually feed back the information that the client has given him in such a way that the client will be further amazed at how much the reader "knows" about him. Invariably the client leaves the reader without realizing that everything he has been told is simply what he himself has unwittingly revealed to the reader.

The stock spiel

The preceding paragraphs indicate that the cold reader is a highly skilled and talented individual. And this is true. But what is amazing about this area of human assessment is how successfully even an unskilled and incompetent reader can persuade a client that he has fathomed the client's true nature. It is probably a tribute to the creativeness of the human mind that a client can, under the right circumstances, make sense out of almost any reading and manage to fit it to his own unique situation. All that is necessary is that the reader make out a plausible case for why the reading ought to fit. The client will do the rest.

You can achieve a surprisingly high degree of success as a character reader even if you merely use a stock spiel which you give to every client. Sundberg (1955), for example, found that if you deliver the following character sketch to a college male, he will usually accept it as a reasonably accurate description of himself: "You are a person who is very normal in his attitudes, behavior and relationships with people. You get along well without effort. People naturally like you and you are not overly critical of them or yourself. You are neither overly conventional nor overly individualistic. Your prevailing mood is one of optimism and constructive effort, and you are not troubled by periods of depression, psychosomatic illness or nervous symptoms."

Sundberg found that the college female will respond with even more pleasure to the following sketch: "You appear to be a cheerful, well-balanced person. You may have some alternation of happy and unhappy moods, but they are not extreme now. You have few or no problems with your health. You are sociable and mix well with others. You are adaptable to social situations. You tend to be adventurous. Your interests are wide. You are fairly self-confident and usually think clearly."

Sundberg conducted his study over 20 years ago. But the sketches still work well today. Either will tend to work well with both sexes. More recently, several laboratory studies have had excellent success with the following stock spiel (Snyder and Shenkel 1975):

> Some of your aspirations tend to be pretty unrealistic. At times you are extroverted, affable, sociable, while at other times you are introverted, wary and reserved. You have found it unwise to be too frank in revealing yourself to others. You pride yourself on being an independent thinker and do not accept others' opinions without satisfactory proof. You prefer a certain amount of change and variety, and become dissatisfied when hemmed in by restrictions and limitations. At times you have serious doubts as to whether you have made the right decision or done the right thing. Disciplined and controlled on the outside, you tend to be worrisome and insecure on the inside.
>
> Your sexual adjustment has presented some problems for you. While you have some personality weaknesses, you are generally able to compensate for them. You have a great deal of unused capacity which you have not turned to your advantage. You have a tendency to be critical of yourself. You have a strong need for other people to like you and for them to admire you.

Interestingly enough the statements in this stock spiel were first used in 1948 by Bertram Forer (1949) in a classroom demonstration of personal validation. He obtained most of them from a newsstand astrology book. Forer's students, who thought the sketch was uniquely intended for them as a result of a personality test, gave the sketch an average rating of 4.26 on a scale of 0 (poor) to 5 (perfect). As many as 16 out of his 39 students (41 percent) rated it as a perfect fit to their personality. Only five gave it a rating below 4 (the worst being a rating of 2, meaning "average"). Almost 30 years later students give the same sketch an almost identical rating as a unique description of themselves.

The technique in action

The acceptability of the stock spiel depends upon the method and circumstances of its delivery. As we shall later see, laboratory studies have isolated many of the factors that contribute to persuading clients that the sketch is a unique description of themselves. A great deal of the success of the spiel depends upon "setting the stage." The reader tries to persuade the client that the sketch is tailored especially for him or her. The reader also creates the impression that it is based on a reliable and proven assessment procedure. The way the sketch is delivered and dramatized also helps. And many of the rules that I give for the cold reading also apply to the delivery of the stock spiel.

The stock spiel, when properly delivered, can be quite effective. In fact, with the right combination of circumstances the stock spiel is often accepted as a perfect and unique description by the client. But, in general, one can achieve even greater success as a character analyst if one uses the more flexible technique of the cold reader. In this method one plays a sort of detective role in which one takes on the role of a Sherlock Holmes. (See the "Case of the Cardboard Box" for an excellent example of cold reading.) One observes the jewelry, prices the clothing, evaluates the speech mannerisms, and studies the reactions of the subject. Then whatever information these observations provide is pieced together into a character reading which is aimed more specifically at the particular client.

A good illustration of the cold reader in action occurs in a story told by the well-known magician John Mulholland. The incident took place in the 1930s. A young lady in her late twenties or early thirties visited a character reader. She was wearing expensive jewelry, a wedding band, and a black dress of cheap material. The observant reader noted that she was wearing shoes which were currently being advertised for people with foot trouble. (Pause at this point and imagine that you are the reader; see what you would make of these clues.)

By means of just these observations the reader proceeded to amaze his client with his insights. He assumed that this client came to see him, as did most of his female customers, because of a love or financial problem. The black dress and the wedding band led him to reason that her husband had died recently. The expensive jewelry suggested that she had been financially comfortable during marriage, but the cheap dress indicated that her husband's death had left her penniless. The therapeutic shoes signified that she was now standing on her feet more than she was used to, implying that she was working to support herself since her husband's death.

The reader's shrewdness led him to the following conclusion—which turned out to be correct: The lady had met a man who had proposed to her. She wanted to marry the man to end her economic hardship. But she felt guilty about marrying so soon after her husband's death. The reader told her what she had come to hear—that it was all right to marry without further delay.

The rules of the game

Whether you prefer to use the formula reading or to employ the more flexible technique of the cold reader, the following bits of advice will help to contribute to your success as a character reader.

1. *Remember that the key ingredient of a successful character reading is confidence.* If you *look* and *act* as if you believe in what you are doing, you will be able to sell even a bad reading to most of your subjects.

The laboratory studies support this rule. Many readings are accepted as accurate because the statements do fit most people. But even readings that would ordinarily be rejected as inaccurate will be accepted if the reader is viewed as a person with prestige or as someone who knows what he is doing.

One danger of playing the role of reader is that you will persuade yourself that you really are divining true character. This happened to me. I started reading palms when I was in my teens as a way to supplement my income from doing magic and mental shows. When I started I did not believe in palmistry. But I knew that to "sell" it I had to act as if I did. After a few years I became a firm believer in palmistry. One day the late Dr. Stanley Jaks, who was a professional mentalist and a man I respected, tactfully suggested that it would make an interesting experiment if I deliberately gave readings opposite to what the lines indicated. I tried this out with a few clients. To my surprise and horror my readings were just as successful as ever. Ever since then I have been interested in the powerful forces that convince us, reader and client alike, that something is so when it really isn't.

2. *Make creative use of the latest statistical abstracts, polls, and surveys.* This can provide you with a wealth of material about what various subclasses of our society believe, do, want, worry about, and so on. For example if you can ascertain about a client such things as the part of the country he comes from, the size of the city he was brought up in, his

parents' religion and vocations, his educational level and age, you already are in possession of information that should enable you to predict with high probability his voting preferences, his beliefs on many issues, and other traits.

3. *Set the stage for your reading.* Profess a modesty about your talents. Make no excessive claims. This catches your subject off guard. You are not challenging him to a battle of wits. You can read his character; whether he cares to believe you or not is his concern.

4. *Gain his cooperation in advance.* Emphasize that the success of the reading depends as much upon his sincere cooperation as upon your efforts. (After all, you imply, you already have a successful career at reading characters. You are not on trial—he is.) State that due to difficulties of language and communication, *you may not always convey the exact meaning which you intend.* In these cases he is to strive to reinterpret the message in terms of his own vocabulary and life.

You accomplish two invaluable ends with this dodge. You have an alibi in case the reading doesn't click; it's his fault, not yours! And your subject will strive to fit your generalities to his specific life occurrences. Later, when he recalls the reading he will recall it in terms of specifics; thus you gain credit for much more than you actually said.

Of all the pieces of advice this is the most crucial. To the extent that the client is made an active participant in the reading the reading will succeed. The good reader, deliberately or unwittingly, is the one who forces the client to actively search his memory to make sense of the reader's statements.

5. *Use a gimmick such as a crystal ball, tarot cards, or palm reading.* The use of palmistry, say, serves two useful purposes. It lends an air of novelty to the reading; but, more important, it serves as a cover for you to stall and to formulate your next statement. While you are trying to think of something to say next, you are apparently carefully studying a new wrinkle or line in the hand. Holding hands, in addition to any emotional thrills you may give or receive thereby, is another good way of detecting the reactions of the subject to what you are saying (the principle is the same as "muscle reading").

It helps, in the case of palmistry or other gimmicks, to study some manuals so that you know roughly what the various diagnostic signs are supposed to mean. A clever way of using such gimmicks to pin down a client's problem is to use a variant of "Twenty Questions," somewhat like this: Tell the client you have only a limited amount of time for the reading.

You could focus on the heart line, which deals with emotional entanglements; on the fate line, which deals with vocational pursuits and money matters; the head line, which deals with personal problems; the health line, and so on. Ask him or her which one to focus on first. This quickly pins down the major category of problem on the client's mind.

6. *Have a list of stock phrases at the tip of your tongue.* Even if you are doing a cold reading, the liberal sprinkling of stock phrases amidst your regular reading will add body to the reading and will fill in time as you try to formulate more precise characterizations. You can use the statements in the preceding stock spiels as a start. Memorize a few of them before undertaking your initial ventures into character reading. Palmisty, tarot, and other fortune-telling manuals also are rich sources for good phrases.

7. *Keep your eyes open.* Also use your other senses. We have seen how to size up the client on the basis of clothing, jewelry, mannerisms, and speech. Even a crude classification on such a basis can provide sufficient information for a good reading. Watch the impact of your statements upon the subject. Very quickly you will learn when you are "hitting home" and when you are "missing the boat."

8. *Use the technique of "fishing."* This is simply a device for getting the subject to tell you about himself. Then you rephrase what he has told you into a coherent sketch and feed it back to him. One version of fishing is to phrase each statement in the form of a question. Then wait for the subject to reply (or react). If the reaction is positive, then the reader turns the statement into a positive assertion. Often the subject will respond by answering the implied question and then some. Later he will tend to forget that he was the source of your information. By making your statements into questions you also force the subject to search through his memory to retrieve specific instances to fit your general statement.

9. *Learn to be a good listener.* During the course of a reading your client will be bursting to talk about incidents that are brought up. The good reader allows the client to talk at will. On one occasion I observed a tea-leaf reader. The client actually spent 75 percent of the total time talking. Afterward when I questioned the client about the reading she vehemently insisted that she had not uttered a single word during the course of the reading. The client praised the reader for having so astutely told her what in fact she herself had spoken.

Another value of listening is that most clients who seek the services of a reader actually want someone to listen to their problems. In addition many clients have already made up their minds about what choices they are going to make. They merely want support to carry out their decision.

10. *Dramatize your reading.* Give back what little information you do have or pick up a little bit at a time. Make it seem more than it is. Build word pictures around each divulgence. Don't be afraid of hamming it up.

11. *Always give the impression that you know more than you are saying.* The successful reader, like the family doctor, always acts as if he knows much more. Once you persuade the client that you know one item of information about him that you could not possibly have obtained through normal channels, the client will automatically assume you know all. At this point he will typically open up and confide in you.

12. *Don't be afraid to flatter your subject every chance you get.* An occasional subject will protest such flattery, but will still cherish it. In such cases you can further flatter him by saying, "You are always suspicious of people who flatter you. You just can't believe that someone will say good of you unless he is trying to achieve some ulterior goal."

13. Finally, remember the golden rule: *Tell the client what he wants to hear.*

Sigmund Freud once made an astute observation. He had a client who had been to a fortune-teller many years previously. The fortune-teller had predicted that she would have twins. Actually she never had children. Yet, despite the fact that the reader had been wrong, the client still spoke of her in glowing terms. Freud tried to figure out why this was so. He finally concluded that at the time of the original reading the client wanted desperately to have children. The fortune-teller sensed this and told her what she wanted to hear. From this Freud inferred that the successful fortune-teller is one who predicts what the client secretly wishes to happen rather than what actually will happen (Freud 1933).

The fallacy of personal validation

As we have seen, clients will readily accept stock spiels such as those I have presented as unique descriptions of themselves. Many laboratory experiments have demonstrated this effect. Forer (1949) called the tendency to accept as valid a personality sketch on the basis of the client's willingness to accept it "the fallacy of personal validation."

The early studies on personal validation were simply demonstrations to show that students, personnel directors, and others can readily be persuaded to accept a fake sketch as a valid description of themselves. A few studies tried to go beyond the demonstration and tease out factors that influence the acceptability of the fake sketch. Sundberg (1955), for example, gave the Minnesota Multiphasic Personality Inventory (known as the MMPI) to 44 students. The MMPI is the most carefully standardized personality inventory in the psychologist's tool kit. Two psychologists, highly experienced in interpreting the outcome of the MMPI, wrote a personality sketch for each student on the basis of his or her test results. Each student then received two personality sketches—the one actually written for him or her and a fake sketch. When asked to pick which sketch described him or her better, 26 of the 44 students (59 percent) picked the fake sketch!

Sundberg's study highlights one of the difficulties in this area. A fake, universal sketch can be seen as a better description of oneself than can a uniquely tailored description by trained psychologists based upon one of the best assessment devices we have. This makes personal validation a completely useless procedure. But it makes the life of the character reader and the pseudopsychologist all the easier. His general and universal statements have more persuasive appeal than do the best and most appropriate descriptions that the trained psychologist can come up with.

Some experiments that my students and I conducted during the 1950s also supplied some more information about the acceptability of such sketches. In one experiment we gave some students a fake sketch (the third stock spiel previously discussed) and told half of them that it was the result of an astrological reading and the other half that it was the result of a new test, the Harvard Basic Personality Profile. In those days, unlike today, students had a low opinion of astrology. All the students rated each of the individual statements as generally true of themselves. The groups did not differ in their ratings of the acceptability of the individual statements. But when asked to rate the sketch as a whole, the group that thought it came from an accepted personality test rated the acceptability significantly higher than did the group that thought it came from an astrologer. From talking to individual students it was clear that those who were in the personality-test group believed that they had received a highly accurate and unique characterization of themselves. Those in the astrology group admitted that the individual statements were applicable to themselves but dismissed the apparent success of the astrologer as due to the fact that the statements were so general that they would fit anyone. In other words, by changing the context in which they got the statements we were able to manipulate the subjects' perceptions as to whether the statements were generalities that applied to everyone or were specific characterizations of themselves.

In a further experiment we obtained a pool of items that 80 percent or more of Harvard students endorsed as true of themselves. We then had another group of Harvard students rate these items as "desirable" or "undesirable" and as "general" or "particular" (true of only a few students). Thus we had a set of items that we knew almost all our subjects would endorse as true of themselves, but which varied on desirability and on perceived generality. We were then able to compose fake sketches which varied in their proportion of desirable and specific items. We found that the best recipe for creating acceptable stock spiels was to include about 75 percent desirable items, but ones which were seen as specific, and about 25 percent undesirable items, but ones which were seen as general. The undesirable items had the apparent effect of making the spiel plausible. The fact that the items were seen as being generally true of other students made them more acceptable.

The most extensive program of research to study the factors making for acceptability of fake sketches is that by C. R. Snyder and his associates at the University of Kansas. A brief summary of many of his findings was given in an article in *Psychology Today* (Snyder and Shenkel 1975). In most of his studies Snyder uses a control condition in which the subject is given the fake sketch and told that this sketch is generally true for all people. On a rating scale from 1 to 5 (1, very poor; 2, poor; 3, average; 4, good; 5, excellent) the subject rates how well the interpretation fits his personality. A typical result for this control condition is a rating of around 3 to 4, or between average and good. But when the sketch is presented to the subject as one which was written "for you, personally," the acceptability tends to go up to around 4.5, or between good and excellent.

In a related experiment the subjects were given the fake sketch under the pretense that it was based on an astrological reading. The control group, given the sketch as "generally true for all people," rated it about 3.2, or just about average. A second group was asked to supply the astrologer with information on the year and month of their birth. When they received their sketches they rated them on the average at 3.76, or just below good. A third group supplied the mythical astrologer with information on year, month, and day of birth. These subjects gave a mean rating of 4.38.

From experiments such as those we have learned the following. The acceptability of a general sketch is enhanced when (1) the reader or source is believed to know what he is doing, (2) the instrument or assessment device is plausible, (3) a lot of mumbo jumbo is associated with the procedure (such as giving month, day, hour, and minute of birth along with a lot of complicated calculations), and (4) the client is led to believe that the sketch has been tailored to his personality. When these conditions are met, the client, and possibly the reader as well, have a strong "illusion of uniqueness"—that is, the client is persuaded that the sketch describes himself or herself and no one else.

Why does it work?

But why does it work? And why does it work so well? It does not help to say that people are gullible or suggestible. Nor can we dismiss it by implying that some individuals are just not sufficiently discriminating or lack sufficient intelligence to see through it. Indeed one can argue that it requires a certain degree of intelligence on the part of a client for the reading to work well. Once the client is actively engaged in trying to make sense of the series of sometimes contradictory statements issuing from the reader, he becomes a creative problem-solver trying to find coherence and meaning in the total set of statements. The task is not unlike that of trying to make sense of a work of art, a poem, or, for that matter, a sentence. The work of art, the poem, or the sentence serve as a blueprint or plan from which we can construct a meaningful experience by bringing to bear our own past experiences and memories.

In other words the reading succeeds just because it calls upon the normal processes of comprehension that we ordinarily bring to bear in making sense out of any form of communication. The raw information in a communication is rarely, if ever, sufficient in itself for comprehension. A shared context and background is assumed. Much has to be filled in by inference. The good reader, like anyone who manipulates our perceptions, is merely exploiting the normal processes by which we make sense out of the disorderly array of inputs that constantly bombard us.

A variety of examples and experiments could be drawn upon to illustrate this point. Let us look at a couple of instructive cases. Dr. Joseph Weizenbaum, of M.I.T., developed a computer program that would enable a human to apparently carry on a conversation with a computer. One version of his program is called "Doctor." The program is designed to "talk" like a Rogerian psychotherapist. Such a therapist is relatively easy to imitate because the basis of Rogers' technique is simply to feed the patient's statements back to him in a paraphrased form. The human, playing the role of the patient, types in questions and answers on a teletype machine. The computer, under control of the program, types replies and questions on the same teletype paper. A typical example (Weizenbaum 1976) of a conversation between a young lady and Doctor follows. (The lady speaks first. The computer's responses are printed in italics.)

> Men are all alike.
> *In what way?*
> They're always bugging us about something or other.
> *Can you think of a specific example?*
> Well, my boyfriend made me come here.
> *Your boyfriend made you come here?*
> He says I'm depressed much of the time.
> *I am sorry to hear you are depressed.*
> It's true. I am unhappy.
> *Do you think coming here will help you not to be unhappy?*
> I need some help, that much seems certain.
> *What would it mean to you if you got some help?*
> Perhaps I could learn to get along with my mother.
> *Tell me more about your family.*
> My mother takes care of me.
> *Who else in your family takes care of you?*
> My father.
> *Your father?*
> You are like my father in some ways.
> *What resemblance do you see?*
> You are not very aggressive, but I think you don't want me to notice that.
> *What makes you think I am not very aggressive?*
> You don't argue with me.
> *Why do you think I don't argue with you?*
> You are afraid of me.
> *Does it please you to believe I am afraid of you?*
> My father is afraid of everybody.
> *What else comes to mind when you think of your father?*
> Bullies.
> *Does that have anything to do with the fact that your boyfriend made you come here?*

Doctor is a quite primitive program as natural-language programs go. It employs a lot of tricks and stock phrases. It has no mechanisms for actually understanding sentences. Instead it seeks out key words that are typed and does some simple syntactical transformations. For example, if the program sees a sentence of the form "Do you X" it automatically prints out the response "What makes you think I X?" When Doctor cannot match the syntax of a given sentence it can cover up in two ways. It can say something noncommittal, such as *"Please go on"* or *"What does that suggest to you?"* Or it can recall an earlier match and refer back to it, as for example, *"How does this relate to your depression?"* where depression was an earlier topic of conversation.

In essence Doctor is a primitive cold reader. It uses stock phrases to cover up when it cannot deal with a given question or input. And it uses the patient's own input to feed back information and create the illusion that it understands and even sympathizes with the patient. This illusion is so powerful that patients, even when told they are dealing with a relatively simple-minded program, become emotionally involved in the interaction. Many refuse to believe that they are dealing with a program and insist that a sympathetic human must be at the controls at the other end of the teletype.

Sociologist Harold Garfinkel has supplied another instructive example (1967). He conducted the following experiment. The subjects were told that the Department of Psychiatry was exploring alternative means to therapy "as a way of giving persons advice about their personal problems." Each subject was then asked to discuss the background of some serious problem on which he would like advice. After having done this the subject was to address some questions which could be answered "yes" or "no" to the "counselor" (actually an experimenter). The experimenter-counselor heard the questions from an adjoining room and supplied a "yes" or "no" answer to each question after a suitable pause. Unknown to the subject, the series of yes-no answers had been preprogrammed according to a table of random numbers and was not related to his questions. Yet the typical subject was sure that the counselor fully understood the subject's problem and was giving him sound and helpful advice.

Let me emphasize again that statements as such have no meaning. They convey meaning only in context and only when the listener or reader can bring to bear his large store of worldly knowledge. Clients are not necessarily acting irrationally when they find meaning in the stock spiels or cold readings. Meaning is an interaction of expectations, context, memory, and given statements. An experiment by the Gestalt psychologist Solomon Asch (1948) will help make this point. Subjects were given the following passage and asked to think about it: "I hold it that a little rebellion, now and then, is a good thing, and as necessary in the political world as storms are in the physical." One group of subjects was told that the author of the passage was Thomas Jefferson (which happens to be true). The subjects were asked if they agreed with the passage and what it meant to them. These subjects generally approved of it and interpreted the word *rebellion* to mean minor agitation. But when subjects were given the same passage and told that its author was Lenin, they disagreed with it and interpreted *rebellion* to mean a violent revolution.

According to some social psychologists the different reactions show the irrationality of prejudice. But Asch points out that the subjects could be acting quite rationally. Given what they know about Thomas Jefferson and Lenin, or what they believe about them, it makes sense to attribute different meanings to the same words spoken by each of them. If one thinks that Jefferson believed in orderly government and peaceful processes, then it would not make sense to interpret his statement to actually mean a bloody or physical revolution. If one thinks that Lenin favors war and bloodshed, then it makes sense, when the statement is attributed to him, to interpret *rebellion* in its more extreme form.

Some recent research that my colleagues and I conducted might also be relevant here. Our subjects were given the task of forming an impression of a hypothetical individual on the basis of a brief personality sketch. In one condition the subjects were given a sketch that generally led to an impression of a nice, personable, friendly sort of fellow. In a second condition the subjects were given a sketch that created an impression of a withdrawn, niggardly individual. Both groups of subjects were then given a new sketch that supposedly contained more information about the hypothetical individual. In both cases the subjects were given an identical sketch. This sketch contained some descriptors that were consistent with the friendly image and some that were consistent with the niggardly image. The subjects were later tested to see how well they recognized the actual adjectives that were used in the second sketch. One of the adjectives, for example, was *charitable*. The test contained foils for each adjective. For example, the word *generous* also appeared on the test but did not appear in the sketch. Yet subjects who had been given the friendly impression checked *generous* just as frequently as they checked *charitable*. But subjects in the other condition did not confuse *charitable* with *generous*. Why? Because, we theorize, the two different contexts into which *charitable* had to be integrated produced quite different meanings. When subjects who have already built up an impression of a "friendly" individual encounter the additional descriptor *charitable*, it is treated as merely further confirmation of their general impression. In that context *charitable* is simply further confirmation of the nice-guy image. Consequently when these subjects are asked to remember what was actually said, they can remember only that the individual was further described in some way to enhance the good-guy image, and *generous* is just as good a candidate for the description as is *charitable* in that context.

But when the subjects who have an image of the person as a withdrawn, niggardly individual encounter *charitable*, the last thing that comes to mind is generosity. Instead, they probably interpret *charitable* as implying that he donates money to charities as a way of gaining tax deductions. In this latter condition the subjects have no subsequent tendency to confuse *charitable* with *generous*.

The cold reading works so well, then, because it taps a fundamental and necessary human process. We have to bring our knowledge and expectations to bear in order to comprehend anything in our world. In most ordinary situations this use of context and memory enables us to correctly interpret statements and supply the necessary inferences to do this. But this powerful mechanism can go astray in situations where there is no actual message being conveyed. Instead of picking up random noise we still manage to find meaning in the situation. So the same system that enables us to creatively find meanings and make new discoveries also makes us extremely vulnerable to exploitation by all sorts of manipulators. In the case of the cold reading the manipulator may be conscious of his deception; but often he, too, is a victim of personal validation.

References

Asch, S. E. 1948. "The Doctrine of Suggestion, Prestige, and Imitations in Social Psychology." *Psychological Review* 55: 250-76.

Forer, B. R. 1949. "The Fallacy of Personal Validation: A Classroom Demonstration of Gullibility." *Journal of Abnormal and Social Psychology* 44: 118-23.

Freud, S. 1933. *New Introductory Lectures on Psychoanalysis.* New York: W. W. Norton.

Garfinkel, H. 1967. *Studies in Ethnomethodology.* Englewood-Cliffs, N.J.: Prentice-Hall.

Snyder, C. R., and R. J. Shenkel 1975. "The P. T. Barnum Effect." *Psychology Today* 8: 52-54.

Sundberg, N. D. 1955. "The Acceptability of 'Fake' versus 'Bona Fide' Personality Test Interpretations. *Journal of Abnormal and Social Psychology* 50: 145-57.

Weizenbaum, J. 1976. *Computer Power and Human Reason.* San Francisco: Freeman.

Sleight of Tongue

Ronald A. Schwartz

> . . . words still manifestly force the understanding, throw everything into confusion, and lead mankind into vain and innumerable controversies and fallacies.
>
> —Francis Bacon, *Novum Organum*

Several years ago Peter Hurkos, a reputed psychic, was extrasensorially perceiving a telephone caller's past, present, and future on a local audience-participation radio program. The listening audience was assured that the caller was a complete stranger to Hurkos. During this telepathic demonstration, Hurkos, who uses a thick Dutch accent, said, "I see a duk" (rhymes with "took"). The caller responded, "Why, that's amazing! Our dog is right here in the room with me." The woman continued to be amazed as Hurkos kept seeing things from afar. Pondering this exchange, it occurred to me that "duk," as Hurkos pronounced it, could sound like a lot of things other than dog, such as: duke, dock, doc, duck, and so on, depending upon the hearer's expectation or orientation. What is heard is what the hearer brings to the stimulus. By presenting such an ambiguous stimulus Hurkos has greatly expanded the domain of possible "hits," because the listener in all probability will be able to extract something from his environment to correspond.

The more closely I listened to Hurkos, the more I realized how little he actually said and how much the callers were contributing to his performance. For example, one exchange went:

Hurkos: One, two, three, four, five—I see five in the family.

Caller: That's right. There are four of us and Uncle Raymond, who often stays with us.

It is curious that the gull unwittingly tries so hard to get duped. He often goes to considerable lengths to make the stimulus intelligible and in accord with something from his experience.

What Hurkos is practicing is magic, but the props for his art are words. As with ordinary magic the art is often subtle and ingenious. Hurkos and other psychics are just full of tricks, and new ones are continually being invented which baffle laymen and scientists alike. Sometimes they even baffle other magicians; but this is less likely, because magicians are usually more adept at discovering the ruse.

Once the unwitting complicity is established the telepathee is psychologically primed for almost any utterance. For example:

Hurkos: I see an operation.

Caller: (no response)

Hurkos: Long time ago.

(Extending the domain. Psychics are partial to operations; almost everybody has had one or is planning to have one.)

Caller: No. We have been very lucky.

Hurkos: (somewhat angrily) Think! When you were a little girl. I see worried parents, and a doctor, and scurrying about.

Caller: (no response)

Hurkos: (confidently) Long time ago.

At this point the psychological tone is such that the observers feel that Hurkos is right and that the dim-witted caller can't even remember her serious childhood operation. The caller feels this way too.

Caller: (yielding) I cannot remember for certain. Maybe you are right. I'm not sure.

Of course he is right. He is so confident and she is so uncertain. As extraordinary as it sounds, the mood is such that participants who disagree with the psychic come off as obstinate or perverse, as though they are deliberately trying to sabotage the effort.

When Hurkos gets going, the assault is truly aggressive. He verbally connects, dodges, bluffs, and weaves, always keeping the opponent off balance. Consider the following ESP demonstration aired on a local TV show. The moderator asked for a volunteer to come up and give Mr. Hurkos a personal object to hold. The volunteer was an attractive female who looked to be in her early thirties. Hurkos was at this time looking directly at the lady, although he said at the beginning of the show that he wanted "to see the people not at all."

Moderator: Well, I think we are ready with our first very attractive guest. Have you ever met or spoken to Mr. Hurkos?

Woman: No.

Hurkos: Oh, my God, your mama had an operation! Did mother have an operation?

Woman: Yes!

Hurkos: And thank the Lord that she is alive! She had a double operation.

(Stop! Just what exactly is a double operation? Two operations? Two organs involved? Two surgeons? Two lumps? You'll see we never get to find out.)

Woman: Yes, you are correct—I know what you mean.

Hurkos: (nodding knowingly) You have something of your mother? (long pause)

Moderator: Do you have anything of your mother's?

Woman: No, I don't.

(Watch out Hurkos. Try nonsense.)

Hurkos: Then why is it to give it no to you?

Woman: (long pause)

Moderator: Do you have something your mother gave to you?

(Hurkos, you are in trouble. The moderator has clarified the question.)

Woman: (long pause)

Hurkos: The wallet.

Moderator: The wallet. Did she give you a wallet?

Woman: (embarrassed and confused) Yes!

Masterful! At first the question implied that the item was on her person, but Hurkos, missing on that, said "the wallet." In this way the domain got expanded by the moderator from "do you have" to "did she give," the latter implying only ownership. The "hit" was possible because the original stimulus was ambiguous and because almost all girls at some time get wallets from their mothers.

The next line advances cleverly. Hurkos now knows that she does not have it on her person, but that she must have a wallet so it should be in her purse left by her seat.

Hurkos: Where is it? I will go get it for you. (laughter) Where is the purse?

Woman: Uh.

Hurkos: If you was my daughter I would spank you. (laughter) The reason why I tell you this—can I have your wallet? It is lousy money.

Woman: Do you know what . . . I took it out of my purse just before I left the house. It was too bulky.

(One in a hundred! Oh well, it doesn't make any difference now. Her original acknowledgment of the gift was so impressive. On to other areas.)

Hurkos: When did Mommy went to the doctor? When?

Woman: Last—

Hurkos: Sept—

Woman: Week.

Hurkos: (loudly) No, no you are wrong!

Woman: Maybe, I might be wrong.

Hurkos: Was it nine days ago, on a Monday?

Woman: It could be.

(Hurkos had lightly guessed September. The gull told the correct date and Hurkos not only switches to her account but makes it seem like she is a little off in her recollection and that he knows precisely when Mommy visited the doctor. The woman acquiesces in the face of his conviction.)

Hurkos: Who is Ann?

Woman: Ann? I have an aunt. My aunt's name is Ann.

Hurkos: Yes. She has trouble with her legs too—left leg.

Woman: Could be.

(At this point the victim is primed for almost anything.)

Hurkos: And I see you married.

(Now or later?)

Woman: Yes.

Hurkos: How is it possible that you fell in through the bed? Is not normal.

Moderator: Did you fall off the bed?

Woman: No.

(If the moderator had not clarified the question the "no" would have been less likely.)

Hurkos: 'Bout a year ago?

Woman: I do not remember.

(That's better, now her memory is faulty.)

Hurkos: I don't see number five, but only see three. There gonna be three. How many are in the family?

Woman: Well, we have two now.

Hurkos: Yeh, that is what you think. (laughter)

(". . . only see three" and "there gonna be three" permits this latitude. Of course now the poor thing probably thinks she is pregnant.)

Moderator: Congratulations.

Hurkos: (loudly) I must say your husband is a very hard worker.

Woman: That is very true.

Hurkos: (loudly) And he acts like a child!

(All wives see this.)

Woman: Yes.

Hurkos: (screaming) This man don't have the time to relax! He always busy. He married with you, but he is married with his job!

(Elaboration of her acknowledgment of her husband's diligence.)

Woman: Very true.

Hurkos: (still screaming) And he, I don't want to see work with a whole bunch of peoples!

(This could mean almost anything.)

Woman: That is right. Very true!

Hurkos: Did you lose your bracelets?

Woman: I did lose a (pause)

(A desperate search is initiated for any item that could be even remotely corroborative.)

Hurkos: Did you got it too?

Woman: I got a pin. Yes, I lost a pin.

(Loss of bracelets confirmed by loss of a pin!)

Hurkos: You are a very good person. You are against the marriage a little bit but that came over.

Woman: That is very true!

(In her strong confirmation Hurkos is given a clue.)

Hurkos: (at the top of his lungs) When you want to break the marriage that time he did not have a chance. When you said, "If I don't, if I don't want him and I lose him—I like him, I am not listening to anybody—I want a want a want, or get the house!" This is correct?

Woman: That is fantastic!

Hurkos: Bye-bye. (huge audience applause)

(Absolute victory from nonsense.)

Moderator: How many percent would you say he is correct?

Woman: Oh, I would say ninety-five percent.

Moderator: (to audience) I just asked her how many percent she thought he was correct and she said almost a hundred percent.

A really impressionable gull can produce a most dramatic effect. Consider this interaction with a middle-aged male.

Moderator: Have you ever met or spoken with Mr. Hurkos?

Man: No.

Hurkos: May I have your glasses, please?

(Symbolic gesture of removing man's sight and giving it to Hurkos?)

Man: (gives Hurkos his glasses) Here.

Hurkos: Sir, (loudly) why did you finish university?

(Here we go again. Hurkos has it both ways. It sounds like he is asking both why the man did and why he did not finish school.)

Man: Well, the war broke out. I had no money to finish it.

Hurkos: But you have brilliant mind. You have a lot of ideas and you can bring those together.

(Most people would agree.)

Man: Yes.

Hurkos: And you was working when you were nine years old, sir.

Man: Well not exactly. I was helping, yes.

(Helpful, indeed.)

Hurkos: One, two, three, four, five—how many people are in the family, sir?

(Again, any one of the above is a "hit.")

Man: Six.

Hurkos: Five besides you.

Man: Five besides me, yeh.

Hurkos: But you are wrong, it is not true. (pause) There were seven in the family, one died very young and are six now. One died very young.

Man: That I don't remember.

(Boggles the mind. Can't recollect losing a member of the family!)

Psychology has something to tell us about why Hurkos and other of his ilk are so effective. A number of psychological principles are operating which contribute to the ESP effect. Specifically, investigation into the artifacts in behavioral research has shown that more dominant experimenters effect greater inducement of their expectation on their subjects than less dominant experimenters. Also, it has been demonstrated that experimental subjects generally try to comply with the demand characteristics of the situation, that is, try to be good subjects. It is likely, therefore, that the excessive compliance to Hurkos we've observed is simply due, in part, to his excessive demand and the subject's natural inclination to comply.

Added to this we have the P. T. Barnum Effect—the willingness of people to accept a general personality description as accurate for themselves. Much of the psychic's output is of this variety. Too, the phenomenon of prestige suggestion—the tendency for prestigious names to compel consent or approval regardless of the merits of their argument—is no doubt operative, for psychics usually get lavish billing.

Most important, there is the well-known effect of stimulus ambiguity on conformity. Conformity studies have shown that people are more inclined to acquiesce when they are uncertain of their perception. This effect, of course, is Hurkos's stock-in-trade. Since perception is a synthesis, the more ambiguous the stimulus, the more the construction is determined by the perceiver. The skilled use of ambiguity is very convincingly precise because the gull is responding to his own internal construction, which naturally seems right or true to him. Ambiguity tends to permit an easy internal confirmation.

It's a good bet that subject complicity is the main explanation for all sorts of wondrous phenomena: clairvoyance, telepathy, precognition, astrology, and so on. Whenever people are called upon to perceive the seer's message, their perception can be influenced and shaped.

The eagerness with which people latch on to the fanciful idea of extrasensory perception to explain the apparent accuracy of the utterance, rather than to the recognition of their own complicity, is the really intriguing aspect of the demonstration. Perhaps recognition of one's own contribution is just too lackluster an explanation to be given much credence. Our persistent delight with fanciful explanations suggests that skilled practitioners of sleight of tongue will enjoy a bright and prosperous future.   ●

Cold Reading Revisited

James Randi

Two years ago in the pages of this journal (Spring/Summer 1977), Ray Hyman gave a quite definitive account of the art of "cold reading" as practiced by those with psychic pretensions. Dr. Hyman, as a psychologist and an advanced student of "mentalist" techniques, outlined the often surprising subtleties that are employed to convince the unwary that marvelous forces are at work. More recently (Fall 1978) Ronald A. Schwartz showed how stage performer Peter Hurkos applied some of the same techniques.

Recently Dr. Paul Kurtz and I were invited to appear on Canadian TV with one Geraldine Smith, a renowned "psychic" in Toronto. She had been extensively and convincingly written up in the Canadian papers as a powerful worker, and we were looking forward to being astonished. I had brought along an object that I thought would have strong "vibrations" for a genuine psychic, and the host of "Point Blank," Warner Troyer, had an article of his own to try for results.

We reproduce here the text of Miss Smith's "reading" for each article and suggest that students of cold reading note carefully the techniques used. The end result is hardly convincing, but that is due to the fact that the owner of each object was careful not to provide any feedback to Smith at any time. The words are transcribed directly from a tape recording provided by the people at Global TV in Toronto. We thank them for their courtesy.

The host decided to try a silver chain bracelet.

Host: Tell me something about the owner of that [handing her a "Medic Alert" bracelet].

Smith: Okay.

Host: A little bracelet. Other than the fact that they need a Medic Alert bracelet.

Smith: [laughs] Okay. First of all I'm getting a very strong gold. Now one thing I should let everybody know is, I also work with what is called "auras," and I've tuned myself into picking up vibrations, colors, from the article that the person has worn—had for some time. Now, first color is gold, and that is an extremely sensitive color. It's also, of course, allergy color too. It's the color of super, super, super nervousness. This person I'm feeling physically, upper, lower back area problems. I'm also seeing some upper stomach area things going on. Do you know this person?

Host: Uh-huh.

Smith: Personally?

Host: Uh-huh.

Smith: All right, there's something in this area here [points to chest]. You see, as soon as I pick up an article, I will physically feel different areas that perhaps have been affected by the person. Feeling this area [indicates chest], feeling this area [indicates forehead], headaches, eyestrain, something like that. Upper and lower back areas. It's also the intuitive color. I would also, I'd have to say that the person who owns it is extremely intuitive, probably clairsentient, which is very clear on the gut-feeling type areas. I'm seeing— there's a separation around this person. Are they in this area now? Because I see them, like, not here.

Host: Physically not here?

Smith: Physically not here. No.

Host: Uh-huh. Well!

Smith: Which—which would make me wonder if the person is [laughs] either dead or if there's been some very, very bad health problems with them, because I'm just *really* feeling sluggish myself. Interesting. I'm seeing the month of January here—which is now—but there would have to be something strong with the person with January as well.

Host: Okay. What color is *my* aura?

Smith: You've got blue, you've got gold, you've got green.

Host: What do those things *mean*? What's an aura?

Smith: An aura is—I perceive it as a color. A color energy vibration that surrounds the person's body. It tells me mental, physical, spiritual—ah— general personality. Green is the color of the communicator. Anybody in the communications area has to have green in their aura. In other words, in any area.

Host: So you can tell me that without looking, huh?

Smith: Yeah. In other words, yeah. In other words, you've got to have good communication in all areas. The one thing I would emphasize very strongly right now is that the green is a little bit blocked in the mental area of your aura, so that means you have been very frustrated in the area of communication on a personal level. It's almost as if you've been talking to the wall. I'm seeing a lot of [laughs], a lot of vibration there. And it's more in a personal area rather than in—you know—a work, business situation.

Host: Okay.

The host, who had been very careful not to tip his hand, following my admonitions to that effect before the program, was the owner of that Medic Alert bracelet. He had not tried to conceal the fact, though he picked it up from a small table to hand it to the psychic. She assumed that it belonged to another person and gave her "reading" with that assumption. Note that she did not tell us the age, sex, or relationship of this mythical person. She tried to "pump" the host by asking whether he knew the person and whether the person was physically there or not. Warner carefully gave answers. Yes, he knew the person. And he answered her question about the whereabouts with another question. I'm proud of him.

Jumping the gun, she guessed the person was absent, and then covered all bases with a classic generalization/cop-out—". . .which would make me wonder if the person is either dead or if there's been some very, very bad health problems." Note that she was only "made to wonder," not *know*, and as phrased, it was also a question that might prompt an answer but failed to do so.

The host was neither dead nor absent. His back, he assured her, was excellent, nothing was wrong in the chest area, and when she tried to add a neck/upper-shoulder area quickly to the reading as he revealed his good state of health, he denied that as well. But she mumbled an encouraging "excellent . . . fine" as he outlined his condition to cover the fact that she was dead wrong. He is neither nervous nor allergic and has no stomach problems. Headaches and eyestrain, he assured us, do not bother him either.

But I must report that our host tumbled for one of the more common tricks of such readings. You see, the victim is allowed and encouraged to read more into the recitation than is already there. Smith had said, "I'm seeing the month of January here," and during his denial of the accuracy of her reading, he admitted that she had determined that his birthday was in January! But she had said nothing about a birthday, particularly *his* birthday, since she didn't even know the bracelet belonged to him! When confronted with all this evidence, Ms. Smith explained: "The thing is, that, to me, in a reading, means quick removal from a situation, which means either leaving this place, leaving the country—quick removal." Perhaps the lady was expressing her own desires of the moment. It certainly didn't make any sense to me.

Looking at Paul Kurtz and myself, she tried another tack. "Understand, I can totally see what the two of you are saying. The thing that I take a little bit further down the line is—my readings—as much as many things can be applied to many people, aren't there a lot of similarities in life? [There was a short stunned pause here, as we all tried to think what she could possibly be trying to say.] You know, we get married or we don't, we're male or we're female, we have children or we don't."

Ms. Smith was a little bit subdued, but smiled bravely after this confrontation. She was yet to be tested with an object that I had brought along specifically for the purpose. It was an object that I had a complete history of, and it was something I had owned for some time. In terms of a psychometry reading, it should have been excellent material. I will reverse the regular procedure in that I will tell you in advance all about the object, and then give you her entire word-for-word "reading" of it. You will then be in the position I was and be able to do your own analysis of her accuracy.

The object was a small bisque-fired ceramic, black in color, of Peruvian origin. It measured seven inches long and was in the form of a bird, with a spout coming from the top. It was a fake—a replica—of a genuine Mochica grave object and had been made by a friend in Lima who is Peruvian by birth but Chinese in ancestry. He is a short man, five-foot-seven or so, heavy, straight black hair, 28 years of age, with totally Chinese features to all appearances. He is single, or was at the time the reading was made. He speaks only Spanish. His business is making accurate replicas of original Peruvian art and repairing ceramics. The ceramic was given to me by him because it had been broken, and I repaired it myself when I returned to the United States. I have a large collection of similar peices, both genuine originals and good replicas. I had brought this one with me to avoid breaking a valuable original, and to get around the tendency of psychics to expound on nonexistent people long extinct, who they can safely claim were associated with such an article. I knew the whole history of the object, from molding to breaking and beyond.

Geraldine Smith took a deep breath and started on her "reading" of this article, while I carefully sat there unblinkingly.

Smith: Okay. The first thing that—actually I'm very quickly being taken over maps, and I'm tuning in very strongly to Mexico, United States— general area there. I'm seeing three *very* strong personalities, two females and a male, and I'll describe them all for you. First of all, the man I'm seeing is approximately five-foot-eight, five-foot-nine. To me, that's short for a male. Very deep brown hair, but receding at the temples. Glasses, quite thick. Obviously very bad eyes, because the focus I'm seeing is very, very strong. I'm seeing kind of a round-neck shirt. It's not the type you have on now. I guess it would be more along the line that he is wearing [she pointed at two of us]. Then I'm going to the two ladies. Oh, I didn't give you an age on the man. He would have to be 45 to 50. Something like that. The ladies I'm seeing, one would be, hmmm, five-foot. Very short. Four-nine, five-foot. The other lady is quite a bit taller. One is very, very, very heavy-set, and shortish curlyish hair, but fluffy. And the other one, the shorter one, is just, well, there's nothing really big about her. I'm seeing these three people very much in connection with this. Do you recognize them at all?

Randi: Now you're asking me something. You're supposed to be *telling* me.

Smith: I mean, *do* you recognize them?

Randi: Do you want now the history of the object? Is the reading finished?

Some things about this exchange will be obvious. She took a guess at an Indian origin, but missed. It could have been North American Indian or Mexican to the uninitiated, but it was not. She threw in three people for a try-on, and I'm still trying to fit in the two women, but I can't. The man's eyesight is excellent—he does not wear glasses. As for height, she's great on that, and the fellow does wear turtleneck sweaters frequently. His hair is very definitely *not* receding—quite the contrary. The age she gave is at least 17 years off.

Now, you should know two other things about this test. First, I told Geraldine Smith very clearly that this was not to be considered a formal test. My reasons should be obvious. It was quite possible for her to have looked into my background and discovered my interest in Peruvian archaeology. She could have visited a museum and come up with the origin of the object easily. Second, she agreed to consider a formal test by the CSICP, to be conducted regardless of the outcome of this demonstration. We have not heard from her since that time.   •

Geller-Type Phenomena

The Nonpsychic Powers of Uri Geller

David Marks and Richard Kammann

Renewed scientific interest in psychic or paranormal phenomena has followed recent experiments conducted with the Israeli stage-performer Uri Geller. We present here results of an investigation of alleged paranormal phenomena produced by Mr. Geller. Geller has submitted to laboratory controlled tests of his ESP ability at Stanford Research Institute with positive results (Targ and Puthoff 1974), but those experiments have been criticized on the grounds that the researchers did not take adequate precautions to prevent the use of ordinary modes of communication. These include the possible use of radio signals sent by an accomplice (Hanlon 1974) and imperfect shielding of target material from Geller's visual field.[1]

A recent visit by Geller to the New Zealand stage-circuit (23 March through 4 April 1975) provided an opportunity to conduct further tests of his powers, subject to his willingness to cooperate. Regrettably Geller was not willing to participate in any further experiments, and it now seems unlikely that he will agree to do so on further occasions. Since Geller and other alleged psychics may avoid study under adequately controlled laboratory conditions, it is necessary to develop techniques of field obser-

---

1. One of the authors (R.K.) visited the S.R.I. laboratories in November 1975 and observed that there was a hole three-to-four inches in diameter between the "shielded room" and the target area. This tubular hole was at least one foot in length and normally provides cables into the shielded room when it is used for EEG experiments. Its location is about three feet above the floor, high enough to look through by bending over or kneeling. It would provide a cone of visible area perhaps three-to-five feet in diameter against the opposite wall of the anteroom. Dr. Puthoff stated that the hole was completely "shielded" with a metal cover (no longer available for inspection) and that cotton gauze was also used to block the duct. Auditory cues may have been possible if Geller unstuffed the cotton in the hole from the inside and listened through it. Perhaps just as important as the four-inch hole is the window between the "shielded" room and the anteroom. It appeared to be a one-way-vision screen (i.e., a reversible mirror) about two feet wide and 1.5 feet high. In November 1975 it was covered by a bulletin board screwed over it in the anteroom. Dr. Puthoff stated that it was even more thoroughly covered by a shield when Geller was being tested.

vation that can help to distinguish between the normal and the paranormal hypotheses.

Our method was to record the apparently essential features of the allegedly paranormal effects observed in the field, and then, by controlled follow-up investigations under similar conditions with nonpsychic subjects, to determine whether or not the same effect could be readily produced using normal sensory and motor functions. In short, we used a delayed control group. While this design is necessarily imperfect in that we could not replicate the exact conditions of the original data, if the follow-up experiments show a striking similarity in detail to the Geller effects, then it is more reasonable and parsimonious to assume that Geller used the same normal modes of communication.

Although Geller declined our invitation to conduct controlled laboratory experiments, he did agree to several informal demonstrations of his powers. We were also able to observe and record a variety of Geller's demonstrations during radio, television, and stage performances. An important advantage of the field-study approach is that it does not involve the intrusiveness and social alienation of the subject that may accompany rigorous measurement procedures and that may, as a number of researchers have recently noted, interfere with sensitive paranormal phenomena (Hasted et al. 1975; Taylor 1975). In all of the cases reported herein, Geller performed voluntarily, without interference by us, and in each case he claimed to be using paranormal abilities.[2]

Perception of graphic material

In one series of trials Geller demonstrated his ability to reproduce pictures that had been drawn and then sealed inside envelopes. A total of nine trials are listed in Table 1; these took place in a number of different locations over an eight-day period.

Trial 1 took place during a recorded interview with Geller in a Wellington, New Zealand, hotel on 24 March 1975. A picture of a ship (Cook's barque, *The Endeavour*) was drawn on white paper in light-blue, ball-point-pen ink (see Figure 1a). The paper was a standard sheet of hotel

---

2. All quotations attributed to Geller are taken from tape recordings made on the occasion. This particular statement is obviously incorrect since James Cook was not a pirate. It is also of interest to note Geller's report of "orange sails" in light of the fact that the orange center of the insignia printed on the paper coincides with one of the lower sails in the drawing when the paper is folded and help up to the light (see Figure 1).

stationery, boldly printed with the hotel's name and address (in black), an insignia, and a border (both orange and dark blue). The upper and lower thirds of the paper were folded over the drawing, which was placed inside a single white envelope.

Table 1. Summary of trials

Trial	Date (day, mo., yr.)	Place	Target	Presented by	Prepared by	Figure
1	24/3/75	Radio studio	ship	D.M.	D.M.	1a and 2a
2	25/3/75	Radio studio	flower	B.E.	R.E.	1b and 2b
3	25/3/75	Radio studio	goblet	B.E.	S.E.	1c and 2c
4	25/3/75	Radio studio	stripes/ grid*	D.M.	A.E.	No drawing
5	30/3/75	T.V. studio	2 bars/ cross*	P.S.	R.K.	No drawing
6	30/3/75	T.V. studio	various/ man*	P.S.	R.K.	No drawing
7	1/4/75	Radio studio	house/ hat*	G.D.	D.M.	No drawing
8	1/4/75	Radio studio	2 bars/ cross*	G.D.	D.M.	No drawing
9	1/4/75	Radio studio	various/ man*	G.D.	R.K.	No drawing

*Figures described in the "x/y" format were so designed that when they unfolded they appeared as described by the x term; but when observed through the folded paper they appeared as described by the y term.

D.M., R.K., and A.E. are research psychologists. B.E., P.S., and G.D. are radio and television announcers.

Two psychologists (D.M. and A.E.) were present during the trial, but the target drawing was known only to D.M. There was a friendly, cooperative atmosphere in which the observers allowed Geller to control the conditions. The trial lasted 11.5 minutes. Geller asked that he be allowed to hold the envelope in his hands, and he did so continuously for the first 8.5 minutes while he talked and answered questions about his ESP ability. Although Geller handled the envelope extensively during this time, he made no attempt to reproduce its contents. He then asked D.M. to go into

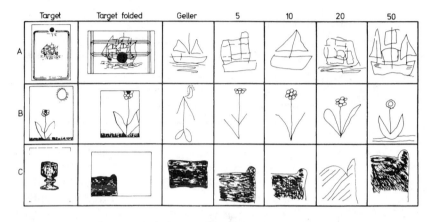

Figure 1. Target drawings presented to Geller and to nonpsychic controls.
The first and second columns show targets as they appeared in unfolded and folded states. Geller's final attempts at reproducing the targets are shown in column 3. Columns 4-7 show the best student reproductions following inspection times of 5,10,20, and 50 seconds. The drawings were presented to the control subjects in the same envelopes used in the trials with Geller under similar lighting conditions. Note that in the case of target B the sun was drawn in light-yellow crayon and is invisible in the folded state. The wine goblet (target C) was totally transformed in shape when folded.

another room and redraw the picture. D.M. was instructed to repeat the drawing a total of three times. Meanwhile A.E. was asked to draw what he could "pick up" and, at one point, to fetch another pen (although Geller had already been provided with one), thus leaving the subject completely unobserved for five to ten seconds.

Trial 1 concluded when Geller reported: "I've got a sailing boat; but many sails and water. Suddenly I saw not white sails but orange sails, and I said, 'If it's not a pirate boat then it's nothing.' "[3] Geller then revealed the drawing reproduced in Figure 1a.

Trials 2 and 3 took place in the studio of a radio station in Wellington on 25 March 1975. Geller was invited to be interviewed on a live "talk-back" program hosted by Dr. B. Edwards (B.E.), and D.M. was invited to observe and record the phenomena which took place.

3. This research was supported by a grant from the University of Otago Research Committee. We thank Linda Addis, Gillian Denny, Brian Edwards, Tony Egan, Martin Fisher, Brent Gracie, Robyn Irwin, Paul Savage, and Roxane Smith for their help and encouragement. We also acknowledge our debt to Uri Geller for his time and patience.

For Trial 2 a picture was drawn in crayon on off-white, lined paper. The picture showed a dark-blue flower, dark-green grass, and a light-yellow sun (see Figure 1b). The drawing was folded twice, first along the horizontal median and then along the vertical median, and placed inside two white envelopes. The envelopes were then sealed. The target was known only to B.E. and his daughter (who had prepared the drawing but was not present during the trial). The trial lasted 85 minutes during which time Geller handled the envelope extensively but also demonstrated various other effects of the psychokinetic type and functioned in his role as guest speaker on the program. Geller produced a series of 29 tentative drawings before submitting a final attempt; these included a number of different images, most of which were triangular in shape with a vertical center line. The final drawing is shown in Figure 1b.

In Trial 3 the target was another crayon drawing on off-white, lined paper (see Figure 1c). This drawing, a wine goblet, was heavily shaded, folded twice, as in Trial 2, and placed inside two white, sealed envelopes. The target was known only to the person who prepared it, B.E.'s wife (S.E.), who was not present during the trial. The trial lasted 18 minutes, during which time Geller handled the envelope. He produced four attempts. His final drawing is shown in Figure 1c. When the envelopes were unsealed and the target drawing was revealed, Geller was obviously disappointed by his lack of success. He stated, "I never get a drawing when it is shaded. I could have got it if it wasn't shaded in."

As indicated in Table 1, for the remaining six trials of the series Geller made no further attempts to reproduce the target material. These further target drawings were either politely refused, ignored, or flatly rejected. Reasons for refusal included lack of time, fatigue, or the inability to produce effects mechanically or on command.

Comparison with nonpsychic controls

On the basis of observations made with the original target drawings presented to normal subjects, we cannot support Geller's claim that in Trials 1 through 3 he used paranormal or extrasensory perception. Although the target drawings were originally believed to be invisible when placed inside the containing envelopes, the experimental evidence suggests that when the envelopes are pressed flat and rotated to certain light orientations the target material can be seen through the envelopes as faint lines.

Forty-eight naive university students (aged 18 to 30 years) were presented with the target drawings of Trials 1 through 3 and two other target drawings for practice trials. They were given the following instructions: "This is an experiment on visual perception. I'm going to give you five envelopes. Inside each one is a drawing done on notepaper. The paper will be folded, and it may be inside another envelope. Copy the drawing inside exactly as you see it. You may hold the envelope up to the light or use any method you like, but you cannot open the envelope. We want you simply to draw an outline of what you see. The drawings are not complicated. There may be a few details in the drawing but they're not important. Just draw the main idea. No guesswork—just draw exactly what you see. The first two envelopes are for practice only." To allow for the special features of the notepaper containing the target for Trial 1, the subjects were told: "The paper in this envelope is folded and imprinted with a hotel name, along with the hotel's insignia and a border. You can ignore these aspects, but draw what you can see of the drawing." The subjects were randomly allocated to four different viewing times—5, 10, 20 or 50 seconds—and the 48 attempts at each target were drawn on three-by-five-inch cards.

In the second phase of the experiment six naive undergraduates judged the accuracy of each set of 48 student responses and of Geller's final drawings. Accuracy was judged against the original target in both its folded and unfolded states. Each of Geller's final drawings was faithfully copied onto a three-by-five-inch card so that it would be indistinguishable from others in the series. The judges used a scale defined as follows: 0 = nothing drawn; 1 = totally inaccurate; 2 = very inaccurate; 3 = inaccurate; 4 = slightly accurate; 5 = moderately accurate; 6 = very accurate; 7 = almost perfectly accurate. Mean interjudge reliability was 0.81.

Mean accuracy ratings obtained by the responses at each observation interval are given in Table 2. The criterion for accuracy in Table 2 is the unfolded version of each target. Geller's performance was significantly better than the average student performance at 50 seconds for both the drawings of the ship and goblet and about equal to the 20-second average for the drawing of the flower.

Unfortunately the total time Geller spent effectively examining each drawing is unknown, as is the effect of the prior experience Geller has had at perceiving drawings in sealed envelopes. In addition, Geller was allowed multiple attempts at reproducing the target material. It seems reasonable, therefore, to compare Geller's final attempts with the best of the student attempts at each inspection interval. These are presented in Figure 1. As

Table 2. Average accuracy of drawings by Geller, by students at four inspection
times, and by best students at each inspection time.

Target	Geller	Average students				Best students			
		5s	10s	20s	50s	5s	10s	20s	50s
1. Ship	4.33	1.65	1.65	1.84	2.83	3.83	4.00	2.83	5.17
2. Flower	3.33	2.48	2.87	3.36	4.23	4.33	4.67	5.00	5.67
3. Goblet	2.50	1.02	1.42	1.36	1.80	2.33	2.33	2.17	2.67

indicated in Table 2, with the exception of the flower drawing, where
Geller's accuracy was actually lower than that obtained by the best student
attempt at even 5 seconds, Geller's responses were within the range of
rated accuracy of the best student drawings. Geller's accuracy was no
greater than that obtained through methods of visual inspection. These
results lead, therefore, to the conclusion that Geller simply perceived the
faint outlines of the folded drawings through the envelopes.

The folded versions of the target material as perceived by transmitted
light are illustrated in Figure 1. The folded version of the goblet is of
special interest since, because of the heavy shading and the positions of the
folds, it bears little resemblance to the unfolded drawing. As shown in
Figure 1c, all responses, including Geller's, matched more closely the
folded versions of the target. Trial 3 was crucial because the goblet
drawing was the only case for which genuine paranormal powers were
necessary to accurately reproduce the target material. Geller's performance did not differ from that of subjects using normal visual perception
and can provide no support for the paranormal hypothesis.

Starting watches

Uri Geller claims that he can fix or start broken watches by a paranormal
method usually known as psychokinesis. Typically he instructs each participant to hold a "broken" watch in his hands and to think, "Work! Work!
Work!"

Seven jewellers that we surveyed estimated that over 50 percent of the
watches brought in for repair are not mechanically broken, but have
stopped because of dust, dirt, gummed oil, or badly distributed oil. In
such cases holding the watch in one's hand for a few minutes can warm
the oil, and additional handling or winding movements can help free the
working parts. These physical procedures are necessarily involved in

carrying out Geller's "psychic" instructions.

During one week six jewellers tried to start stopped watches that were brought in for repairs by such holding-and-handling methods before opening the case for inspection. Of 106 watches attempted, 60 (57 percent) started working. Forty-five of the 46 watches that did not start were judged by the jewellers in their "post-mortems" to have been impossible to start without opening the case to correct a mechanical defect.

In order to determine whether the "Geller effect" was greater than an equivalent holding-and-handling physical effect, two surveys were conducted. First, each member of a class of psychology students was asked to find one person who had attempted to start a watch with the aid of instructions Geller had given on TV or radio and to record the results. The students were also told to try to find a person with a broken watch and to carry out a fixed series of holding-and-handling procedures. The students were given detailed printed instructions that emphasized that they must have no prior knowledge of the results obtained by the possible subjects.

Second, these same two testing procedures were employed in a random-sample telephone survey. If the telephone respondent had not tried to start a watch by following Geller's instructions, he or she was asked whether there was a broken watch in the house, and if so, to bring it to the telephone. The experimenter then described the handling procedures and engaged the respondent in a nonpsychic topic of conversation during the holding period. The holding-and-handling procedures in both surveys consisted of a sequence of: (1) holding the watch in a loosely clenched fist for three minutes; (2) gentle rotations; (3) winding. The results are shown in Table 3, where it can be seen that the "Geller effect" was not superior to the simple physical effect.

Table 3. Percentages of watches that started following Geller's instructions and ordinary procedures.

Method	Geller's instructions		Physical procedures	
	Number attempted	Percentage success	Number attempted	Percentage success
1. Student survey	13	38.4	32	68.9
2. Telephone survey	30	53.3	29	44.8
Totals	43	48.8	61	57.3

What little evidence is available from interviews of audience members at a public stage-demonstration in Dunedin, New Zealand, suggests that Geller's presence in the room does not enhance his effect. Out of a sample of 12 watches given to Geller on stage, four started. Out of 17 watches held in the hands of audience members, three started.

At the several private demonstrations and public performances by Uri Geller we also observed a wide variety of other "Geller effects," including metal bending. On analysis every one of these cases had design features which would allow the use of normal (nonpsychic) procedures to achieve the effect. However, as analysis of these cases depends to a large degree on our accuracy as observers (albeit with audiotape recordings and film), they are not reported here.

Conclusion

Parsimony dictates the choice of normal explanations for the phenomena described here. Geller's procedures allow him to use ordinary sensory channels and ordinary motor functions. While we cannot with our data refute the Stanford Research Institute experiments, we question whether it is credible that Geller uses normal sensory-motor means outside the laboratory but switches to paranormal means inside it.

References

Hanlon, Joseph 1974. *New Scientist* 64: 170.
Hasted, J. B. et al. 1975. *Nature* (London) 254: 470.
Targ, Russell, and Harold Puthoff 1974. *Nature* (London) 251: 602.
Taylor, J. G. 1975. *Nature* (London) 254: 472.

New Evidence in the Uri Geller Matter

James Randi

In October of 1977, I received a letter from Tel Aviv, Israel, signed by Mr. Yasha Katz. It was an invitation to get from him the "truth" about Uri Geller. Katz had been Geller's manager for two years. I had previously met and spoken with Mr. Katz in New York concerning details of my book *The Magic of Uri Geller*, and in that book I declared that I felt Mr. Katz to be innocent of any wrongdoing in regard to promoting the Geller myth. It was my opinion that Katz was a dupe, and not one of the dupers. Our conversation at that time showed him to be very naive and an experienced apologist for Mr. Geller, but not necessarily a confederate in the deceptions.

These matters, however, are pretty well predictable. Those who are first taken in often are subsequently asked to cooperate in the fakery, for a variety of plausible reasons. Psychics have for generations pressed investigators to accept the "powers" they exhibit as ephemeral, spontaneous, and not always available, particularly under pressure or under "negative" influences such as skepticism. Quite honest disciples are convinced, in light of the miracles they have already witnessed, that occasionally cheating must be allowed—to maintain a reputation and to pay for the groceries. And seldom do the disciples fail to go along with this suggestion. But now, said Katz, he was prepared to tell all he knew about the psychic superstar.

It was inevitable that Yasha Katz would arrive at this point in his relationship with Geller. He had obtained a contract early in the Geller rise to fame that entitled him to a percentage of all engagements booked outside of the United States; and since he was in on all the press conferences and attended all the shows, he had to become aware of the methods of deception Geller was using. Still, a strong conviction remained that Geller did have genuine psychic powers. Katz discussed some of these things with me during that secret meeting in New York City, and I recorded the meeting in detail.

Two things had assured him that Geller had real psi-powers. One evening, he told me, he had attended a theater with Geller and a few other friends. Awaiting the arrival of a car after the show, he witnessed a "teleportation." When Geller had taken his seat in the theater, he'd called attention to the fact that there was an armrest missing from the seat he'd chosen. Now, while the two waited for the car to come, Katz was to witness a miracle. He heard a sound above his head, and looked up in time to see an object falling to the ground at his feet. It fell in a small puddle, and Katz retrieved it. Lo! it was a vinyl-covered armrest —doubtless from that same theater!

On another occasion, Katz had left Uri alone in their New York apartment and gone to the street, several floors down, to buy a newspaper. Upon his return, he discovered that a huge "planter" formerly located inside the apartment, was now standing outside in the hall. He excitedly awakened Uri, who was inside, supposedly sleeping soundly, and together they struggled the thing back into the room. And, said Katz, it was impossible that Geller could have put the thing there by normal means, since it was so heavy that even with both of them lifting it Uri strained himself in the process.

Both these events had convinced Yasha Katz that, regardless of the "feats of clay" that Geller exhibited, he did have *some* genuine abilities. Today, his opinion is rather different. Let us examine both these convincing claims.

In the first case, Katz gave me two different versions. During the recording we made in Israel, he embellished the story to say that he had looked up and had seen the armrest "levitated, floating in the air." Then it had fallen at his feet into the puddle. ". . . I removed it, and it was perfectly *dry*, not a drop of water on it.'' Really? I picked up a glass of water and poured part of it on the armrest, which Katz had saved as a relic all those years. The water ran off it immediately, and it was perfectly dry. I looked at Katz. He was smiling uncomfortably. As for the ''levitation,'' I reminded Yasha of what he had said just previously about Geller's methods of ''materialization''—but we'll come to that item shortly (point 6).

As for the planter that had spirited itself out of the apartment, Yasha was adamant. He had referred to it in our New York interview as ''like cast iron'' but this was a reference to its mass, not the material. My recording of this talk does not make that clear. Actually, as Yasha showed me, it was made up of heavy glass blocks, and weighed about 120 pounds or so. For Katz had it sitting right there in his Tel Aviv apartment! He'd brought it with him to Israel.

''You *can't lift* that planter!'' said Katz. I stood up to try. ''No, no! Never mind!'' he exclaimed as I began removing potted plants from it. He came over to me and tried to discourage me from trying this simple experiment. I heaved, and up it came. I replaced it, and put the plants back in position. Now, if I, at forty-eight years and weighing 155 pounds, can *lift* this thing, surely Uri Geller could *pull* it out into the hall a few feet while Yasha Katz is out of the building buying a paper!

On my recording of this event, there is an abrupt change of subject by Mr. Katz. I could not bring him back to the planter episode again.

My expenses for the Israel visit were picked up by RAI-TV, the Italian television network. Earlier in 1977, I'd contacted Piero Angela, a leading journalist in Italy, and he had set out to do a lengthy documentation of parapsychological work around the world. At first believing that he would have a fairly convincing story to tell about the scientific work being done, since he'd read the books and found it difficult to disbelieve them, he'd had a rude awakening after visiting the laboratories and interviewing the scientists involved. He'd found evasion and poor rapport everywhere. When I notified him of Katz's desire to confess his involvement, I called Angela and was immediately told that he was interested. Though the hour-long segment on Geller (there are five hours of this documentary) was already edited and ready to be broadcast, Angela felt the new information might be valuable enough to redo the segment. And it was.

First, I had a private recording session with Katz. From it, I made a list of main points he told me, and the next day, Katz sat before the cameras and gave details of these events. He also expressed serious doubt about his previous positive attitude about Geller, since a demonstration I'd given the night before had shaken him up a bit. There was hardly an intact spoon in the place, and mutilated keys were everywhere in the apartment.*

Here is a list of Katz's points, prepared from the tape and written up within hours of the interview. I have omitted several items from this list, since they are very personal in nature, and have little to do with the question of whether Geller is genuine or not:

Yasha Katz, interviewed by James Randi in Tel Aviv, December 1977, gave details of his association with Uri Geller, the "psychic," whom he worked with as manager for two years.

1. Katz says that Geller brainwashed him by performing impromptu tricks for him, even after Katz was aware that Geller resorted to tricks to fool his audiences. Such tricks were designed to keep Katz believing that Uri had *some* powers.

2. Katz was instructed to sit in the audience and give Geller special signals to indicate what word had been written on a blackboard out of Geller's sight. He signaled, using both his cigarette and his hands. This trick was presented by Geller as a genuine demonstration of ESP.

3. Geller took notes on unusual models of cars and license plates used by persons in the audience, and later revealed these details to his audience. On one particular occasion, Katz noted that Geller watched a newspaper reporter in California get out of a Porsche, and later he stunned the reporter by describing his car and the license plate number.

4. Geller stationed Katz near the box-office to note details about certain people entering the theater, which were relayed to Geller so he could reveal them from the stage as if he were using ESP.

5. Katz stooged for Geller to "teleport" a spoon for an *Express* reporter in Paris. He threw the spoon into the air to the ceiling.

6. Geller frequently, says Katz, threw objects from behind his back, over his shoulder, to make them "materialize," and Katz witnessed it many times.

7. Once, in Palm Beach, Katz caught Geller sneaking the lens cap off a

*Katz also introduced me to a young Israeli who conducted "telepathy" demonstrations with his eight-year-old sister. Katz was convinced it was real, since the two "never spoke to one another." The boy, a member of the local Israeli Magic Club made no claims of genuine powers. He said that Katz was just "determined to believe" and could not be dissuaded from wanting to manage another "real" psychic . . .

camera to perform his "psychic photo" trick.

8. When Geller was to appear in Birmingham, England, word arrived backstage that the front row was packed with magicians. Geller, says Katz, turned white and refused to go out on stage. He talked Katz and Werner Schmid, the producer, into making up a story about a bomb threat, and the two told this to the press. Geller claimed to the reporters and audience that Katz and Schmid would not let him appear, but that he wanted to do so. They left town with a police guard and the show was canceled.

9. Katz often caught Geller doing an ESP trick wherein he would try to guess what the subject had written; then when the subject's target was revealed, he would quickly duplicate it on his own pad secretly, and show it to the subject with great elation.

10. Visiting Sylvia Fine, Danny Kaye's wife, Geller broke a special piece of heirloom jewelry. Katz clearly saw Geller take it between his hands when she was not looking, and break it; he then claimed it had broken in a supernatural manner.

11. In London, Geller wanted to impress a publisher who was trying to get him to sign a contract. He arranged with Katz to leave a hotel telephone off the hook and concealed near the publisher, so that he, Geller, could listen in several rooms away. Then he burst into the room and "psychically" revealed details of their conversation.

12. Geller has been featuring an old trick recently in which four persons stand about a seated subject and lift him with their fingers. It is supposed to be a genuine "levitation." Katz showed this trick to Geller in Germany, as a gag. It is a common parlor-trick.

13. As an example of methods used on the media, Katz says that on an ABC morning TV show from San Francisco, he was instructed by Geller to go into the producer's office when Geller distracted the staff and determine what was in a sealed envelope. It was a drawing of a flag. Katz informed Geller, and Geller acted out, in his usual manner, the "ESP" impressions he was getting from the sealed envelope when he appeared on the TV show.

14. Last year, reports Katz, a reporter from the *National Enquirer* came to Katz for a story about Geller; but since he could only give her negative opinions and facts, she abandoned the story, saying her editor had instructed her to get only a positive story, regardless of the truth.

15. Katz says Geller and his confederate, Shipi Shtrang, now have become citizens of Mexico, since Mexico has no extradition treaty with Israel. Shipi left Israel without reporting for military service, and Geller got him false medical papers in New York from a cooperative doctor. Now that Shtrang and Geller are wanted for questioning in Israel, Geller is taking refuge in Mexico with Shtrang.

16. Geller's and Shipi's passports were arranged by president-to-be Portillo, at the urging of his wife, who is a devoteé of parapsychology and a good friend of Geller. The residency requirement was waived.

In light of this evidence, can there be any remaining "proof" of Geller's authenticity? He has consistently refused to appear for testing at reputable laboratories, though he has promised to show up. He will be tested only by those who already have a firm belief in such matters. The record is clear: Geller, as the others who claim these powers, ignores legitimate offers from interested scientists who can perform adequately controlled tests. His abilities are not only unproven—they are mythical.

Examination of the
Claims of Suzie Cottrell

James Randi

Suzie Cottrell is a 20-year-old woman from Meade, Kansas. Some months before she came to my serious attention, and that of the CSICP, she had appeared on the "Tonight Show" with Johnny Carson on NBC-TV, doing a card trick for the persons present and apparently fooling Carson, who has a reputation as a magician from his early days in the entertainment business. Carson had told me previously that he was not about to allow anyone on the show who claimed paranormal powers unless that person would submit to controls. One such person, Mark Stone, had approached the staff of the Carson show asserting that what he demonstrated was the real thing. That night, Carson wiped him out by implementing a simple change in the conditions of the trick, and Stone failed. Johnny told me that Stone's manager had insisted that he be presented as the real thing and that he was willing to be put to the test. He was.

It was not too difficult to see why Carson had been fooled by Suzie Cottrell. He's a sharp and observant man, but a pretty 20-year-old farm girl was too much for him when she performed a rather obscure card trick that Carson was unlikely to be familiar with. Besides, I have heard a story concerning Jimmy Grippo, a highly skilled card worker in Las Vegas who is said to have touted Cottrell to Carson and to have provided a demonstration of her talents for Johnny in advance of the NBC appearance.

Suzie's father contacted Paul Kurtz of CSICP, asking that his daughter be tested by us. But it was carefully specified that my own offer of $10,000 must not even be mentioned, to avoid putting a crass commercial flavor on the procedure. After some negotiations, the date of March 16, 1978, was agreed upon, and Kurtz designated Martin Gardner and me as those responsible for designing and controlling the procedure. We prepared a testing area by installing a videotape camera and outlining with white tape the exact area that the camera covered. This was to be the limited test area in which all action had to take place. Should Suzie's hands or any of the cards wander out of the test area, the experiment was to stop immediately and return to ground zero, all materials being confiscated and held for examination. We had a good number of decks of cards, all labeled and fresh. And, most important, we were going into the fray with a definitive statement that had been read and understood and agreed to by Suzie Cottrell and her group.

A word about the last circumstance would be proper at this point. In my 35 years of looking into these matters, I have found that the most common reason to fail to come to any firm conclusion in such testing procedures is the lack of a firm understanding of the conditions and parameters from the beginning. Thus, I insist that the subject know in advance just what is expected, agree in advance that conditions are satisfactory for the production of whatever miracle is to be shown, know exactly what will be accepted as proof, and finally agree to abide by the decisions reached under these conditions. This way, secondguessing and weak rationalizations for failures are not acceptable.

We told Cottrell that from the moment she entered the test room she would be under video surveillance, and she agreed to that. We asked if all was well—the "vibrations," the weather, the time of day, and so on, and she agreed that things were just dandy. She'd had a press conference in which she announced just how nervous she was, and we wanted to eliminate that alibi as a reason for any failure that might be noted. But she assured us of her excellent state of mind, and we were all ready to go. After Suzie and her parent signed the agreement, we entered the test room.

A TV camera was perched on a tripod, fixed upon the marked-off table. There was, unfortunately, an ABC-TV camera crew there as well; and in spite of my admonitions not to block our camera set-up, they did so repeatedly, but fortunately not at any moment of import. The subsequent story was not used by ABC in the documentary on the paranormal that they were preparing, which was broadcast on December 5, 1978. The reason for this was not made clear; but I feel that since Cottrell failed utterly, it was the kind of thing they refer to as a "non-story." That's the way it often is—successes are glorified, and the far more numerous failures and exposures are discarded.

A reel-to-reel videotape machine sat on the large table, off to one side. Psychologists Irving Biederman and James Pomerantz from the State University of New York at Buffalo were there, since we needed their careful observations and record-keeping for future analysis of what we were about to see. Martin Gardner and I sat behind the chair in which Cottrell was to be seated, one to each side, about five feet away. Several other SUNY people were there to act as a subject-pool for Cottrell to choose from.

We questioned Cottrell about her abilities. She said she could name all of the cards in a face-down deck "some of the time." We expressed understandable dissatisfaction with the estimate, and she then estimated her success rate as 48 out of 52, on the average. On several other feats that had been claimed for her in the papers, and during interviews, she gave similar estimates of success rates. But the one demonstration, the Carson stunt, was the one that she said was her best. She was "almost invariably" successful on that one.

Briefly, the routine referred to consisted of her predicting (in writing) which card would be selected from a spread-out, face-down deck. Now that is about the way it would be described by the unwary and, essentially, that description is correct. In addition, the deck used is a new, shuffled deck not supplied by her. The choice is quite free, and the prediction is written down well in advance, clearly and boldly, and the paper is retained by another person, not her.

What you have just read is what the popular pres has seen fit to use as a description of the famous Cottrell routine. Again, it is quite correct—as far as it goes. But, as the reader will discover, there are a few twists in there that make it far less than convincing as a psychic marvel.

But back to the lab. While the videotape machine near her remained on the ready but not running, Cottrell was told that she might perform her special routine without any controls being imposed. But she also was made to understand that this would not be part of the test, only a warm-up for her. She agreed, but not until her father was ushered out of the room. He made her nervous, Suzie told us; and the father averred that this was usually the case and that he was accustomed to being put out of the room when Suzie worked. There was a certain amount of chuckling at this comment, but Gardner and I exchanged glances. Something was becoming rather evident now, and the picture was a bit clearer.

We gave her a deck of cards. She noted that it was "plastic covered" stock, and she said she didn't much like that kind of card to work with. Immediately, I offered to withdraw the deck and give her a standard Bicycle poker deck, but she decided to use the one offered. It was important, you see, not to allow her any later excuse for failure. Even such a factor as a deck of cards she did not feel comfortable with could have provided adequate excuse (in terms of the usual parapsychological reasoning) for failure.

She shuffled and cut the deck many, many times, commenting at one point that she was "not very good at this sort of thing," with a cute smile. Gardner and I did not smile, recognizing a put-on at first flush, and she dropped the inept act from then on. She fussed endlessly with the cards, until the moment of truth came. Cottrell asked for a piece of paper, wrote something on it, and continued. Gardner and I observed the necessary moves being made, then watched as she suddenly banged the deck on the table and began spreading them wildly about on the surface of the table with both hands in a circular fashion, rearranging them repeatedly and "patting" them into seemingly random patterns. She pushed them into various configurations while she spoke, asking that her subject, a young woman she'd chosen to sit opposite her at the table, reach into the spread and select "any five cards." As each card was slipped out of the spread, Suzie would rearrange them somewhat, seemingly at random. At one point, however, she would stop doing this and back away from the table.

The subject had five cards facedown before her. Cottrell then eliminated some until one was singled out (more about that later) and announced it was the required card. And, in the four times she performed for us, she was right three times! Odds of this being done are a whopping 1 in 36,000—assuming, that is, that no trickery is being used.

But the trickery was quite obvious to Gardner and me. The routine was one well known to card magicians. It has been in print and available to the public for many years. Cottrell's fussing with the cards was quite haphazard until the point where she gathered the pack up and straightened it out. At that time—the "moment of truth"—she glimpsed the top card as it fell into place. This was the card she then wrote upon the piece of paper as the one that would be chosen. Further shuffling became more orderly, and each shuffle and cut retained the top card in place securely. Suddenly, the deck was slammed on the table top and a rapid spreading commenced, with the curious circular motion. But, we noted, the top card almost immediately got tossed off to one side, close to Cottrell and near her arm, which would cover it during the fussing about. It would be picked up into the motion by one thumb, slid about a bit, then dropped off again in a different position, but always in full control and available.

Finally, Cottrell introduced the chosen card into the main mass of cards, arranging it all by patting and pushing so that the one important card lay in a position where it was most likely to be chosen. As the selection process took place, however, additional patting served to push the desired card into a spot immediately within reach of the subject.

After all, she had 5 chances in 52 of having the required card chosen. And the greater mass of cards were kept in motion, or under her reach, so out of contention. But how was she able, even with the correct card among the 5 selected ones, to force the final selection of that one from the five?

Here is where the demonstration really falls apart. Cottrell used almost *any* means to arrive at the correct one-out-of-five, and seldom the same means. Suppose the needed number was 4. She would say, for example, "Name any two numbers between 1 and 5." If the answer was, "1 and 3," she would brightly suggest that $1 + 3 = 4$. A winner! Of, if the named numbers were 1 and 2, she would promptly elimiante cards number 1 and 2, and narrow it down to the remaining three cards, telling the victim to choose "another number," and cashing in if it were now 4, otherwise continuing to play about until only 4 was left. Or, she could make a 2 into a 4 by counting from the opposite end of the row of cards. The variations are endless, and Cottrell knows them all.

On the Carson show, seeing the ace of spades on the bottom of the deck as Ed McMahon shuffled it, Cottrell had asked *four* people at the table to select *five* cards each! Here she had 20 chances in 52 of her force-card being selected as she manipulated them about. She saw McMahon take the correct card, and then, with her back turned, she asked them to choose "the highest card" in anyone's hand—and of course, it turned out to be the ace of spades in McMahon's hand! How can we be so certain of that? Because, in this day of home video recorders, several people taped the performance, and the ace of spades even showed up on the bottom of the deck for the home viewers. Also, viewers saw Cottrell kneeling down at the table to get a better peek.

Which brings us to a point that certainly must have troubled the astute reader who has been following the protocol and implementation thereof in this account. I said that Cottrell had agreed to being recorded on videotape from the moment she entered the test room—but I then referred to the video recorder as "ready but not running" *after* we had already entered the room. True, *that* machine was *not* running—but the actual machine of record, which was *in the next room, was* running and recorded all of Suzie's "informal" uncontrolled tests in great detail, unknown to her. Thus she made no effort to hide her methods, figuring we were falling for the trick.

Later, of course, we were able to verify our observations very easily and were able to see the peek at the top card, the false cuts and false shuffles, and the final manipulation of the crucial card. Frank Garcia, the renowned expert on card techniques and gambling trickery, has seen this tape and certifies that the observations made by Martin Gardner and myself are *quite* correct.

Well, Suzie Cottrell, not realizing that we had the cat in the bag already, agreed that it was time for us to get down to serious business. We started up the dummy videotape machine, and Suzie tried her trick several times more. Now, however, there was a slight change in the procedure. We cut the desk just before she spread the cards—and she scored *zero* from then on in. For the reader's information, her chances of getting no cards right in another set of 4 attempts is 92.5 percent in a fair test.

In another test, she asked that each of two different subjects draw 5 cards from the deck and comapre their hands to see if they had "corresponding" cards, that is, one would have the same number and color of card as the other (jack of hearts/jack of diamonds, 2 of clubs/2 of spades, etc.) and in this experiment—which she herself suggested—she scored zero.

She tried to predict which persons would select the highest cards in a simple dealing of the pack. She tried it 80 times—again, her idea, not ours, but strictly controlled by us. She failed 80 times.

In another set of 104 tries at predicting cards to be turned up, she fell somewhat short of her estimated 92.3 percent success rate. She scored zero. A following set of 40 predictions failed 100 percent. In a series designed to test her ability to determine the suit of a card by psychic means—an ability that she had claimed—she obtained 22 percent where 25 percent would be expected by chance. And so on.

In a press interview later, Suzie complained that I had been "setting up mental blocks," though I offered to leave the room and the building, if she wished, and she had declined my offer.

But there are two things I have not revealed to you. In the document that Suzie Cottrell signed before beginning the tests was a set of statements that she agreed to, explicitly. She certified that she had not in the past used any form of trickery, subterfuge, or sleight-of-hand, and that she would not on this ocasion use any trickery, subterfuge, or sleight-of-hand, and that she would not feel compelled by the pressure of the situation to resort to any such methods. Yet she performed, in magician's parlance, the top-peek, false three-way cuts, top-retaining shuffles, and the Schulein force, followed by multiple-choice forces. There's not a great deal left in the lexicon of the card-sharp.

Second, there came a pause in the middle of the testing procedure where Suzie asked for a break. I had just been shuffling a deck for her to use, and I placed it on the table, within camera range, of course, to await her return to the room. Shen she re-entered, she saw the videotape reels not moving, and she sat at the table while the hubbub went on about her. On the video tape, we see her hands pick up the deck. She blatantly looks at the top card, then replaces the deck in position. Then, in a loud voice she announces that she is "ready." You bet she was, and until I reached forward at the last moment to cut the cards before she recommenced the tests, she felt an imminent success in the air.

Suzie Cottrell does tricks with cards. The methods are standard conjuror's methods and not in any way psychic miracles. Her *mise en scene* is an appealing one: she says she wants to develop her powers to help autistic children and that she wants no money and has no interest in becoming a professional psychic. After the encounter in Buffalo, she need not specify the latter intention. However, it seemed fairly obvious, from certain aspects of the examination, that her father was unaware of the trickery and believed that Suzie had genuine powers.

As she left the session, in tears, Suzie Cottrell turned to one of the psychologists who had controlled the tests: "I'm going to forget what you did to me today," she cried, "and if you ever have any autistic children, I'll help them all I can, in spite of what you've done!" Martin Gardner and I turned away in dismay. And Irving Biederman, the psychologist, summed it up for the press this way: "On the basis of the tests, one cannot discriminate between Suzie Cottrell and a fraud." Amen.

A Test of 'Psychic' Photography

Christopher Scott
and Michael Hutchinson

We were asked by Granada Television to participate in tests it was to conduct in London on August 8-10, 1978, with Masuaki Kiyota, a 16-year-old Japanese who had been featured in part of an NBC-TV program on the paranormal. It is claimed that Masuaki can project mental images onto unexposed film under very tight conditions.

Granada proposed to start the tests very informally under whatever conditions the young man desired. Further tests would be progressively more difficult, first of all under the control of Professor John Hasted in a laboratory at Birkbeck College, and finally under the skeptically tight conditions of the writers. We would therefore not be required to participate until the third day. In the meantime we were kept informed of progress.

Through a change in plans, the very first day was spent in the laboratory of Professor Hasted. Whatever the protocol, the results were negative. The monotony of the day was broken by the production of what Hasted hailed as a "whitey," a term invented by American parapsychologist Jule Eisenbud. This was a photograph which, when peeled, was completely white, whereas it should have been black, as Masuaki

The "Trafalgar Square" photo

works without operating the camera shutter. When Morris Smith, the technical representative from Polaroid, was asked for his comments, he pointed out that a light-sensitive diode—which had been used for the previous experiment—had been removed, leaving a hole through which light could pass. When the rest of the pictures from the pack were removed and peeled, they too were "whiteys," to no one's surprise.

At the start of the second day, Masuaki managed to produce a photograph of poor quality, with the foreground completely underexposed and the rest of the picture blurred. In the center was a grey "pole" which could have been Nelson's Column in London's Trafalgar Square, although it was not clear enough for anyone to be absolutely positive. A building on the right showed up well enough to be compared with an original, if this could be identified. We later spent considerable time in and around Trafalgar Square but failed to find an angle from which the picture could have been taken. However, most people who saw the photograph initially thought that it *was* Trafalgar Square.

The second photograph, even poorer in quality, was said to show six guardsmen in file, from the right to the left. Questions were then raised as to whose camera and film had been used for these shots. It transpired that both had been given to Masuaki the previous evening and had been with him overnight. How many shots had been left on the film was not known, but "Trafalgar Square" was on frame seven and the "Guardsmen" on frame eight. While the former could have been the result of a normal shutter release exposure without pulling the photograph from the camera, the latter could not. Morris Smith stated that it was not possible to tell if the film pack has been tampered with.

The rest of the day was spent conducting tests using a camera and film supplied by Morris Smith. The location was changed from the previous day's laboratory to a room in the White House Hotel—where the first two photographs were produced—and later to the local park. Relatively tight conditions produced no results. But once more, after being allowed to have a camera and film to himself—this time for a two-hour lunch break—Masuaki achieved some success. What actually happened was later to cause a fierce argument between the interpreter on one side and the television presenter and his director on the other.

The interpreter was an Anglo-American journalist named David Tharpe, who had lived for some years in Japan. He had written at least one article about Masuaki, had been present at the young psychic's home during some successful experiments, and had been proposed by Masuaki (or his advisers) as interpreter for the London trip.

In an interview at the end of the test, Tharpe said that the most interesting photograph taken during the two and a half days was the one taken after the lunch break on the second day. He said that the camera and the film had not left his sight or that of his girlfriend, a Japanese-American who had been in London for some days. She was not part of the "official" group from Japan, which consisted of Masuaki, Tharpe, and Professor T. Miyauchi, who is an expert on "nengraphy," which is Japanese for "thoughtography." He didn't take any part in the Granada experiments.

According to Tharpe, the first two photographs were black, but the third produced a picture. This "picture" looked like a black square partly washed out by a white circle, leaving two black corners in the shape of triangles. Tharpe said that Masuaki had asked him how to symbolize Switzerland. His answer had been, "By the Alps, or the Matterhorn," subsequently explaining that the shape of these was triangular. The photograph was therefore captioned "The Matterhorn."

The version given us by the producer on the morning of our arrival was more detailed. The session after lunch took place in the hotel room. Masuaki said he wanted to work without the film crew—there were too many observers around, he said—so the camera and lights were switched on by Gordon Burns, the program presenter. Also present were Jeremy Fox (the producer), David Tharpe, and his girlfriend. Granada suggested the use of another Polaroid camera, and a fresh film pack was loaded in front of the movie camera. But, said the producer, Masuaki wanted to use the camera that had been with him over lunch. He drew attention to this camera and the case that had held it. This was a small black case with a hinged lid, and was only just large enough for a Polaroid camera.

With both hands Masuaki placed the camera in the case and closed the lid. After several seconds, the producer told us, Masuaki took the camera from the case and produced the "Matterhorn" photograph. This was the first photograph, not the third as claimed by Tharpe, who also seems to have forgotten that, for no apparent reason, Masuaki had placed the camera in the case.

We have tried, without success, to find out what the developed movie film showed. However, the circumstances do not convince us that the photographs were evidence to prove the case for "thoughtography," for Masuaki could have operated the shutter release while placing or removing the camera from the case.

As previously mentioned, the afternoon brought no further results. Masuaki thought that he might be able to produce some photographs that evening if a small party were held in his room. This the television team arranged, and the party was held. But still Masuaki was unable to live up to his claims.

When we arrived on the third morning we were told that Masuaki had again been left with a camera and film overnight. It was confidently expected that—as on previous occasions—this would lead to an early success.

The first photograph to be peeled was black. But the second had a gray impression of the Eiffel Tower with a lamppost in the foreground and slightly to the right of the tower. After inspection, Morris Smith again said that there was no evidence to show whether or not the film pack had been tampered with. The serial number on the photo confirmed that the film was from the batch supplied by Polaroid for the tests. This had also been the case with the other "successes."

Further tests using fresh film were not successful. After about an hour a short break was called. While few people were paying attention, Masuaki pulled a shot from the camera. On it were a number of vague light smudges running down the picture. These, it turned out, can be produced by pulling the film out of the camera in a jerky manner, as Morris Smith effectively demonstrated. Further tests using fresh film

Image of Royal Albert Hall exposed onto Polaroid film by Mike Hutchinson without a camera, using a color transparency and a "pinhole projector."

were not successful.

Morris Smith told us that, while in theory it is possible to take a film pack to pieces, expose each frame to the light, and reassemble it, he wasn't exactly sure how. (Obviously this is not something that Polaroid needed to do.) The main problem is that complete darkness is necessary. The film is very fast (3000 ASA, compared to around 100 ASA for normal film) and even a "safe" light used in a conventional darkroom would fog the film. He and one of the writers (M.H.) decided to take the pack to pieces and reassemble it. M.H. tried to reassemble it with his eyes closed, but found that when trying to slide the final piece on, it fouled the film, and if he continued would have folded it back. Opening his eyes, he realized that this final piece need not be slid on but could be snapped in place without difficulty.

It still remained to be discovered how an impression could be made on each frame without showing signs of tampering. We knew that the film could be folded back without showing signs that it had been. Perhaps at this stage a contact print could be made. Alternatively, it might be possible to make a crude pinhole "projector" using a slide viewer. If pointed at the folded film, a quick press on the slide illuminator might create enough light to produce a poor quality picture.

To test this theory, M.H. carried out experiments in his apartment hallway. This was the darkest place he could find. Even so, a small amount of light came through glass panels at the top of five doors. This light was just enough to see by when his eyes became dark-adapted, but did not fog the film in any noticeable way.

The front lens of the slide viewer was covered with a piece of cardboard that had a hole in it made by a small darning needle. The pack was taken to pieces and the film was unfolded. One end of the film is fixed to the pack by a staple. Taking care not to tear the film from the staple, it was folded back with one hand. The other hand pointed the "projector" toward the film at a distance of six to eight inches. An exposure of two or three seconds is all that is needed. In this way, up to eight frames can be exposed.

In M.H.'s experiment the "projector" was hand-held. While this did not cause much "shake" on the prints, it did make it difficult to expose the full frame every time. However, it would be quite simple to make a small stand that would both steady and aim the projected image correctly.

The size of the subject on the print can be varied by altering the distance of the projected image. This means that sometimes only part of the original slide will be reproduced. The positioning of the main subject can also be altered by different angles of projection. For example, two of Masuaki's previous successes were aerial photographs of the Statue of Liberty. While the view and the shadows were identical, the Statue in one picture was in a different position of the frame.

These variations are to be expected when using a pinhole "projector." Similarly, variations of exposure also occur. Could it be that, after seeing the underexposed photographs of "Trafalgar Square" and the "Guardsmen," Masuaki decided to give a second or two extra to the "Eiffel Tower," which turned out to be overexposed?

Conclusions

Under fairly tight conditions Masuaki Kiyota was unable to fulfill his claim to project mental images onto film. The only time he achieved any success was after the film had been in his possession—and not under any control—for at least two hours.

Unfortunately, no steps were taken to ensure that the film pack could not be tampered with. In fairness to the television team, this only seems important with hindsight. Given the benefit of this report, we hope that future experimenters will either seal the pack or, preferably, prepare it in such a way that any signs of tampering would be obvious.

Finally, a few words of praise for the television team. They started the assignment believing in Masuaki and hoping for good results. Their initial enthusiasm and confidence gave way to understandable skepticism when it became evident that Masuaki couldn't perform under tight conditions as claimed. However, they didn't let this sway them in their professional approach, nor encourage them to loosen the controls. Whether the negative results will lead to a final program being broadcast remains to be seen. •

The Extraordinary Mental Bending of Professor Taylor

Martin Gardner

No one can say that John G. Taylor, professor of mathematics at Kings College, University of London, is not a brilliant and colorful personality. He was born in 1931 at Hayes, Kent, the son of an organic chemist. After getting his doctorate at Cambridge University, he taught mathematics and physics at a number of colleges in England and the United States, including a stint as professor of physics at Rutgers University. His technical papers (more than a hundred) display a wide range of interests that include pure mathematics, particle physics, cosmology, and brain research.

There is another side to Professor Taylor that I can best characterize as that of a ham actor who thrives on crowd adulation and personal publicity. When in the United States, he studied acting at the Berghof Herbert Studio, in Manhattan, and for a while was "sex counselor" for *Forum* magazine. In England, his constant appearances on radio and television shows made him such a celebrity that in 1975, when the respected British magazine *New Scientist* conducted a poll of readers to determine the world's top twenty scientists, Taylor made the list. The magazine's cover ran his picture alongside Archimedes, Darwin, Einstein, Galileo, Newton, and Pasteur!

Taylor also enjoys writing popular books about science, of which his best known was the international best-seller *Black Holes* (1973). It is not a bad introduction to black-hole theory, but toward the end of the book Taylor indulges in lots of freaky conjectures. He thinks it quite possible, for example, that Earth was visited in the distant past by extraterrestrials, who may have come in spaceships driven by "black-hole power generators." Saturn, he tells us, is the most likely planet that "high-gravity aliens" could have used as a way-station in their explorations of our solar system.

In his last chapter, Taylor considers the possibility that we have souls that are structured forms of energy capable of moving from one body to another. The universe, he reminds us, has two possible destinies. It may expand forever to die the familiar thermodynamic "heat death," or it may go into a contracting phase and eventually be crushed out of existence by a black hole. In either case, no matter will be left "which could realistically be said to be worth having a soul." However, the universe may bounce back from the big crunch. "The only chance of immortality then is in an oscillating universe. Even in that, everlasting life will not be of the usual form but one in which there may be no relation at all between one cycle and the next due to the enormous re-scrambling of matter in the collapsed phase. It could well be that souls will have to cast lots as to which of the variety of bodies they will inhabit in subsequent lives. That is, of course, unless the hand of God intervenes, his wonders to perform."

There is one other possibility of immortality. If one fell into a black hole, says Taylor, he might emerge in a parallel universe. This, however, has a big shortcoming. If two "close friends" fell into different holes, they could find themselves in separate universes with no possibility of reunion. "So there is always the chance that the immortality gained by falling through a rotating black hole may be a very lonely one."

In view of such quirky speculation, it was not surprising that in 1973, when Taylor appeared on a BBC television show with Uri Geller, he was so stunned by Geller's magic that he became an instant convert to the reality of ESP and PK. Geller did his familiar trick of duplicating a drawing in a sealed envelope. "No methods known to science can explain his revelation of that drawing," wrote Taylor with his usual dogmatism. The professor's jaw dropped even lower when Geller broke a fork by stroking it. "This bending of metal is demonstrably reproducible," Taylor later declared, "happening almost wherever Geller wills. Furthermore, it can apparently be transmitted to other places—even hundreds of miles away."

"I felt," said Taylor in his most often quoted statement, "as if the whole framework with which I viewed the world had suddenly been destroyed. I seemed very naked and vulnerable, surrounded by a hostile and incomprehensible universe. It was many days before I was able to come to terms with this sensation."

Although Taylor was supremely ignorant of conjuring methods, and made not the slightest effort to enlighten himself, he at once set to work testing young children who had developed a talent for metal bending after seeing Geller on television. Taylor's controls were unbelievably inadequate. Children, for example, would put paper clips in their pocket and

One of the youngsters that Taylor, in *Superminds*, claimed could bend metal.

later take one out twisted. Nevertheless Taylor was persuaded that hundreds of youngsters in England had the mind power to deform metal objects. Curiously, Taylor never actually *saw* anything bend. One minute a spoon would be straight, later it would be found twisted. Taylor named this the "shyness effect." Metal rods were put inside sealed plastic tubes and children were allowed to take them home. They came back with the tubes still sealed and the rods bent. One boy startled Taylor by materializing an English five-pound note inside a tube.

So certain was Taylor that his high I.Q., combined with his knowledge of physics, gave him the ability to detect any kind of fraud that he rushed into print a big book called *Superminds* (published here by Viking in 1975). It will surely go down in the literature of pseudoscience as one of the funniest, most gullible books ever to be written by a reputable scientist. It is even funnier than Professor Johann Zöllner's *Transcendental Physics*, inspired by the psychic conjuring of the American medium Henry Slade. Taylor's book is crammed with photographs of grinning children holding up cutlery they have supposedly bent by PK, tables and persons floating in the air during old Spiritualist seances, glowing ectoplasmic ghosts, psychic surgeons operating in the Philippines, Rosemary Brown displaying a musical composition dictated to her by the spirit of Frederic Chopin, and numerous other wonders.

Not the least peculiar aspect of Taylor's volume was his argument that all paranormal feats, including religious miracles, are explainable by electromagnetism. "The Geller effect is a case in point. Will it ever turn out that the miracles of Jesus Christ also dissolve in scientific speculation. ...This book has presented the case that for one modern 'miracle,' the Geller effect, there *is* a rational, scientific explanation. This explanation is also claimed to allow us to understand other apparently miraculous phenomena—ghosts, poltergeists, mediumship, and psychic healing. What, then, of other miracles? Can they too be explained by these newly discovered powers of the human body and mind, and the properties of matter broadly described in the book?"

After writing *Superminds*, of which let us hope he is now super-ashamed, Taylor slowly began to learn a few kindergarten principles of deception. When the Amazing Randi visited England in 1975, Taylor refused to see him; but Randi managed to call on him anyway, disguised as a photographer-reporter. You'll find a hilarious account of this in Chapter 10 of Randi's Ballantine paperback, *The Magic of Uri Geller*. Taylor proved to be easier to flimflam than a small child, and his "sealed" tubes turned out to be so crudely sealed that Randi had no trouble uncorking one and corking it again while Taylor wasn't looking. Randi even managed to bend an aluminum bar when Taylor's attention was distracted, scratch on it "Bent by Randi," and replace it among Taylor's psychic artifacts without Taylor noticing.

Another crushing blow to Taylor's naive faith in Geller was a test of the "shyness effect" by two scientists at Bath University. They allowed six metal-bending children to do their thing in a room with an observer who was told to relax vigilance after a short time. All sorts of bending at once took place. None was observed by the observer, but the action was secretly being videotaped through a one-way mirror. The film showed, as the disappointed researchers wrote it up for *Nature* (vol. 257, Sept. 4, 1975, p. 8): "*A* put the rod under her foot to bend it; *B, E* and *F* used two hands to bend the spoon ... while *D* tried to hide his hands under a table to bend a spoon."

Slowly, as more evidence piled up that Geller was a charlatan and that the "Geller effect" never occurs under controlled conditions, Taylor began to have nagging doubts. After several years of silence, he suddenly announced his backsliding. Of course he didn't call it that. Instead, he and a colleague at Kings College wrote a technical article for *Nature*, "Can Electromagnetism Account for Extrasensory Phenomena?" (vol. 276, Nov. 2, 1978, pp. 64-67; also Skeptical Inquirer, Spring 1979, p. 3.)

In *Superminds*, after considering all possible ways to explain psi phenomena by known laws, Taylor concluded that only electromagnetism offered a viable possibility. The *Nature* paper reinforces this view. Electromagnetism, the authors decide, "is the only known force that could conceivably be involved." They then report on a series of carefully controlled tests of ESP and PK using talented subjects. No psi phenomena occurred. When controls were eased, the phenomena did take place but the experimenters could not detect a whiff of electromagnetic radiation. Their conclusion is that all the phenomena they investigated, metal bending in particular, have normal explanations.

More was to come. In *Nature* (vol. 279, June 14, 1979) the same authors published a sequel to their first paper. In this sequel, titled "Is There Any Scientific Explanation of the Paranormal?" they again stress the fact that "on theoretical ground the only scientific explanation [for psi forces] could be electromagnetism." Their conclusion is that neither electromagnetism "nor any other scientific theory," including quantum mechanics, can explain dowsing, clairvoyance, or telepathy. "In particular there is no reason to support the common claim that there still may be some scientific explanation which has as yet been undiscovered. The successful reductionist approach of science rules out such a possibility except by utilization of energies impossible to be available to the human body by a factor of billions. We can only conclude that the existence of any of the psychic phenomena we have considered is very doubtful."

Now it is pleasant for skeptics like me, who also regard psi phenomena as possible but "very doubtful," to welcome Taylor back to our ranks. But surely his reasons are as shaky as those that converted him to the paranormal six years ago. The history of science swarms with observed phenomena that were genuine but had to wait for centuries until a good theory explained them. A lodestone's magnetism was sheer magic until the modern theory of magnetism was formulated, and even today no physicist knows why the acceleration of electrical charges inside atoms causes magnetic effects. It is not even known why electricity comes in units of positive and negative charge, or whether magnetic monopoles exist as theory seems to demand.

Kepler correctly decided, on the basis of confirmable correlations, that the moon causes tides; but in the absence of a theory, even the great Galileo refused to believe it. One could add hundreds of other instances in which a phenomenon was authenticated long before a theory "explained" it. On this I find myself in full agreement with J. B. Rhine and other parapsychologists who regard the lack of a physical theory as no obstacle whatever to the acceptance of psi.

Science cannot absolutely rule out the possibility of anything, but it can assign low degrees of probability to extraordinary claims. In my view, which is the view of most psychologists, the classic psi experiments are more simply and plausibly explained in terms of unconscious experimenter bias, unconscious sensory cuing, fraud on the part of subjects eager to prove their psychic powers, and, on rare occasions (such as those recently disclosed about S. G. Soal), deliberate fraud on the part of respected investigators.

The central point is this. When science assigns a low degree of credibility to an extraordinary claim, it does so by evaluating the empirical evidence. Geller and the spoon-bending children are indeed frauds, but the reasons for thinking this have nothing to do with the fact that the supposed "Geller effect" is unsupported by an adequate physical theory. It is because the conjuring techniques for fraudulently bending metal are now well known, and because the metal invariably refuses to twist whenever the controls are commensurate with the wildness of the claim. •

Downfall of a Would-Be Psychic

Donald H. McBurney and Jack K. Greenberg

Steve Shaw bends spoons, reads minds, and performs other amazing feats. He has lived in England, South Africa, and Australia, besides western Pennsylvania. Would I be willing to go on television to test him? I (D. McB.) was talking with Susan Michaels of the "Pittsburgh 2day " show, which is broadcast over station KDKA. Susan was obviously taken by Steve's story and his naive personality. Sound like anybody familiar? If you guessed Uri Geller, give yourself an A+, because Shaw says he developed his powers after seeing Geller on TV. I explained to Susan, who **is the assistant producer of the show,** that I was a psychologist with a skeptical interest in ESP as a hobby, and that a magician was needed to do a test properly. Nevertheless, she wanted me because of my credibility as a professor and also because she felt, wisely, that Steve would balk at the presence of a magician.

After some negotiation, she agreed that magician Jack Greenberg could hang around the set, posing as a crew member. Jack and I set up the conditions for the spoon-bending, which essentially consisted in having Susan obtain six stainless-steel spoons, carefully marking them, and keeping them locked up until just before the show. No one else was to know their whereabouts. Steve was not to touch the spoons before the show, although he was to be permitted to see them from a few feet away. After many conversations with Susan and Jack, we were ready to go. Both Steve and his agent called me in a veiled attempt to find out what I knew about such things. Steve wanted to see me the night before the show. I agreed to meet him for lunch the next day at 11:30 A.M., just before the show.

Just like the wolf meeting the three little pigs in Farmer Brown's garden, Steve showed up at the station much earlier than scheduled. But

Jack, anticipating this, was there by 9:30 A.M. He spooked around as Steve badgered everyone in sight for a look at the spoons. Much to the credit of the people at KDKA, Steve got nowhere.

Our lunch was as predictable as the McDonald's fare we ate. He tried to probe my degree of sophistication. I detected that he had done some homework on me (as I had done on him). Just before the show, he asked hosts Patrice King and Wayne Van Dyne to go out into the hallway with him. Each of them was asked to think of a word and write it inside a circle on a paper. After ostentatiously folding the paper and tearing it up (the center-tear fold), he had a sudden call of nature. Jack dove into the men's room right behind him to find him stuffing paper into his pocket.

The show itself was something of an anticlimax. He "managed" to choose which film canister out of a dozen or so had a nickel in it and they chatted about how he discovered his powers. This canister trick was so obvious that the hosts and the producer detected how he did it, but they kept mum. We felt that foiling him on this trick would lead to his passing on the spoons, and after all of our work we wanted him to try this. I came on after a commercial break and talked a bit with the hosts. When asked whether I thought Shaw had supernatural powers, I said that many people bent spoons. Some call themselves magicians and others, mentalists, The difference is in their fees. Shaw said, when I asked him, that he would not use magic in bending the spoons. He was absolutely correct, because the spoons did not bend. He gave up after a minute or two, obviously shaken. He also decided not to try to bend the three-inch masonry nails that Susan had provided for him. After the show, when he was performing for two members of the audience, we noticed that he had his own three-inch masonry nails, of a slightly different cast from the ones that had been provided. Not surprisingly, one of these had a pronounced bend in it by the time we came upon the impromptu show.

I felt sorry for Shaw. After all the trouble Jack, Susan, and I had gone to, I was hoping that something interesting would happen. I felt a little guilty about tricking him, even though he would have tricked us if we had let him. He returned to Washington, Pennsylvania, and to his job mopping the floors of the local hospital, having wasted his chance to impress the largest Pittsburgh TV audience for that time-slot. He, like many other magicians, will try to make the big time by copying his self-proclaimed inspiration, Uri Geller. Perhaps he should have studied Geller's career more closely, because most of what I suggested in the way of controls was borrowed from Randi's book *The Magic of Uri Geller*.

There is no doubt that this episode will be replayed many times, as long as some people will want to believe in the paranormal and others are willing to gull them.   ●

Stories of Life and Death

Psychology and Near-Death Experiences

James E. Alcock

> *...and if I die before I wake,*
> *I pray thee, Lord, my soul to take.*
>
> —From a children's prayer

It seems easy enough for us to accept the ephemeral existence of a bolt of lightning or a cannon's roar. At one moment it doesn't exist, a moment later it does, and in yet another moment it is dissipating into nothingness. We don't ask, "Where is it now?" It is nowhere and it has gone nowhere. It just doesn't exist anymore. Neither do we wonder about the continued existence, in some form, of those garden vegetables whose growth we so carefully guided and nourished until we ended their lives in order to lengthen our own. Mosquitoes and mice, kumquats and kangaroos, they too, at least in the view of most people brought up in the Western tradition, must meet their end without promise of a flora-fauna heaven to receive their eternal souls. It is quite a different matter, however, when we ourselves are faced with shuffling off our own mortal coils. Whether suffering the loss of a loved one, or pondering our own certain demise, it is emotionally difficult to accept that a lifetime of learning, living, and loving, of struggle and self-improvement, leads but back to dust whence we came. For many, life would seem devoid of meaning if death really marks the end of our existence.

During the social evolution of human behavior, religions gradually developed that helped to assuage these age-old existential and ontological anxieties about living and dying. No one knows, of course, how religion began; but one can imagine that our cave-inhabiting ancestors, as their reasoning skills slowly grew, must have puzzled over such things as death and the appearance during sleep of the forms of people whose bodies, they knew, had long since provided a snack for some wandering beast of prey. We may not be far wrong if we speculate that, at the same time they gave magical (supernatural) explanations to the mysterious workings of nature, they explained their dreams of the dead by believing that the dead were still alive in some form. Not only would that have accounted for their experiences, but presumably it would have been comforting with regard to their own existence as well.

At any rate, one way or another, religions developed that taught that the "soul" survived bodily death in some form, and this belief is still a central tenet of almost all religions. Such a belief is very functional for most people, since it is anxiety-reducing. To give it up without having anything to take its place is a very serious matter indeed. Imagine the consternation of those among nineteenth-century scientists who also happened to be devoutly religious as the rapid rise of post-Darwinian materialistic thought threatened to explain away, in mechanistic terms, all the mysteries of human existence. So it was that in 1882 a group of British scholars, anxious to find a "scientific" basis for their religious beliefs by means of finding empirical evidence of the postmortem existence of the soul, organized themselves into the first formal parapsychological group, the Society for Psychical Research. Thus, as Rogo (1975) points out, parapsychology was born and bred in a survivalist milieu.

Until very recently, interest among parapsychologists in survival research had waned considerably from the early days, and there has been a great divergence of opinion among parapsychologists about the importance of such research. Joseph Banks Rhine, the acknowledged "dean" of American parapsychologists, believes that survival is an issue that can never be resolved and, thus, that time shouldn't be wasted trying to resolve it. Gardner Murphy, too, believes that no evidence can ever prove postmortem existence (Rogo, 1975). Yet we are now in the midst of a resurgence in survival research, spurred perhaps by the deep existential uncertainty of our times, as well as by the will of James Kidd. Kidd left a fortune to be given to any person or group that could find proof that the soul continues to exist after death; and following a legal battle among 130 contestors, the American Society for Psychical Research was awarded the money ($270,000), which it immediately began to channel into research projects dealing with the question of survival (Osis &

Haraldsson, 1977).

Today, there are a number of researchers studying survival and matters related to it. Some, like Ian Stevenson at the University of Virginia, claim to have experimental evidence for the existence of a body other than the physical one, which, even during life, can at times separate itself from the body of flesh and blood. Others, as we shall see, have concerned themselves with the reported experiences of people who have been near death, whose hearts have stopped beating for a short time but who were eventually resuscitated. Some such people believe that they were transported to another level of existence during what they consider to have been a period of death. (Incidentally, why anyone would consider temporary cardiac inactivity to indicate a period of death is difficult to understand in this day and age.) Yet other researchers claim to have evidence of reincarnation from their studies of subjects hypnotically "regressed" to memories of some allegedly prior existence.

My aim is to examine these claims of evidence of survival and to present information about relevant "normal" psychological phenomena that would seem to account for all such evidence.

The Experience of Dying

According to Raymond Moody (1975, 1977), the experiences of patients who have survived "death" provide compelling evidence of the continued postmortem existence of the soul. People from widely divergent backgrounds and belief systems experience very similar events following their deaths, he says, and this communality of experience points to the *reality* of the experience. While there is considerable variation on the theme, the "theoretically complete model experience," according to Moody, is as follows: At the moment of greatest physical discomfort, the patient hears the physician pronounce him dead, and he hears an uncomfortably loud ringing or buzzing sound. He feels himself being drawn rapidly through a long tunnel. He notices that he has a new body with powers very different from the old one, and may even see his old body lying on the bed with the resuscitation team gathered around it; but his vantage point is outside and above his body. He catches sight of dead relatives and friends and encounters a "being" of very bright light, a "loving, warm spirit." This spirit helps him to panoramically review the events of his past life. He is overwhelmed by feelings of love, joy, and peace. He has a vision of all-encompassing knowledge, the wisdom of the ages. Finally, he comes to some kind of barrier, but is made to turn back and go, reluctantly, back into his body. Following his resuscitation, he is emotionally very moved and is no longer afraid of death.

Moody is a psychiatrist and is well aware that certain physiological conditions might bring on such experiences, but he takes great pains to argue that no physiologiccal or psychological factors could account for all the data. However, he appears to ignore a great deal of the scientific literature dealing with hallucinatory experiences in general, just as he quickly glosses over the very real limitations of his research method. He argues that, since obvious possible causes of hallucinations can't account for his data (a conclusion he draws more or less by fiat), therefore it must be a matter of genuine psychic experience. This argument is of the classic Abbott and Costello form, where one of them "proves" that the other person "isn't here":

> "Are you in London?"
> "No."
> "Are you in Paris?"
> "No."
> "Are you in Moscow?"
> "No."
> "Well, if you aren't in London, Paris, or Moscow, you must be some-where else."
> "Yah, I guess so."
> "Well, if you're somewhere else, you can't be here."

If it's not due to psychological, neurological, or pharmacological factors (that can be identified), then it has no natural explanation. The explanation must lie somewhere else, beyond our "ordinary" world. This kind of reasoning is common in the survivalist literature.

We must wonder, too, about the reliability of the memories of Moody's respondents. They related their death experiences to him usually long after they occurred, and generally after having heard Moody speak on the subject. Even then, he finds a wide spectrum of experience, some people reporting only one or two of the key elements, others reporting most of them. Also, he says, many people report no experience at all. It is interesting to note (Moody, 1975, p. 26) that the respondents "uniformly characterize their experience as inexpressible," despite their ability to describe them in considerable detail. Moody admits that people have reported similar experiences in situations where no jeopardy to life was involved. Mystical and drug experiences can be quite similar, he admits. To account for this, Moody begs the question and speculates that during such non-death experiences, the mechanism that releases the soul at death may be prematurely triggered.

Karlis Osis and Erlandur Haraldsson are both psychologists who obtained Ph.D.'s for parapsychological research. Like Moody, they have been interested in the experiences of the dying. Their research is described in their recent book *At the Hour of Death* (1977). According to them, a typical death experience involves:

1. a period of exaltation shortly before death,
2. a much higher than normal rate of hallucinatory experience just before death, and
3. a "vision," usually while the patient is fully conscious and not sedated. The vision is generally that of a dead friend or relative, who typically is described by the patient as being there to take him into death.

This, if true, would be both interesting and comforting. However, can we have any confidence in the veridicality of the patients' reports? The authors in this case didn't even interview the patients, but instead sent out large numbers of questionnaires to physicians and nurses (in the United States) or personally handed out the questionnaires to medical staff (in India). They argue that these trained observers are likely to be more accurate in their accounts of what the patients reported than would be the patients themselves. This is difficult to accept, of course, since what they were really doing was asking for anecdotal reports about the observer's impressions of what a patient was experiencing in a situation that occurred in all likelihood years before. But to compound the weakness of this work, the researchers used a questionnaire that could, even under the best of circumstances, yield nothing but confusing data. *All* death experiences observed by the practitioners were lumped together. For example,

Did the patients see hallucinations of persons not actually present? No. of cases _____

Did the patients, or anyone else, identify the hallucinatory person or persons seen as: (a) someone living, No. of cases _____; (b) someone dead, No. of cases _____; etc.

Please describe one characteristic case of such a hallucination of persons.

Except for questions asking about a "characteristic case," it would be impossible to even sort out the respondent's impressions of what he thought the patients saw. If twenty people saw hallucinations and ten saw living persons and ten saw dead persons and ten saw landscapes, which visions were associated with which? The other items on the questionnaires compounded the difficulty of sorting out these impressions.

An additional weakness is that only 20 percent of the people in the American sample chose to respond at all. The authors try to sidestep that problem. They argue that, if bias was introduced at this point, one could check it by comparing reports from those respondents who replied quickly with those of respondents who were more tardy, on the assumption that tardy respondents were more typical of those medical people who were less inclined to accept the afterlife hypothesis. Since no difference was found, they concluded that the 20 percent who did respond were no different in their beliefs, attitudes, and biases than the 80 percent who didn't. They dealt with other possible sources of bias in a similar fashion. What can one say?

Thus, the data is by its very nature unreliable and uninterpretable. All one can conclude from it is that *some* medical people recall *some* patients' having had death "experiences." An analysis of the "characteristic cases" they describe might be useful in generating hypotheses for further research, but nothing more. They seem to have hoisted themselves on their own petard: their defense of the objections they themselves raise is so weak as to be embarrassing.

They also report some data on near-death experiences, the resuscitation type. Strangely, however, their reports differ considerably from those of Moody. Moody emphasizes the "tunnel experience"; they don't report it. Even when discussing Moody's findings, they don't mention the famous tunnel. Why, one wonders? Similarly, they were unimpressed with reports of panoramic recall, another typical Moody experience. One reason for this, they tell us (p. 24), was that such panoramic recall also occurs in other situations, not just in patients who are near death—Penfield, for example, was able to elicit such experience via brain stimulation (Penfield & Jasper, 1954). Again, while Moody tells us that inexpressibility of experience was typical, Osis and Haraldsson tell us that they found only three such cases and take this as more evidence that the experiences were real, since "ineffability is said to be one of the main characteristics of mystical experiences and psychedelic trips."

Elisabeth Kubler-Ross is another famous proponent of the "scientific'' basis for life after death, claiming to be 100 percent certain of it. In a recent television interview,* she even told of her own out-of-body experiences and her encounter with a spirit-form that assured her that whatever critical abuse she had suffered because of her writings on the subject was only intended as a test of her mettle and that the time has come for the great enlightenment of mankind with regard to immortality, and apparently she has been chosen to carry the message. When questioned about her conclusive scientific evidence, she said it had not yet been published.

It serves no purpose to go any further in this regard. However, since responsible people are being swayed and seduced by the life-after-life literature, in order to persuade them to give serious pause to whatever they may read in the future, it is worthwhile to review briefly some of the literature relevant to ''extraordinary'' experiences.

Mystical Experience

Psychologists have not given much attention to what has variously been called ''mystical experience,'' ''ecstasy,'' the ''illumination,'' the ''great awakening,'' and so on. Materialists often react only with quiet amusement to people's reports of having felt the presence of God or having experienced Cosmic Consciousness. Since the sixties many young people have been seeking such metaphysical or ''transcendent'' states through drugs or meditation, with their interpretations of the ''reality'' of their experiences dependent greatly on their preexisting belief systems. It is important for us to realize that such states, having been reported by mystics and saints throughout recorded history, in all likelihood represent a normal, though uncommon, ''state'' of consciousness that can be experienced by virtually everyone, given the proper conditions. A ''transcendental'' experience, such as that brought on by meditation, for example, can be just as easily brought about by a simple relaxation procedure that parallels, except for the metaphysical trimmings, the meditation procedure followed by the ''transcendental'' meditators (Benson, 1975). However, since most people aren't aware that such experiences form a part of the spectrum of ''normal'' human experiences, when they have such an experience it is usually compelling enough

---

*''Man Alive,'' Canadian Broadcasting Corporation, October 2, 1978.

to push them toward a metaphysical explanation.

By way of analogy, imagine a young woman who is raised in a rather strange family where she is shielded from all knowledge of sexual function. One day, while reading a book of inspirational verse, for some reason she experiences spontaneous orgasm. How would she interpret such an event? How would she describe it to others? Ecstasy? Ineffable experience? Communion with God? In absence of a normal explanation, a metaphysical one might be very appealing. Lest this example seem too arbitrary, remember that sexual arousal, although apparently not recognized as such, appears to play a part in demonstrations of religious excitation in various parts of the world (Sargant, 1973).

So, as Scharfstein (1973) urges, mystical experience is important and deserves careful study, even if we reject out of hand any metaphysical claims for it. Many features of these experiences, called "peak experiences" by Maslow (1976), are like those of the "near-death" experiences described earlier. For instance, a loss of fear of death is a common consequence, and such experiences often leave the individual with the feeling of having perceived the universe as an "integrated and unified whole," whatever that means (Maslow, 1976).

Peak experiences can be so "wonderful" that they parallel what the person thinks to be an eager, happy dying, a "sweet death" (Maslow, 1976). (This brings to mind the common description of a particularly intense female orgasm as a "little death.") Accounts of peak experiences often contain statements like "I felt that I could willingly die," or "No one can ever again tell me that death is bad." The individual seems to gain a sense of immortality, and then "knows" that life continues after physical death (Greeley, 1974). Why people think of death at times of such ecstasy is an interesting question in itself.

Interesting, too, is the fact that some biofeedback researchers (e.g., Lynch, 1973) have found that subjects who erroneously believed that they were in an "alpha" state reported having highly moving and meaningful emotional experiences. One researcher (Doxey, 1976) himself attests to the near ecstasy involved in the phenomenon: when testing his equipment, using himself as subject, he was led to believe that he was in an alpha state and experienced something akin to a peak experience. Analysis of EEG recordings later revealed that he had not been in an alpha state but that, because of an error in the setting up of the equipment, the light that was meant to signal an alpha-state was malfunctioning. This underlines the important role that expectation plays in such experiences. A dying person, because of the emotional stress and because of cultural beliefs about death, may be more likely than others to be open to such experiences. Relevant, too, are the findings of Schacter and Singer (1962): Subjects given an injection of epinephrine (a stimulant), but led to believe they had received a vitamin injection that would have no arousing effects, interpreted their arousal as being either angry or happy, depending on the cues in the situation in which they found themselves. Other subjects, *informed* of the arousing side-effects of the injection, experienced no such emotions. It was concluded that the same state of physiological arousal can lead to different "emotional states" and that people choose the label for their emotion depending on the surroundings. If one attributes the arousal to a drug, there is no need to label the emotion at all. It is simply "due to the drug." (Incidentally, this finding does *not* demonstrate that joy and anger have the same physiological basis in natural situations.)

The importance of this cannot be overstressed: our interpretation of emotional experience depends greatly on our expectations and on our "attributions" (i.e., to what do we attribute the cause of the emotional experience).

Out-of-Body Experiences

Quite apart from the issue of body and soul on the deathbed, there is considerable belief among most, but not all, parapsychologists that something like a soul (call it "astral body" or whatever) can spontaneously leave the body during sleep, or even while awake. Some believe that a kind of cord, silver in appearance, links the physical and metaphysical bodies during the out-of-body experience ("OBE"). At any rate, during the OBE, the person sees himself from a vantage point above his physical body. Some researchers have presented evidence that, they argue, gives empirical support to the existence of OBEs: people during sleep (and, so far as I can determine, *not* under visual observation) have supposedly been able, through the agency of their "astral body," to "read" a random number written on a piece of paper placed on a shelf beyond their reach (e.g., Tart, 1968). However, there is considerable debate about the validity of such experiments, even within parapsychological circles. John Palmer (1978), himself a professor of parapsychology, concludes that hypnagogic sleep (to be discussed below) accounts for the overwhelming majority of reported OBE experiences, and that "the OBE is neither potentially nor actually a psychic phenomenon. It is an experience or mental state, like a dream or any other altered state of consciousness" (p. 21). In line with this view, proponents of the OBE interpretation admit that OBEs are almost always experienced when going to sleep or waking up. These are precisely the moments when hypnagogic (and its relative, hypnopompic) sleep are most likely to occur.

Palmer (1978) argues that in both Western and Eastern cultures death is typically assumed to involve the separation of body and soul and that, regardless of an individual's personal view on the subject, it is likely to be on his mind as he lies dying. The stress associated with the possibility of imminent death may bring on such an "experience."

Psychologists and psychiatrists are familiar with reports of OBEs, but they view them as a kind of psychological reaction to stress/conflict. Reed (1972) describes this process, which he calls "ego-splitting": "As well as feeling unnaturally calm, the subject ... feels as though he is actually *outside himself.* The sensation is one of being suspended outside one's body, at some vantage point from whence one can calmly observe and hear oneself in the third person" (p. 125).

It is interesting to note that in even everyday memory processes there is a hint of this kind of phenomenon. Imagine yourself as you sat having dinner last evening. What do you "see"? Virtually everyone visualizes a scene they have *never* perceived: a scene that includes themselves, from a vantage point outside the body. That memory, like most memories, is a construction, a partial invention. Again, think of yourself lying on the beach, and what do you "see"? Not the world around you as viewed through your own eyes, as you originally saw it, but in all likelihood you again see yourself as part of the scene.

Now suppose you have just awakened from a deep sleep and someone asks you to recall the scene in your bedroom just *before* you dozed off. Would you not again "see yourself" lying there? So it is not surprising that resuscitated patients from time to time recall having "seen" themselves. The difference, of course, is that their memories are influenced by the (possible) state of "mystic experience" that occurred at the same time.

Hypnagogic Sleep

Hypnagogic imagery is a curious phenomenon (but *not* a metaphysical curiosity, except to those who wish to make it that) experienced by most people at one time or another in their lives during the period between wakefulness and sleep. (A similar phenomenon, labeled "hypnopompic" sleep, occurs, sometimes, between sleep and wakefulness, as a person is waking up.) It involves dreamlike fantasy, sometimes mixed with elements of reality; but unlike a normal dream, the person experiences it as "real." If you've ever been "half-awake," heard the telephone or the doorbell, got up, and realized that you were mistaken, that is a simple form of the hypnagogic (or hypnopompic) phenomenon. It's quite common. (A 1957 study cited by Reed [1972] found that 63 percent of a group of 182 students reported having had a hypnagogic experience, and a somewhat smaller number a hypnopompic one.)

Hypnagogic imagery is characterized by a relatively consistent pattern of sequential development (Schacter, 1976):

1. Colors, lights, geometric forms.
2. Vivid and detailed images of faces and objects. Unlike the reported death experiences, however, the faces are almost always those of strangers. (Yet, if one thinks that one is dying, and is preoccupied with death, it would not be surprising if the hypnagogic imagery reflected thoughts about those who have "gone before," would it?)
3. Landscapes and scenes of unusual grandeur and beauty.

Some people argue that hallucination of one's own image is very rare in hypnagogic sleep, but Foulkes and Vogel (1965) found that subjects did indeed see themselves in their imagery and that their participation increased as they moved into deeper levels of the hypnagogic state.

If hypnagogic sleep accounts for OBE experiences, as Palmer (1978) argues, and if the dying patient, possibly preoccupied by thoughts of death and of the dead, passes through a physiological state like that of hypnagogic sleep, then the reported death experiences don't appear to be extraordinary. This possibility is at least strong enough that one is surprised by the lack of concern for (or knowledge of?) they hypnagogic state shown by most survivalist writers.

Hallucinations

While we often think of hallucinations (imagery so powerful that one is certain that it is real) as having an infinite variety of themes and contents, such is apparently not the case. As far back as 1926, Kluver (cited by

Siegal, 1977), at the University of Chicago, found that mescaline-induced imagery was constructed around four constant form-types ("form constants"): (1) grating or lattice, (2) cobwebs, (3) tunnel/funnel, (4) spirals. These forms are characterized by vivid colors and intense brightness (perhaps brought about by the failure of inhibiting mechanisms in the visual system). Kluver discovered that these same form constants were typical of imagery brought about in a wide range of hallucinogenic conditions, from hypnagogic and hypnopompic sleep to delirium to dizziness to sensory deprivation. More recent experiments (Siegal, 1977) using cannabis found drug-induced imagery to have two stages. The first is characterized by Kluver's form constants, the second by more complex imagery that can incorporate the form constants but which can include memory images of familiar people and objects. One might again expect almost infinitely diverse images in the complex stages, but this wasn't the case. (Using LSD, 72 percent of subjects experienced the same form constants, and 79 percent experienced quite similar complex images. Seventy-two percent saw religious images and 49 percent saw human and animal images.) The typical imagery in the cannabis study was characterized by a bright light in the center of the field of vision: the location of the point of light created a *tunnel-like perspective* (a la Moody?). Subjects reported viewing much of their imagery in relation to the tunnel, and the other imagery seemed to move relative to the tunnel. (Subjects given placebos reported only black and white "random" form imagery.) Further studies showed that, whether the cannabis, mescaline, or LSD, after ninety minutes, most of the forms viewed were *tunnel-lattice* in nature. More complex imagery appeared usually only well after the shift to the tunnel-lattice. Common reported images involved childhood memories and scenes associated with strong emotional experiences that the person had undergone (similar to *panoramic review*?). They often reported an aerial perspective of themselves, and feelings of dissociation from the body. The imagery was also influenced by environmental stimuli. During the peak period, they asserted that the images were real.

Siegal concluded that "the experiments point to underlying mechanisms in the central nervous system as the source of a universal phenomenology of hallucinations" (p.132) and points to recent electrophysiological studies that "confirm that hallucinations are directly related to states of excitation and arousal of the central nervous system, which are coupled with a functional disorganization of the part of the brain that regulates memorizing stimuli.''

It seems, then, that there is no doubt that ordinary hallucinations can contain virtually all of the elements described by Moody, and that preoccupation with dying is likely to make more likely the notion of the separation of the soul from the body and the meeting with loved ones who have died. Again it is surprising to see virtually no references to hallucination research (apart from cursory attempts to discredit the possibility that ordinary hallucinations are involved) in the life-after-life literature. Moody (1977) agrees that sensations of being drawn down a tunnel are often reported by persons being placed under anesthesia, expecially when ether is used. But he argues that since very few of his subjects were under the influence of drugs, and since few had had any neurological problems, these experiences are not equivalent to drug-induced or other hallucinations. Again, he uses the Abbott-Costello logic—they hadn't had drugs, they had no neurological problems, therefore it was psychic. As mentioned earlier, Moody's trump card is to argue that, even if other types of hallucinations are similar to the death experiences, it is quite possible that the other hallucinations *themselves* are a manifestation of the premature release of the soul.

Hypnotic Regression

One more tool in the survivalist's armamentarium is hypnotic regression to lives past. Although you'd never know it on the basis of the literature in this area, there is considerable controversy in modern psychology about what ordinary hypnosis, not to mention regression, really does. Some argue that it puts the subject into a special state of consciousness; others argue that it leads only to a state of heightened suggestibility, where the subject and hypnotist tacitly (''unconsciously''?) collaborate to bring about the ''experience'' of hypnosis.

As for age-regression (even within the *current* lifetime!) there is mounting evidence that, while the subject may focus on long-dormant memories, no real ''regression'' (e.g., in cognitive function) actually occurs. While that is still the subject of some debate, the regression to memories of *past* lives is what's relevant for us at this point, and it goes without saying that past-life regression is not accepted as veridical by the vast majority of psychologists.

One of the most famous cases was the classic Bridey Murphy, the subject of the best-selling book *In Search of Bridey Murphy*. The story began in 1952, when a Colorado housewife by the name of Virginia Tighe (called Ruth Simmons in the book) was hypnotically "regressed" and began to describe her previous life as Bridey Murphy in Cork, Ireland. The fact that she spoke at this point in an uncharacteristic Irish brogue lent credence to her "experience." She told a very descriptive story about life and people in Ireland, even though investigators hired by the hypnotist-author, Morey Bernstein, could find no record in Ireland of a Bridey Murphy who lived in Cork (Gardner, 1957).

Dr. Ian Stevenson, a University of Virginia parapsychologist and a renowned "expert" and prolific writer on the subject, remains very impressed with the Bridey Murphy case. In fact, in an article (Stevenson, 1977) in the new *Handbook of Parapsychology*, he states that it still "has not been improved upon in the many books since written for the general public that have reported experiments with hypnotic regression" (p. 636). Yet, while Stevenson extols the merits of the Bridey Murphy case, he remains unimpressed by what others consider to be the undoing of Virginia Tighe (Gardner, 1957). In 1956, a *Chicago American* reporter visited Virginia's hometown and discovered that her background was far from being unrelated to her stories of Bridey Murphy. The reporter interviewed a Mrs. Anthony Corkell, who had lived across the street from Virginia for many years and whose Irish background and stories of Ireland apparently fascinated the young girl. Moreover, Virginia had been active in her high school drama program and had learned several Irish monologues, which she had delivered in what her former teacher referred to as a heavy Irish brogue. She even had an Irish aunt who had also entertained her with stories of Ireland. And, oh yes, about the lady across the street, Mrs. Corkell: her maiden name was Bridey Murphy! Despite this, and despite the assessment by Irish experts of her descriptions of Ireland as artificial and contrived, survivalist parapsychologists such as Stevenson reject this evidence and continue to consider this case to be a strong one, and the Bridey Murphy book is still undergoing regular reprintings.

In all likelihood, the material generated by the hypnotized subjects is a fictional blend of information and stories that they had once learned but had "forgotten." Stevenson and others naturally disagree with this interpretation. However, a study by Zolik (1958) lends credence to this view. Zolik hypnotized subjects and "regressed" them to "past lives," and then induced posthypnotic amnesia; and at a later time rehypnotized them and questioned them about characters and events from their earlier "past-life" accounts. He found that subjects had constructed the past-life stories from events of their own lives combined with events and people from books and plays. Surely those who were not impressed by this would at least have used Zolik's approach to examine the reports of their subjects. But no. None of this has had any effect on the confirmed survivalist. For example, Dr. Wambach, the author of a recent book on past-life regression (Wambach, 1978), has admitted to me that the reports she has gathered from hypnotized subjects who had been "regressed" to past lives do not in any way differ from those of unhypnotized subjects. She interprets this as evidence that one can have access to past-life material *without* being hypnotized.

Assessment

Each and all of the various characteristics of the "death" experience have been found to occur, alone or in a combination, in various "normal," non-death circumstances, such as those associated with emotional or physical stress, sensory deprivation, hypnagogic sleep, drug-induced hallucination, and so on. We know that the nervous system can process these experiences, even if we can't always predict when the experiences will occur. The famous "tunnel," the very bright light, the visions of others, the sense of ineffability, the out-of-body experiences, and the subsequent loss of a fear of death, etc., are, at the very least, not *unique* to any postmortem existence. Thus, unless one accepts either that postmortem reality mimics these earthly experiences that some people have from time to time, or that, as Moody suggests, these experiences of the living are brought about by a premature and temporary release of the soul, the reports of people who have been near death pose no demand for metaphysical interpretation.

It is clear that the "scientific," "objective" evidence for life after death is very unimpressive indeed. However, survivalist researchers are undeterred by such criticisms of their work; and it is abundantly apparent that, evidence or no evidence, they are, most of them at least, thoroughly convinced of their own immortaility. Moody admits this directly, as does Kubler-Ross (1977) when she describes death as the "peaceful transition into God's garden." I have no argument with people's theology or philosophy. What is bothersome, however, is the necessity these people feel to try to provide "objective" evidence to support their beliefs, and their attempts to fool the layman with their claims of scientific rigor and exactitude. Survival research is based on belief in search of data rather than observation in search of explanation. It is an extension of individual and collective anxiety about death. Already such research has yielded a palliative vision of death as a grand, beautiful transition to a newer and better life. Gone are the worries about hellfire and damnation of old.

If one were to believe Moody and others, why not abandon this often frustrating earthly existence and dispatch oneself forthwith to the wonderful world beyond? Even Moody doesn't want to encourage that, and he tells us that those who have survived death report that they had the "feeling" that those on the "other side" take a dim view of those here on earth who try to speed their admission to paradise. (The early Christians had a similar problem; for they too promised a wonderful life hereafter, and many of their converts, not too well taken care of in their earthly lives, chose to go directly to the next life without delay. It is hard to build a social movement if the recruits keep killing themselves, and so suicide quickly became a heinous sin for Christians.) Despite Moody's discouragements about suicide, there are bound to be those who are enthused enough by his reports to go ahead with it anyway. I have already heard of one woman whose child was killed and, having read a book like Moody's, attempted suicide in order to rejoin her child.

At any rate, we should not, in our irritation at both those who disseminate survivalist pseudoscience and those who so quickly swallow it whole, overlook the fact that some dying people do have "mystical" experiences just as some living people do. Remember that Mesmer was uniquely successful at treating hysteria; but when the scientists branded him a fraud because they were able to prove that magnets weren't essential to his treatment (contrary to his belief that they were), he was put out of business; and as a result there was no one around who could cure hysteria. We shouldn't overlook the phenomenon just because we reject the explanation. Even while seeing no reason to resort to metaphysics to explain it, we should nonetheless study it in its own right. A few medical researchers (e.g., Noyes, 1972; Noyes & Kletti, 1976) have gathered reports of near-death experiences that are quite similar to some of those described here but see no need to involve metaphysical explanations. We need more such research. It would be a pity to leave it all to the psychics.

References

Benson, H. 1975. *The Relaxation Response.* New York: Avon.

Currie, I. 1978. *You Cannot Die.* New York: Metheun.

Doxey, N. 1976. Personal communication.

Foulkes, D. and G. Vogel 1965. "Mental Activity at Sleep Onset." *Journal of Abnormal Psychology* 70: 231-243.

Gardner, M. 1957. *Fads and Fallacies in the Name of Science.* New York: Dover.

Greeley, A. M. 1974. *Ecstasy—A Way of Knowing.* Englewood Cliffs, N.J.: Prentice-Hall.

Lynch, J. J. 1973. "Biofeedback: Some Reflections on Modern Behavioural Science." In *Biofeedback: Behavioural Medicine,* ed. L. Birk, pp. 191-203. New York: Grune & Stratton.

Maslow, A. 1976. *Religions, Values, and Peak-Experiences.* Harmondsworth: Penguin. (Originally published in 1964.)

Moody, R. 1977. *Reflections on Life After Life.* Atlanta: Mockingbird Books.

_____. 1975. *Life After Life.* Atlanta: Mockingbird Books.

Noyes, R. 1972. "The Experience of Dying." *Psychiatry* 35: 174-183.

Noyes, R. and R. Kletti 1976. "Depersonalization in the Face of Life-Threatening Danger. *Psychiatry* 39: 19-27.

Osis, K. and E. Haraldsson 1977. *At the Hour of Death.* New York: Avon.

Palmer, J. 1978. "The Out-of-Body Experience: A Psychological Theory." *Parapsychology Review* 9: 19-22.

Penfield, W. and H. Jasper 1954. *Epilepsy and the Functional Anatomy of the Human Brain.* Boston: Little, Brown.

Reed, G. 1972. *On the Psychology of Anomalous Experience.* London: Hutchinson.

Rogo, D. S. 1975. *Parapsychology—A Century of Inquiry.* New York: Dell.

Sargant, W. 1973. *The Mind Possessed.* London: Heinemann.

Schacter, D.L. 1976. "The Hypnagogic State: A Critical Review of the Literature." *Psychological Bulletin* 83: 452-481.

Schacter, S. and J. Singer 1962. "Cognitive, Social, and Physiological Determinants of Emotional States." *Psychological Review* 69: 379-399.

Scharfstein, B. 1973. *Mystical Experience.* Baltimore: Penguin.

Siegal, R. K. 1977. "Hallucinations." *Scientific American* (October): 132-140.

Stevenson, I. 1977. "Reincarnation: Field Studies and Theoretical Issues." In *Handbook of Parapsychology,* ed. B. B. Wolman, pp. 631-663. New York: Van Nostrand.

Tart, C. T. 1968. "A Psychophysiological Study of Out-of-the-Body Experiences in a Selected Subject." *JASPR* 62: 3-27.

Wambach, H. 1978. *Revisiting Past Lives.* New York: Harper & Row.

Zolik, E. S. 1958. "An Experimental Investigation of the Psychodynamic Implications of the Hypnotic 'Previous Existence' Fantasy." *Journal of Clinical Psychology* 14: 179-183.

The Case of The Amityville Horror

The Amityville Horror. by Jay Anson. Prentice-Hall, Englewood Cliffs, N.J., 1977. 201 pages, $7.95.

Reviewed by Robert L. Morris

This book claims to be the true account of a month of terrifying "paranormal" events that occurred to a Long Island family when they moved into a house in Amityville, New York, that had been the scene of a mass murder. Throughout the book there are strong suggestions that the events were demonic in origin. On the copyright page, the Library of Congress subject listings are "1. Demonology—Case studies. 2. Psychical research—United States—Case studies." The next page contains the following statement: "The names of several individuals mentioned in this book have been changed to protect their privacy. However, all facts and events, as far as we have been able to verify them, are strictly accurate." The front cover of the book's dust jacket contains the words: "A True Story."

A close reading, plus a knowledge of details that later emerged, suggests that this book would be more appropriately indexed under "Fiction—Fantasy and horror." In fact it is almost a textbook illustration of bad investigative journalism, made especially onerous by its potential to terrify and mislead people and to serve as a form of religious propaganda.

To explore this in detail, we first need an outline of the events that supposedly took place, as described in the text of the book, plus a prologue derived from a segment of a New York television show about the case.

On November 13, 1974, Ronald DeFeo shot to death six members of his family at their home in Amityville. Shortly after, DeFeo was sentenced to six consecutive life terms, despite a plea for insanity by his attorney because DeFeo claimed to have heard voices in the house telling him what to do.

In the early middle of November 1975, George and Kathy Lutz were shown the house by a realtor who, at the completion of the tour, told the couple of the house's history. They nevertheless agreed to buy the house, since the price was good.

On December 18, the Lutzes and their three children moved in. Their house was blessed by a friend, Father Frank Mancuso, in the afternoon. Earlier that day Mancuso had lunched with four friends, including three priests who had advised him of the house's history and had suggested he not go. During the blessing, Mancuso heard a strong, masculine voice say, "Get out!"

Twenty-eight days later, on January 15, 1976, the Lutzes moved out, leaving their possessions behind. During these twenty-eight days the Lutz family, so the story goes, was beset by a wide variety of unusual events. Some were physical: a heavy door was ripped open, dangling on one hinge; hundreds of flies infested a room in the middle of winter; the telephone mysteriously malfunctioned, especially during calls between the Lutzes and Mancuso; a four-foot lion statue moved about the house; windows and doors were thrown open, panes broken, window locks bent out of shape; Mrs. Lutz levitated while sleeping and acquired marks and sores on her body; mysterious green slime oozed from the ceiling in a hallway; and so on. Some phenomena were experiential: Mrs. Lutz felt the embrace and fondling of unseen entities; Mr. Lutz felt a constant chill despite high thermostat temperatures; the Lutzes' daughter acqured a piglike playmate; the Lutzes saw apparitions of a pig and a demonic figure; the children misbehaved excessively and the family dog slept a lot and avoided certain rooms; marching music was heard; et cetera.

During this same period of time, Father Mancuso is said to have experienced unusual phenomena also, although his only contact with the Lutzes afterward was an occasional phone call (some calls got through, although most did not). A few hours after his blessing of the house, the hood of his car smashed back against his windshield, tearing loose a hinge, and the door flew open. Then the car stalled. In the following month, Mancuso was beset by a series of illnesses, including sores on his hands and the flu, and a strong unexplained stench emanating from his room at the rectory following a votive Mass on behalf of the Lutzes.

Approximately two weeks after they moved out, the Lutzes met William Weber, an attorney representing Ronald DeFeo, through a mutual friend. A week later, on February 5, Weber stated on a local TV news program (described in the prologue) that he hoped to prove that some force capable of influencing human behavior (including his client's) existed at the Amityville house, that he had commissioned scientists to rule out certain kinds of physical phenomena, and that it would then be turned over to a group of psychic researchers. Two weeks later the Lutzes held a press conference in Weber's office, at which time they announced that they intended to keep the house for a while but not live in it, and were awaiting the results of an investigation by parapsychologists and other professional occult researchers.

On February 18, according to the epilogue, a group of people spent the night at the Amityville house and conducted informal investigations, including three seances. Included were a clairvoyant, a demonologist, two psychics, two parapsychology field investigators, and a local TV news crew. Several reported unusual subjective impressions, but that was all. In March the Lutzes moved to California and posted their house for sale.

Since the book's publication, additional information has emerged. According to Curt Suplee's book review in the *Washington Post* (Dec. 9, 1977), the Lutzes were at about this time advised by a friend to sell their story to Prentice-Hall. A Prentice-Hall editor put the Lutzes in contact with a New York writer of documentary scripts, Jay Anson. Working from tapes provided by the Lutzes, plus some interviews with Father Mancuso (and local police officials, according to Anson's afterword), he turned out the book in three to four months. The Lutzes and Anson share the book copyright, although Anson retains the movie rights exclusively. The book has become a national best-seller in hardback and is scheduled for paperback release. The Lutzes have emerged from seclusion and have appeared on television.

According to an article in *People* (Feb. 13, 1978) by Burstein and Reilly, the Amityville house is now owned and occupied by a new family, who report no unusual phenomena save for extensive harassment by tourists and the curious. The new family "have sued the Lutzes, Prentice-Hall, and Anson for $1.1 million in damages—and are trying to enjoin them from characterizing their story as true."

This is the picture as it has been presented to the public. The basic problem is that Anson appears to have made only meager attempts to assess the truthfulness of the Lutzes' story, contrary to his own claims in the book. His only listed sources of information are joint tapes made by the Lutzes, plus additional interviews with Father Mancuso and local police officials. He does not claim to have talked directly with the Lutzes or to have questioned them in any way. Father Mancuso is a poor witness because he set foot in the house only once, and that immediately after having been warned by other priets to stay away. Local police officials were not directly involved in any of the phenomena. The most interesting witnesses would have been others who reported unusual feelings in the house (to see if they corroborated the Lutzes' descriptions of their experiences) and the repairmen called to fix damage done to the house, who would have commented on the nature and extent of the actual damage done. No interviews were described with the scientists mentioned by DeFeo's lawyer or with the parapsychologists from the Psychical Research Foundation. Since I am a former employee of the PRF, I know both investigators. One of them, Jerry Solfvin, had indeed talked to Anson at some length by phone. He described the PRF's involvement in a letter to me as follows:

> We *didn't* carry out an investigation there—just an informal visit on my part, and a collecting of the Lutzes' reports (after they moved) by George K. The case wasn't interesting to us because the reports were confined to subjective responses from the Lutzes, and these were not at all impressive or even characteristic of these cases. All in all, the family moved out rather quickly (about a month after moving in) and refused to return, making further investigation less appealing to us.

In addition to his failure to collect (or at least to include) interview data from the most important witnesses, Anson never (apparently) visited the house himself to check on the damages described, collect impressions of his own, or do investigative journalism of any thorough sort. Thus Anson's statement in the afterword (p. 197), "To the extent that I can verify them, all the events in this book are true," is patently false. It should read. "To the extent that I *bothered to* verify. . . ."

The flaws of this book as evidence for the "paranormal" can be further seen by considering some of the basic problems of spontaneous cases investigations in parapsychology, in general.

1. *Witnesses may be totally unaware of factors involved in the production of certain phenomena.* For instance, the flies that suddenly appeared in one of the rooms (pp. 29, 45) may have hatched from eggs in something that was being stored in the room (at the time it was mainly a storage room), perhaps even something rather recently purchased. On one occasion following a levitation by Mrs. Lutz, it was noticed that she had deep lines on her face, making her look very old, which soon disappeared. These lines may have been produced by wrinkles in the material she was sleeping on, a common phenomenon that could easily be misinterpreted given the circumstances. Once such an event is past, it becomes essentially impossible to recreate the original circumstances completely, such as to ascertain the absence of such "hidden factors," although often a partial recreation can enable us to ascertain their presence.

2. *Witnesses may have faulty perception of what actually happened.*

Father Mancuso had been warmed away from the house because of its history by the priests with whom he had just lunched, and he was nervous in the house. There are many ordinary mechanical noises that contain some or all of the acoustical frequencies found in human voices, and he may have perceived one such sound (familiar to the Lutzes and therefore ignored by them) as a human voice saying the short sentence, "Get out" (p. 17). Also, on page 191 we have the following: "Then, still in his dreamlike state, George saw Kathy levitate off the bed. She rose about a foot and slowly began to drift away from him." If George was still in a dreamlike state, he could easily have greatly misinterpreted his wife's bodily position and movements. On pages 96-97, an incident is described in which George finishes putting out a fire in the fireplace, then notices two "unblinking red eyes" in the window, eyes which could easily be only the reflection of dying embers. They then went outside and in the light of a flashlight noted clovenhoofed piglike tracks leading around the house. Yet such tracks could easily have been produced by melting ice from the eaves of the house into the snow below, in which case the "tracks" would be expected to follow the contour of the house.

3. *Witnesses may have a faulty memory of the events that happened.* This problem, of course, potentially arises throughout the book. On page 29 is the statement, "Kathy's bathroom door was at the far end of her bedroom," which is contradictory to the floor plan on page 10. On page 81 George Lutz describes having been told by the Amityville Historical Society that "the Shinnecock Indians used land on the Amityville River as an enclosure for the sick, mad, and dying." Yet Curt Suplee's recent inquiries of this and another historical society reveal that the Massapequa Indians lived near Amityville but that the Shinnecocks lived nowhere around Amityville. Apparently either Lutz was misinformed or he misremembered, getting his tribes confused.

4. *Witnesses may give biased or faulty impressions of their own memories.* In making their tapes, the Lutzes would have been highly motivated to remember and report selectively those incidents which made the most interesting listening and the best case. Any tendency to exaggerate could not be teased out by comparing the Lutzes' separate versions for consistency because they made the tape together and were free to evoke a consensus before speaking. For instance, on page 26, it is said of the Lutz children, "Ever since the move, they seemed to have become brats, misbehaved monsters who wouldn't listen, unruly children who must be severely punished." In other places (for example, pp. 35, 54), one is given the impression that the Lutz children started to misbehave only after moving to the new house, and that the Lutzes were only now having discipline problems. Yet on page 86 we read, "Later in the afternoon was

the second time Danny and Chris threatened to run away from home. The first had been when they lived at George's house at Deer Park. He had restricted them to their room for a week, because they were lying to him and Kathy about small things. They had revolted against his authority: both boys refused to obey his orders, threatening to run away if he also forced them to give up television. At that point, George called their bluff, telling Danny and Chris that they could get out if they didn't like the way he ran things at home." The following paragraph states that the children had run away from home but had returned and, "For a while, they stopped their childish fibbing. . . ."

In a similar vein, the family dog, Harry, is described in detail on page 165 as being ill at ease in the basement, finally running from it, and later (p. 180) as refusing to enter the basement. Yet on page 174, brief mention is made of the fact that Harry spent the night in the cellar, with no mention of Harry's behavior at that time. In addition to such selective emphasis and exaggeration, there are instances throughout the book in which the Lutzes describe themselves as unusually calm in the face of what would appear to be tremendously unnerving events.

5. *Witnesses may deliberately fabricate events.* Although it is difficult to separate straight intentional fabrication from some of the possibilities mentioned earlier, there are some suggestive examples. On page 81, Lutz describes having learned about the Shinnecock Indians from the Amityville Historical Society. Yet according to Curt Suplee, the Society not only claims that Lutz's information about the Shinnecocks is completely false, they also claim never to have heard from him. It is possible, of course, that they are the ones with poor memories.

A more telling example involves the Lutzes' descriptions of the weather. A comparison of the weather as described in the book with the weather data my staff and I have extracted from microfilmed copies of the *New York Times* and the *Los Angeles Times (N.Y. Times* was unavailable at my library for December) reveals some rather major discrepancies. Three examples will suffice.

During the days in December covered in the book, the *Los Angeles Times'* daily record shows only one twenty-four-hour precipitation above .18 inches; on December 27, the temperature in New York City ranged from a low of 33°F to a high of 53°, and 1.97 inches of rain fell. Yet on page 64, the weather on December 27 is described as follows: "The weather was bright and clear, the temperatures hovering in the low teens." Perhaps the Lutzes made an honest mistake and were off by a day. On page 57, we read of December 26, 5:30 p.m.: "The roads were reported to be icy from the recent snow, however, and it was a Friday night." December 26, 1975 was a Friday night, so the Lutzes are correct here. The description of December

28 makes no mention at all of the weather.

On pages 83-85, a heavy snowfall is described for December 31, with the snow starting at about 4:30 a.m. and continuing past 10:00 a.m., with a local radio station predicting the Amityville River would be completely frozen by nightfall. The *New York Times* describes light rain that day, with a temperature of 37°F at 8:30 a.m. and a high of 44° at 12:50 p.m. No such snowfall is reported for the adjoining days, and since it was December 31, the likelihood of a confusion over dates is rather low.

Perhaps the strongest example concerns a torrential rainfall that supposedly occurred on January 13, forcing the Lutzes to spend an extra night in their dreaded house. Confusion about dates seems especially unlikely because it was the last day the Lutzes spent in the house, and because the rainstorm had a tremendous impact upon their behavior (and occurred at just the right moment in the story, from a dramatic stand-point). On page 178, we read: "The rains and wind picked up in intensity, and by one o'clock in the afternoon, Amityville was hit by another storm of hurricane strength. At three, the electricity went out, but fortunately the heat remained in the house. George switched on the portable radio in the kitchen. The weather report said it was 20 degrees and that sleet was pelting all of Long Island. Since the radar showed an enormous low pressure system covering the entire metropolitan area, the weatherman could not predict when the storm would subside." On page 179, we read: "By six in the evening, the storm still hadn't slackened. It was as though all the water in the world was being dumped on top of 112 Ocean Avenue." On page 180, we read: "Torrents of water were still smashing against the house, and he somehow knew they wouldn't be allowed to leave 112 Ocean Avenue that night. He picked Kathy up in his arms and took her to their bedroom, noting the time on the kitchen clock. It was exactly 8 p.m." On page 181, we read: "At one o'clock, George felt he was freezing. Because of the noise of the storm raging outside, he knew there was no hope of heat in the house that night from the oil burner." The storm is described as having stopped shortly after.

The *New York Times* for January 14 and 15 gives a different picture of the weather during the day and evening of January 13 and early morning of January 14. The temperature rose above freezing at 11 a.m. on the thirteenth and rose steadily throughout the day and evening. Precipitation from 7 a.m. to 7 p.m. was 0.01 inches. From 7 p.m. to 7 a.m. the following day, there was 0.39 inches of rainfall. This was the only rainfall of any extent that occurred January 10 through January 14. One could argue that Amityville may have experienced its own storm so limited that it did not reach New York City or get mentioned in the weather news of the area. However, the radio forecast in the book described a large low-pressure area over the entire metropolitan area. Also, such "micro-storms" are by their nature of short duration; the storm in the book lasted over twelve hours. Such discrepancies between the weather described in the book and the weather as actually recorded seem, on the surface at least, to be more than mild exaggerations for dramatic purposes. The last storm, especially, was described as influencing the Lutz family's behavior over a considerable period of time. I should mention that the reason I troubled to look up the weather to begin with was my curiosity about the weather report's description of rain and sleet occurring in 20-degree weather.

6. *The investigator may not collect enough information to assess the reports of witnesses.* An example of this is Anson's (the investigator, for present purposes) apparent failure to corroborate the Lutzes' accounts by interviewing other key witnesses or by double-checking some of their factual statements, such as the weather descriptions.

7. *The investigator may collect and disseminate incorrect information.* Anson describes the parapsychologists as from the Psychical Research Institute (pp. 194-195), whereas the correct name is Psychical Research Foundation. The TV news reporter states that the Lutzes moved in on December 23 (p. 2), whereas a couple of pages later it is clearly stated that they moved in December 18.

8. *The investigator may selectively report the data he has collected.* Anson describes the fact that parapsychological investigators were on the scene, but presents no information about their opinions. Yet Jerry Solfvin told me that he had talked extensively with Anson by phone and Anson was aware of the PRF's response to the case. Why was nothing said of this in the book? Perhaps Anson did indeed interview other key witnesses but isn't discussing this material because there was nothing exciting to report.

9. *The investigator may deliberately elaborate upon the details of the case.* Throughout the book, Anson fills in details of dialogue and the Lutzes' behavior that would seem to go beyond their capacity to recall particulars, including very trivial events and dialogue. Also, much of the distortion or elaboration that was placed at the Lutzes' doorstep in the above paragraph may in truth have been the product of Anson's imaginative typewriter. One would need access to the original tapes to resolve where the responsibilities lie.

10. *The investigator may deliberately fabricate events to buttress a case.* Once again, it may have been Anson who created the dramatic weather discrepancies rather than the Lutzes, a question that could be resolved mainly by access to the tapes.

11. *The investigator may interpret the findings in a biased way, thus misleading the reader.* Anson appears to have a bias in favor of occult interpretations of the phenomena in the case. Sometimes it shows up in a

minor way, as on page 81: "John set up residence within five hundred feet of where George now lived, continuing his alleged devil worship." Apparently Anson regards the worship as more than merely alleged. Sometimes it is more general. On pages 198-199, Anson sounds impressed by the fact that many of the phenomena reported by the Lutzes also occurred in other reports of "hauntings, psychic 'invasions,' and the like"; yet if one considers the phenomena reported in other studies of anomalous events, one finds that such a tremendous variety of events have been reported that such parallels are almost inevitable.

In his afterword, pages 199-201, Anson quotes in detail the demonological interpretation of one of those who examined the case, yet gives no comparable space to the PRF investigators or others whose interpretations might stick closer to the information at hand. Although he states that we are not obligated to accept such psychic interpretations, he also states (p. 201) that any others would involve "an even more incredible set of bizarre coincidences, shared hallucinations and grotesque misinterpretations of fact."

12. *It is very difficult to assess the role of coincidence in any assessment of spontaneous cases.* For example, much was made of the static on the telephone lines between the Lutzes and Father Mancuso. Yet we have no knowledge of how likely such an occurrence of static would be by chance without knowing whether the lines were in general filled with static for one reason or another. Coincidences do happen and when we notice them we tend to impute meaning to them.

For all these reasons, *The Amityville Horror* has relatively little value as documentation of a real set of anomalous events. On the surface, it looks as though various problems, including inconsistency, exaggeration, and distortion, are abundant, and there is suggestive evidence of fabrication. If the Lutzes and Anson are to continue to maintain that this is a true story, they are obligated to clarify the discrepancies mentioned. If they can do so, fine and dandy. If not, then the public should be informed loudly and clearly that this book and any further representation of it in the media should be regarded as entertainment only. As it stands, the cover of the book would appear to constitute false advertising and should be handled in the same way as false advertising is handled in analogous cases.

Betty Through the Looking-Glass

The Andreasson Affair. By Raymond E. Fowler. Prentice-Hall, Englewood Cliffs, N.J., 1979, $8.95.

Reviewed by Ernest H. Taves

> *Why, it's turning into a sort of mist now, I declare! And certainly the glass was beginning to melt away. . .In another moment Alice was through the glass, and had jumped lightly down into the Looking-Glass room.*
>
> Through the Looking-Glass, 1872

> *And we are coming to some kind of a glass—mirror, or glass. And they are going through it! We are going through it—through that mirror!*
>
> The Andreasson Affair, 1979

Walking through glass was far from the strangest thing that happened to Betty Andreasson the night of January 25, 1967. This episode was but a minor fragment of a rich and varied narrative we shall return to shortly. First, however, a brief account of our protagonist.

Mrs. Andreasson was a housewife and homemaker living in South Ashburnham, Massachusetts. Ordinarily she lived there with her husband

and their seven children. However, events had altered that arrangement. Betty's husband, James, had been seriously injured in an auto accident the preceding December 23 and was still hospitalized. To help run the household in James's absence, Betty's parents were living in the house.

Betty was a deeply religious fundamentalist Christian. She spent much time with the Bible and there had been a history of "odd events" occurring in the family prior to 1967. What had allegedly happened on that January evening in 1967 was the abduction of Betty aboard an alien spacecraft, whereupon began a nightmarish odyssey. This abduction was (years after the event) subjected to a year-long scrutiny by a team of investigators. Raymond Fowler describes the abduction as "probably the best-documented case of its kind to date." Accordingly it behooves us to examine his account of this incident.

At about 6:30 P.M. on January 25 there was a power failure at the Andreasson house. Betty, her parents, and the seven children were there. Some of the children had been watching television. Betty had been finishing up in the kitchen. Shortly after the lights went out, a pulsating pink light appeared outside the kitchen window. A number of alien creatures entered the house. And there the matter rested until 1974, when Betty submitted an account of the experince to the *National Enquirer* in response to that publication's solicitation of firsthand UFO accounts. (That tabloid awards an annual prize of up to $20,000 for the best UFO-incident report, and a prize of $1 million for "positive proof" that earth has been visited by extraterrestrial spacecraft.) The *Enquirer* expressed no interest. In 1975 Betty learned of J. Allen Hynek's Center for UFO Studies, and she wrote to him of her experience. Hynek shelved the letter for a time before sending it on to the Humanoid Study Group of MUFON—the Mutual UFO Network. MUFON mounted an investigation, into which Fowler was later drawn.

The principal activity of the investigation consisted of 14 sessions during which Betty was hypnotized. On these occasions Betty was taken, in hypnotic state, back to the events of the night of January 25. Each hypnotic session was followed by a debriefing period, during which the investigators discussed the elicited material with Betty.

It will be remembered that at the beginning of the investigation, ten years after the event, Betty's memory of the occasion consisted of a power failure, a pulsating pink light, and the entry into her house of a number of alien creatures. The elaborate narrative that follows was elicited during hypnotic sessions conducted April through July, 1977.

Four large-headed humanoid entities entered the house, by walking or gliding through the unopened kitchen door, which was made of wood. Their heads were shaped like inverted pears. They wore shiny, dark-blue uniforms, with bird emblems on their left sleeves. Their hands, which bore three digits, were gloved. One of the entities, apparently their leader, was somewhat taller than the others. He identified himself as Quazgaa. It occurred to Betty that the uninvited visitors might be angels, and she offered them food. After some confusion she gave Quazgaa a Bible. He gave her a small, thin blue book in exchange. Quazgaa waved his hand over the Bible, causing other copies to appear, which he gave to his companions. Betty looked briefly into the thin blue book; the first three pages were white and luminous. Further on in the book she saw strange designs she couldn't understand. Quazgaa then asked if Betty would follow them. After some turmoil she did.

(It should be interpolated at this point that, when the investigation began, none of the six younger children had any memory of the alien intrusion and that Betty's mother would have nothing to do with the inquiry. Betty's father said he had seen strange characters outside the kitchen window. Betty's oldest daughter, Becky, who also underwent hypnosis, to a degree confirmed parts of Betty's memory of the early part of the evening. From this point on no family member shared in Betty's adventures.)

They leave the house, passing through the closed kitchen door. Betty is astonished to see a strange oval-shaped object settled in her backyard. Quazgaa makes it partly transparent, and Betty sees things in the craft that she had seen pictured in the blue book. They float or glide a few inches above the ground toward the strange vessel.

Now they board the craft, and Betty is taken through different parts of it, including a cleansing station, where she is engulfed in brilliant light, and a changing room, where she removes her clothes and dons a white examination garment. This leads, naturally, to being taken to an examining room. Here a needlelike probe is thrust up her nose. "I heard something break like a membrane or a veil or something—like a piece of tissue or something they broke through." When the probe is withdrawn Betty notes that it now contains upon its end a small round BB-like object with tiny points on it.

A similar probe is thrust through her umbilicus. She must be measured, she is told, for "procreation." The nasal and navel probes are both decidedly unpleasant. "I don't want any more tests! Get this thing out of me!" Finally it is done, and Betty returns to the changing room to dress in her familiar clothes.

But the odyssey has just begun. Now Betty, accompanied by two gnomelike creatures, leaves the spacecraft by floating along and through a long black tunnel. They emerge into a curious compartment containing eight glasslike chairs. She sits on one of these, is covered over with glass or plastic, and is immersed in a fluid. There are tubes, two nasal and one oral, for life support. There are soothing vibrations. Betty is told she is going to be given something to drink. She feels a thick syrup seeping into her mouth through the connecting tube.

"It is a—about a spoonful or so they are giving me through the tube, and it tastes sweet. Tastes good. Oh! This feels good! Oh, so relaxing. (Sigh) And it tastes good. . .it feels good on me."

Later the fluid drains from the chair. The two gnomes reappear, now hooded in black, and the three of them proceed into another long tunnel. They come to a glassy wall, or mirror. They pass through it and emerge into a region where almost everything is red. They see many lemurlike beings, scary creatures, crawling upon the surfaces of buildings. They continue and pass through a membrane into a region where everything is green. This is a welcome realm of beauty. Here Betty and her companions soar over a pyramid, which is crowned by a "feminine-male" head. They see, in the distance, a city of stark beauty, where bright light reflects from crystalline structures like giant prisms. Clearly, Betty has traveled a far piece from the backyard in South Ashburnham, Massachusetts.

Comes now an encounter that confounded the investigators when they heard of it, and in respect of which Fowler confesses persisting bafflement. Betty is taken through these crystal structures to confront an enormous bird. There is much light and heat, and Betty cries out for help and writhes in agony. In due course the temperature drops. Betty squints her eyes open. The bird is gone. A small remaining fire dwindles into embers and ashes, from which emerges a "big fat worm." She has witnessed an enactment, or a vision, or whatever, of one version of the phoenix legend.

A voice then speaks to Betty, and here she has a profound religious experience, for the voice is the voice of God. She did not see God, she said later, but she heard his voice. The consternation of the investigators was considerable as they faced what seemed to them an unlikely admixture of fundamentalist Christianity, on the one hand, and UFOs and the concept of extraterrestrial life, on the other.

Betty and the two creatures then retrace their earlier path, returning to the room with the glasslike chairs. The earlier experience is repeated. "Oh, that feels good. . .Oh, this is so good!"

Back to the ship, then, in the yard in South Ashburnham. In the spacecraft she says farewell to Quazgaa. He tells her that secrets have been locked into her mind, and that her race will not believe her for a long time. The beings have come to help man. They love the human race.

Quazgaa remains on the ship while two of the aliens escort Bettty back to her house. It is now 10:40 P.M. She has been away 3 hours and 40 minutes. Betty asks about the blue book. She is told she may keep it for a time, that it contains formulas, riddles, poems, and writings, "for man to understand nature." Betty's parents and the children, who have been in a state resembling suspended animation, are restored to normal consciousness and everyone goes to bed.

The basic narrative, as outlined above, was obtained during the first eleven hypnotic sessions. Three additional sessions followed. During the first of these Betty spoke for a time in an unintelligible tongue, interspersed with phrases like "Signal Base 32," and "Star Seeso." The entities communicated indirectly with the investigators through Betty and also directly with Betty: They are from a planet far, far away, not in our galaxy. They have been visiting Earth since the beginning of time. They can see the future. Many other humans have been taken aboard their craft, but only a few have "gone to the fullness." In the trip outside of the spacecraft Betty had not been taken to their home planet but to "the high place"— coordinates unspecified.

In another session the investigators tried to find out more about the thin blue book. Betty described what she saw in it: "One comma-dash like a curleque of some kind. A sweeping under in a circle, and then two lines close to each other. . .a zero with a dot and some kind of a line on an angle going through that with a little flag-type thing on the line."

The sessions ended when Betty moved to Florida. Before then she and James had been divorced.

A further incident should be related: When Betty moved to Florida she met one Bob Luca, who had also had a UFO experience. (She and Bob were later married.) Bob was subsequently interviewed by MUFON investigators. On the evening of the day of the interview Bob telephoned Betty to tell her about it. Their conversation was interrupted by an angry male voice, speaking unintelligibly, with clickings and tones. Betty thought that the voice was that of a fallen or evil angel and that she was caught in a supernatural battle between the forces of good and evil. The next day two of Betty's sons were killed in an auto accident.

So much for the basic narrative. It provides a wealth of material for our consideration. What are we to make of it? We might begin with this: The aliens are capable of extragalactic flight. Exceeding the speed of light poses them no problem. They have crossed extragalactic space to observe us since the beginning of time. Our time, that is. They love the human race. They are here to help us. Granting this, we must be incredulous at their manifest inability to communicate meaningful intelligence. Never, surely, in the history of space travel have aliens come so far for so long to communicate so little.

Let us invoke Occam's Razor.[1] This powerful directive suggests that we consider which is the simpler, more reasonable, more rational explanation of this exotic adventure: (1) Betty was taken aboard an extragalactic spacecraft by aliens who have been visiting Earth since the beginning of time but haven't been able to effect meaningful communication with man. (2) Betty recalled, or relived, in hypnosis, a dream or fantasy (or a number of them) that had meaning and utility in terms of her life history and her emotional needs.

With respect to hypnotic regression it must be emphasized that reports obtained from hypnotized subjects do not necessarily correspond to external, objective, verifiable reality. The hypnotist who worked with the Andreasson team, Dr. Harold Edelstein, is an experienced professional; both he and the investigators were convinced that Betty believed that what she brought forth in the hypnotic sessions was true. And there is no reason to challenge that. But whether the events reported correspond to any reality outside of Betty Andreasson is a different question altogether. Indeed, Dr. Edelstein said he could not say how much of Betty's story was "real" and how much was "imaginary." Similarly, Dr. Benjamin Simon, who conducted hypnotic sessions with Betty Hill, did not believe that that other Betty had been taken aboard an alien craft. "Absolutely not," he said. Hypnotic regression is a useful psychotherapeutic tool and the legitimate uses of hypnosis are many, but it is necessary to emphasize that the subjective realities thus elicited may or may not have any correspondence with objective reality.

A minor point, before proceeding to considerations of more importance: It will be recalled that within 24 hours of the interrupted telephone call between Betty and Bob Luca two of Betty's sons were killed in an automobile accident. Before this, shortly before Betty moved to Florida, her father died of cancer and complications. Before that one of the investigators died of a heart attack. There is no suggestion in Fowler's text that the deaths of either Betty's father of the investigator were in any way associated with the investigation. Regarding the death of the two sons in the accident, Fowler states that, though the investigation of that event is "confidential," a "logical reason" for the accident was found. Fowler notes ominously, however, that the auto accident "brought to four the death toll of people who had been associated with our investigation." Then, having just stated that they had found a logical reason for the auto accident, Fowler cannot resist temptation: "The question still remained, though— was the accident a coincidence?"

How did the question remain? This is a cheap shot, yellow journalism, a tired attempt to inject mystery into a situation where there is none. The Curse of Tutankhamen lives on. Betty's narrative is rich enough without the addition of such tarnished trappings.

J. Allen Hynek, in an introduction, writes that even if the entire series of adventures was "the result of some complex psychological drama played in concert" this account would still be "a fine case study in abnormal psychology." Astronomer and UFOlogist Hynek errs in this psychological appraisal. I shall show how the investigation failed and shall suggest how such investigations in the future might be improved.

How, then, does *The Andreasson Affair* fail? Sex is as good a point as any at which to begin. It will be obvious to most readers that the narrative provides a wealth of sexual symbolism and imagery. Phallic symbols appear in Betty's drawings, and the aliens penetrate her orifices. There is displacement here, as might be expected, the umbilicus being implicated rather than the more convenient (medically speaking) and nearby vagina—and not for the first time in abductee literature. And there is much more, including, for example, and in more than one instance, the theme of return to innocence—to, if you will, the womb, as in the immersion in the fluid-filled chair. The feeding in this chair, which felt so good to Betty, is susceptible of multiple interpretation, including breast feeding and adult oral sex, but in the absence of a proper history we don't know what to make of it. Fowler's text is as free from references to sex as the most repressed Victorian novel. A psychosexual history is essential, however, to the scholar attempting to understand the significance of Betty's experience.[2]

Of many reasonable alternative hypotheses to that of the extragalactic spaceship let us set forth one: On the night of the power failure the weather was foggy and misty. With the house lights suddenly out, the headlights of a passing car illuminate the fog, producing the appearance of a light outside the kitchen window. The car bounces, and the light is seen to pulsate. Later the power comes back on and conditions in the house revert to normal. Betty goes to sleep and converts the stimulus of these events into a richly structured dream. For the most part the dream is forgotten, though there remains a vague memory of alien visitation. Later, in hypnosis, the dream is brought back. But the data that would help us understand the dream are not provided.

We must surmise, for example, that the Andreassons had an active sex life—seven children born between 1937 and 1963. We know that on the evening in question Betty had been deprived of her husband for more than a month because of his hospitalization. What was the effect of this upon Betty? What was her state of mind in late January? What was their relationship like? There is a fleeting reference to marital problems, and we know they were divorced between the night of the dream and the time of its recall; but we are told almost nothing else of their relationship. The investigation is seriously flawed, in Fowler's acccount, by the absence of any attempt to ask the obviously indicated questions.

We have also the business of the nasal probe entering an orifice and penetrating a membrane, a veil. Granted that long-distance psychoanalysis is hazardous, some things are relatively obvious; and the chances are that we should wonder here about loss of innocence or initation into adulthood. Indeed, *veil* is a common enough lay term for hymen. But we can't say more about the meaning of this part of the tale because the indicated inquiry was not made—or at least not reported. The point of this kind of inquiry is that the greater the extent to which the elicited material can be shown to make sense in terms of the history and psychodynamics of Betty Andreasson, the lesser the need to seek exotic explanations.

The same lack of inquiry arises in other areas as well. Consider theology. We have seen that religion was an important part of Betty's life. Three aspects of her narrative are of particular interest from a theological point of view: (1) During her trip she was from time to time comforted by the laying on of an alien hand. (2) In the later sessions she began to speak in an unintelligible tongue. (3) She received, from an entity she at first thought might be an angel, a book containing important messages for man but written in unintelligible symbols. Here are three striking parallels with the Mormon religion: The founding of that church was based upon the alleged finding, by Joseph Smith, of the "Golden Bible," a book of metallic plates, given by an angel; the plates were covered with incomprehensible writing that Joseph "translated" by means we needn't go into here. And the concepts of speaking in tongues and the laying on of hands have been important parts of Mormon doctrine from that church's beginning. Was Betty familiar with this history? We need to know the answer to this and other questions, but no information is provided.

Turning from theology to literature we find the same problem. Walking through a glass or mirror into a strange place is a concept Betty shares with Lewis Carroll. Had Betty read *Alice in Wonderland* or *Through the Looking-Glass?* These books have had an influence upon and have been well-remembered friends of generations of girls and women (and boys and men) of all ages. Was Betty familiar with them? Fowler's account leaves the question unanswered.

There is a similar paucity of information even in respect of the episode of the phoenix, which so unsettled the investigators. We gather from Fowler's account that Betty's sister consulted a reference work to look up the phoenix after the session in which Betty spoke of it, but the serious student wants to know if Betty was familiar with the phoenix legend before that session. What was the extent and nature of Betty's awareness and knowledge of the phoenix legend prior to confronting the big bird in the hypnotic session? The answers are not available.

And how about science fiction, and the science-fiction movies? It can be deduced from internal evidence that Betty was familiar both with science fiction and with that genre of illustration. But apparently the investigative team did not consider this a matter worthy of inquiry. There is, in fact, in Fowler's account no evidence whatever that it ever occurred to the researchers at any point to look into Betty's life history for possible sources of the imagery they obtained under hypnosis, through they had a veritable gold mine of material to work with. It is as if they were all hypnotized in their own turn by the concept of the extragalactic spacecraft sitting in the backyard.

Thus the investigation is flawed throughout by a failure to ask obvious questions, answers to which would illumine the meaning of the body of data they collected. Clearly they had no interest in seeking or considering alternative hypotheses. As a serious investigation into an unusual happening *The Andreasson Affair* is a failure.

How to avoid such failure in future investigations? The most obvious suggestion is to use more knowledgeable investigators. To the extent that an investigative staff is comprised solely of participants in NICAP, CUFOS, MUFON, and the like, they might consider whether their collective expertise is equal to the job at hand, whether it might not be a good idea to bring in help from the outside, whether more broadly based backgrounds might not be useful to them. Specifically, in any case similar to the present one the help of an experienced and impartial psychoanalyst is mandatory. Where literary allusions and parallels present themselves, they should be pursued and not ignored. Alternative hypotheses, though perhaps eventually repudiated, should be considered.

Back to Occam's Razor. Are we to believe on the basis of Betty's vivid narrative that an alien spacecraft was present in her backyard that night? I suggest that it is more likely that *The Andreasson Affair* relates the history of a missed opportunity to conduct a moderately interesting study in not-so-abnormal psychology.

Notes

1. Essentia non sunt multiplicanda praeter necessitatem. The number of entities should not be increased unnecessarily.

2. In the course of the investigation Betty was given a psychiatric examination by a psychiatrist who chose to remain anonymous. The doctor said that Bettty displayed no symptoms of active thought disorders or obvious psychiatric problems. This *pro forma* examination does not approach the kind of psychoanalytic inquiry the case required.

All illustrations from the book. By permission of publisher.

Rhythms of Life

Biorhythms: Evaluating a Pseudoscience

William Sims Bainbridge

One of the most interesting current popular fads is a pseudoscience called *biorhythms*. It asserts that the life of every person is shaped by three precise biological rhythms that begin at birth and extend unaltered until death. The biorhythm theory has gained credence with many thousands of people, but is also interesting as a potential research tool for examining the general characteristics of pseudosciences. The theory is unusual in its conceptual clarity. It is not only amenable to empirical test, but also useful for exploring factors that lead people to accept a theory despite the lack of true scientific evidence. This article examines biorhythm theory, uncovers some sources of its persuasiveness, then reports on nine statistical studies that failed to support it.

To inform myself fully about biorhythms, I purchased several books and calculating devices, then enrolled in a biorhythm course taught by Mr. Benjamin Steele (pseudonym) at the experimental college attached to the University of Washington, where I teach. After this course I became a member of the Northwest Biorhythm Association, headed by Mr. Steele. He taught that biorhythms was "our newest science," although "the scientific world just hasn't recognized it yet." He was proud to say, "People into biorhythms are forerunners of a new science and a new age." Steele is also very much involved in astrology, but does not advertise this fact. "I like to keep biorhythms away from astrology, because there's so much skepticism where astrology is concerned." Over the years he has explored several religious denominations, but was never entirely satisfied until he discovered biorhythms: "I can believe that these things are real." Other teachings left him with doubt. "The difference here is, it's observable. I can see it.'' So far, Steele has only proved biorhythms in his own life, but says, "At the present time, I am involved with several biorhythm research projects.''

"The essence of the biorhythm theory is that from the moment of birth all human beings are programmed by nature to have cyclical ups and downs," Steele explained. "The theory suggests that man's behavior is characterized by three innate cycles. The cycles include a 23-day physical cycle, a 28-day emotional or sensitivity cycle, and a 33-day intellectual cycle. The physical cycle influences tasks of a physical nature: physical strength, endurance, energy, resistance, and confidence. The emotional curve takes on increased importance in situations of high emotional content: sensibility, nerves, feelings, intuition, cheerfulness, moodiness, and creative ability. And the intellectual cycle is of particular importance in pursuits requiring cognitive activity: intelligence, memory, mental alertness, logic, reasoning power, reaction, and ambition.''

Each cycle begins at the moment of birth, then oscillates up and down with absolute precision for the entire life. "When our cycles are 'high,' we're most likely to be at our best. When our cycles are 'low,' the opposite is true. But beware of 'critical days'—they occur whenever our cycles are changing. These are unstable days on which we're easily distracted and most prone to accidents." This is the theory. Mr. Steele teaches his students three methods of calculating biorhythms: hand calculation, the use of prepared tables and charts, and the use of the Biolator calculator.

The Biolator is a $30 electronic calculator manufactured by the Casio company. This device is so convenient to use that it makes testing biorhythm theory a very simple job. To perform a calculation, the user simply punches the target data into the Biolator, then the date of birth of the person in question, then presses the BIO button. The Biolator immediately flashes three two-digit "guide numbers" on its display, one for

Table 1
Interpretation of Calculations Using the Biolator

Cycle	Days in Cycle	High Days	Low Days	Critical Days
Physical	23	2-11	13-23	1 & 12
Emotional	28	2-14	16-28	1 & 15
Intellectual	33	2-16	18-33	1 & 17

each of the three rhythms. In operation, it counts the number of days from the person's birth to the target day, separately divides this number by 23, 28, and 33, and displays the remainder from each of these three long divisions. The guide number indicates how far into a new full cycle the target day is. For example, 01-12-24 means that the target day is the first day of a physical cycle, the twelfth day of an emotional cycle, and the twenty-fourth day of an intellectual cycle. Casio provides a little chart that tells the user whether each guide number indicates a high, low, or critical day. In all my calculations for this article, a total of about 2,500, I employed a Biolator and followed Casio's reasonable conventions for rounding off half-days. The only limitation imposed by the Biolator is that all dates must fall between January 1, 1901, and December 31, 1999.

Believing Biorhythms

It is easy to see why Steele himself believes the theory: his self-esteem has been boosted by involvement in biorhythm work. Steele used to play professional football, a grand but uncertain career. Steele's coaches continually found fault with his performance, and he couldn't seem to satisfy them. He happened across a copy of George Thommen's book *Is This Your Day?* This paperback is "the biorhythm Bible," and has sold over 100,000 copies. It explained to Steele that the ups and downs in his performance were not his fault, but caused by unalterable natural rhythms. Then his football career ended. For a while he held an administrative job at our university. When this job also terminated, he found solace in biorhythms. He began teaching the experimental-college course, then launched the Northwest Biorhythm Association. Calling himself a "biorhythm engineer," he claimed professional status, hoping to achieve honor and income in a wonderful new career.

Biorhythm theory is ideal material for a pseudo-profession. The technique of calculation has to be learned, but is relatively simple. There already exists a body of biorhythm literature, containing many subtle variations on the basic theory, permitting the pseudo-professional to acquire a rich fund of specialized "knowledge." Biorhythms are numerologically very efficient. A tiny input of real information (the birth date) produces a high output of pseudo-information (daily biorhythms). The three numbers 23, 28, and 33 have no common factors, so the pattern of cycles does not repeat exactly for $23 \times 28 \times 33 = 21,252$ days, a little over 58 years.

Many people seem interested in making money from biorhythms.

Several publishers have brought out books to compete with Thommen's. Mr. Steele earns $7 from each student in his course, and sells personalized biorhythm charts for $4. A "Biotimer" in Delaware charges $5 for an annual chart. A New York service charges $7.95, while Biorhythm Profiles, in Texas, asks $7.50. The co-founder of one of these enterprises admitted: "We have conducted no extensive research ourselves, being occupied with other concerns. My partner was admitted to the bar last year, and I am working toward my Ph.D. in linguistic anthropology at the University of Texas. We founded the business in July 1973 after reading an article on the subject in *Science Digest*, which makes us one of the oldest firms in the country." I interviewed a young computer consultant at our university who has invested around $1000 devising a computer program to generate really nice-looking biorhythm charts, drawn by a four-color automatic plotter, showing the rhythms as three tall, interlacing sine waves. He says he has no opinion about the validity of the theory, but believes he can make a lot of money selling the charts. This fellow first learned about biorhythms in a *Newsweek* article that proclaimed: "Americans are flocking to the new fad. Lester Cherubin, president of Time Pattern Research, Inc., says his company has sold 100,000 biorhythm printouts ($10 to $20 each) in the three years" (Cowley, 1975). The basic computer program for cranking out biorhythm charts was published in *Byte*, a popular computer magazine (Fox and Fox, 1976).

It is obvious why the biorhythm seller wants to sell. Why does the buyer buy? The advertising on the biorhythm books at least expresses the publishers' theories about the appeal of this pseudoscience. One carries the subtitle, "How to Understand and Predict the Cycles in Your Mind and Body That Hold the Key to Success and Happiness." Another urges the prospective buyer: "Take advantage of your body's natural cycles and lead a happier, more successful life." Another pledges to be "quick and easy," permitting an otherwise bewildered reader to "understand and interpret the life cycles that determine how you feel." The blurb on Thommen's book makes a similar pitch: "Tune in to your natural rhythm—and discover a richer, more secure way of living—with the one book you need for *all* the days of your life!"

The typical biorhythm book attempts to prove its case through many examples showing how biorhythms seem to explain events in the lives of famous persons. Aviator Charles Lindbergh died on a critical day in his emotional rhythm. When President Ford pardoned Richard Nixon, he was low in both the emotional and intellectual rhythms. Marilyn Monroe killed herself near critical days in both of these cycles. Swimmer Mark Spitz won seven Olympic gold medals near physical and emotional highs. I suspect that the authors of these books calculated rhythms for many famous people at important days in their lives and reported only the ones that appeared to support the theory. But these chosen cases also exemplify another favorite tactic of pseudoscientists when they want to appeal to the public: they concern exciting stories about famous people. The average reader, not used to thinking in statistical terms, will be impressed by a collection of vivid, personalistic images.

The books have a nice excuse for failing to prove their claims: they are merely popularizing the findings of other men. Biorhythm theory was supposedly invented and thoroughly tested by three European researchers near the beginning of this century: Wilhelm Fliess, Hermann Swoboda, and Alfred Teltscher. Fliess was a close friend of Sigmund Freud. Indeed, the Japanese manufacturer of the plastic Biomate calculator incorrectly asserts, "The science of Biorhythm, established by Professor Sigmund Freud, a leading psychiatrist of the early 20th century, has been studied by scholars worldwide to become a proven science."

Unfortunately for us, the chief "scientific evidence" cited by Thommen and the other biorhythm authors is entirely unavailable for our inspection. Not a single one of the relevant sources listed in Thommen's bibliography is available in the library of our university. We have Freud's letters to Fliess, but not the replies or anything else Fliess wrote. According to Cohen, "By and large, Fliess' medical and scientific colleagues ignored his book. Some of them even attacked his ideas and the statistics upon which they are based" (Cohen, 1976, p. 20). Whatever the merit of Fliess's work, we cannot inspect it directly. Although Thommen says he corresponded with Swoboda at length, he never saw much of the basic evidence:

> In one of his letters, Dr. Swoboda indicated that eight trunks of research documentation that he had stored in the vaults of the University of Vienna fell into the hands of Russian troops during the occupation of Vienna in 1945. This loss was a bitter blow to Swoboda. (Thommen, 1973, p. 7)

It may have been a bitter blow to Swoboda, but it was a great advantage for the progress of biorhythms. Proponents can cite a wealth of evidence that no one can check. The 33-day intellectual rhythm rests entirely upon the research of Alfred Teltscher, but Thommen himself admits:

Unfortunately, my own search abroad brought to light no original documentation, scientific paper, or book of his, and so my knowledge of Teltscher's work is based on secondhand reports and on articles that discussed his findings. (Thommen, 1973, p. 18)

The biorhythm books cite not only unavailable studies but also irrelevant publications. Conventional scientists use the term *biorhythms* to refer to any roughly periodic changes in a biological organism. The best known are the *circadian rhythms*, approximately a day in length. In many species these rhythms persist even when the organism is screened from the environmental influences of day and night. Natural daily and annual cycles are imprecise and are set by the rotation and revolution of the earth, not by the organism's moment of birth. These biological rhythms exist, but do not exhibit the characteristics claimed for pseudoscientific biorhythms. Authors of the books tend not to make this clear. Arbie Dale, for example, supports the theory in his book *Biorhythm* with an appendix listing supposedly confirming studies. In fact, not a single one of the thirty-one publications listed that date from after World War II has anything at all to do with the 23-28-33 day.

To get perspective on public acceptance of biorhythms, I conducted a research project with the help of undergraduate students at my university. Pseudoscience is one of the topics included in the introductory Sociology of Deviance course I teach two or three times a year. I always perform experiments with my students, both to show what real science is like and to inoculate them against pseudoscience. At one class, in November 1976, I ran the movie *In Search of Ancient Astronauts*, which spreads Erich von Däniken's quack theory. The next day I lectured for fifteen minutes presenting biorhythm theory as if it were an entirely respectable social-scientific perspective. Then I gave the class a questionnaire. After they had filled it out, I debriefed them, using the evidence presented later in this article to debunk biorhythms. The following day I dealt with von Däniken. The questionnaire allowed me to develop several hypotheses about pseudoscience.

The first question asked was: "Do you tend to agree or disagree with the biorhythm theory just presented in class?" Of course many students may have believed the theory simply because it was presented by an intellectual authority, their professor. But I had previously presented several sociological theories in the same manner and then argued against them, so the students were presumably used to the hypothetical statement of theories before they were demolished by evidence. Thus I had some reason to be shocked by the result. Only 5.3 percent of the 113 students felt "the theory is almost certainly true," but a huge 61.9 percent said it "is probably true"! Fifteen percent felt "the theory is probably false," while only 3.5 percent were sure it "could not possibly be true," and 14.2 percent had no opinion. That is, 67.2 percent accepted the theory, while only 18.5 percent rejected it, a ratio of nearly four to one!

In another part of the questionnaire, I asked students to judge six theories, biorhythms among them, and indicate on a scale of zero to 100 percent the "chance each theory is true." On the average, biorhythm

Table 2
Students' Acceptance of Six Controversial Theories

Average "Chance Theory is True"
(Expressed as a Percent)

Theory	All Students	Deeply or Moderately Religious Students	Students Who Are Indifferent or Opposed to Religion
Darwin's theory of evolution	69.8%	62.7%	78.8%
The theory that there is intelligent life on other planets	69.2%	67.1%	71.8%
The theory that "extrasensory perception" (ESP) exists	64.4%	67.1%	61.0%
Biorhythm theory	61.8%	61.7%	61.8%
Von Däniken's theory about ancient astronauts	50.9%	51.0%	50.8%
The theory that miracles actually happened just as the Bible says they did	44.1%	61.1%	22.6%
Number of Students =	113	63	50

theory scored 61.8 percent, only 8 percent behind the score of Darwin's theory. One common hypothesis contends that pseudoscience is really a vestige of traditional religious belief. With this in mind, I asked students how religious they considered themselves: deeply, moderately, largely indifferent to religion, or basically opposed to religion. Religious students were more likely to favor belief in biblical miracles and in ESP and to reject Darwin's theory. It is possible that belief in an immortal soul, or in the possibility of God and his angels speaking to men's minds, makes acceptance of ESP more likely. Religious students were not even the slightest bit more likely to accept von Däniken's theory or biorhythms. This suggests that religiosity is not tied to gullibility, but merely to specific attitudes toward some specific theories.

At the end of the questionnaire I included a page designed to explore the way people respond to their own personal biorhythms. Early in the course, most students gave me their birth dates. Now I told them I had calculated their biorhythms for the day of the questionnaire, and indicated on the last page whether each of the respondent's three rhythms was "high," "low," or "critical." For each one, the student was asked, "On the whole, is this calculation of your rhythm an accurate description of the way you are today?" He was supposed to answer by checking either a "yes" or "no" box.

I actually gave each student only "high" or "low" indications according to the flip of a coin—perfectly meaningless, random "rhythms." However, I believe I convinced almost every student that the rhythms were correctly calculated.

How did students respond to these fake personal biorhythms? Only 5.6 percent of the 108 students checked three "no" boxes, rejecting all three rhythms. Twelve percent checked one "yes" box and two "no" boxes, while 37 percent checked two "yes" boxes and one "no" box. A big 45.4 percent checked all three "yes" boxes! We would expect the student to accept exactly half of these flip-of-the-coin fake rhythms. "High" was supposed to mean "better than average," and "low" was "worse than average." Thus, the average number of "yes" responses should be 1.5 out of the 3.0 maximum possible. In fact the average was 2.24! Students were about as likely to accept "high" fake biorhythms as "low" ones. These results suggest that people do not carry an accurate scale of their personal conditions in their heads. They do not have a well-calibrated "average" against which to compare statements about whether they are currently above or below average. Thus, they are susceptible to spurious influences when asked to make such judgments (Bachrach and Pattishall, 1960; Schachter and Singer, 1962).

Religious students were not more likely than nonreligious students to accept the fake biorhythms. The sex of the student did not make a significant difference here either. One item in the questionnaire asked, "Were you already familar with biorhythm theory before today?" Only 4.6 percent said they were already "very familiar," but a full 53.8 percent said they were "somewhat familiar." Apparently many students have heard about this pseudoscience, but few, as yet, have become deeply involved in the fad. Those previously familiar and previously unfamiliar were equally likely to accept their fake rhythms.

One variable was highly associated with the tendency to accept the fake personal biorhythm: whether the student had been relatively convinced by the lecture on the theory. Those who gave 0, 1, or 2 "yes" responses to their fake rhythms on the average estimated the "chance" biorhythm theory is true at 56.3 percent, while those who gave three "yes" responses, accepting all three fake rhythms, gave an average response of 66.7 percent. We know that differential acceptance of the theory comes first, causally, because the last page of the questionnaire was sealed by another piece of paper until after all students had completed the rest of the questionnaire. Only after completing all the other questions did a student see his fake rhythms and evaluate them. One can imagine the following vicious cycle. Some people are open to the theory. A book on the theory, written in an appealing and authoritative style, makes them think it is probably true. They calculate their personal rhythms, and are particularly likely to accept them. Thus gullibility may operate in a complex way. A gullible person not only may accept a theory because it is stated authoritatively, but also may misperceive in a way to find "independent evidence" for the theory. Such a person may say, "Sure, I thought it was a pretty good theory to start with. But, look! Now I have tested the theory on my own experience and proved it was true."

Testing Biorhythms

Data suitable for testing biorhythm theory scientifically can be found in almost any large bookstore. Several sports encyclopedias contain not only information about wins and losses but also the birth dates of sports stars. A. James Fix was able to test the theory using data from *Baseball Digest* (Fix, 1976). I used information in *The Encyclopedia of Golf* and *The*

Table 3
Four Biorhythm Sport Studies

Arnold Palmer

Percent of 243 Winning
Days on Biorhythm Highs

Cycle	Expected	Actual	z scores and significance
Physical	43.5%	30.0%	4.245 (1/10,000)
Emotional	46.4%	45.7%	0.219 N.S.
Intellectual	45.5%	47.3%	0.563 N.S.

Women Golfers

Percent of 229 Winning
Days on Biorhythm Highs

Cycle	Expected	Actual	z scores and significance
Physical	43.5%	51.5%	2.442 (1/69)
Emotional	46.4%	46.3%	0.030 N.S.
Intellectual	45.5%	44.5%	0.304 N.S.

Men Golfers

Percent of 143 Winning
Days on Biorhythm Highs

Cycle	Expected	Actual	z scores and significance
Physical	43.5%	46.9%	0.820 N.S.
Emotional	46.4%	41.3%	1.223 N.S.
Intellectual	45.5%	42.0%	0.840 N.S.

Baseball Pitchers

Percent of 95 No-Hit
Games on Biorhythm Highs

Cycle	Expected	Actual	z scores and significance
Physical	43.5%	37.9%	1.098 N.S.
Emotional	46.4%	49.5%	0.606 N.S.
Intellectual	45.5%	43.2%	0.451 N.S.

Baseball Encyclopedia. Biorhythm theory holds that athletes are in best form and most likely to win on high days in their three cycles. In four of my studies I calculated athletes' biorhythms for a large number of days on which they won, and inspected the frequencies of highs in all three rhythms.

Biorhythm books by both Thommen and Cohen cite the record of golfer Arnold Palmer as evidence in favor of the theory. Palmer won the British Open at the beginning of July 1962, when he was high in all three rhythms, and lost the PGA two weeks later when he was low in all three. Selected examples cannot prove probabilistic theories, but such theories are susceptible to statistical test. I tabulated biorhythms for all 243 days of the American golf tournaments Palmer won from 1955 through 1971. Only one of the three rhythms, the physical, shows what appears to be a significant departure from the incidence that would be expected by chance. Palmer won far *less* often on physical high days than expected by chance, the opposite of what biorhythm theory predicts! Should we now become supporters of an *anti*-biorhythm theory, the exact opposite of Thommen's? No. The estimate of statistical significance used, the z test, is really not appropriate here. It assumes that each day in our sample is an isolated case, entirely independent of each other day. But golf tournaments cover several days. Palmer won only 59 different tournaments, so the 243 days are not 243 independently selected cases.

Here the error works against biorhythm theory, but sometimes a similar error may inappropriately convince someone the theory is true. A person may feel that a whole series of days exactly fits his rhythms, counting the days as individual pieces of evidence even though they are linked together and there may have been only a couple of actual changes over the period. The Palmer study turned up another artifact. Only one of the 243 winning days came on an emotional "critical day"—0.4 percent versus the expected 7.1 percent of the time. This appears to be statistically highly significant, but the following argument shows that the finding is spurious. Arnold Palmer was born September 10, 1929, a Tuesday. The emotional cycle is 28 days long, exactly four weeks. Therefore, Palmer's emotional critical days always fall on Tuesdays. Golf tournaments are scheduled to cover weekends, and seldom last through Tuesday. Spurious finding indeed! People may find much in the course of their emotional lives that appears to support the theory because the emotional rhythm is synchronized with the cultural rhythm of the week (Melbin, 1960).

I performed a similar study looking at the biorhythms of 31 women golfers on the last days of 229 tournaments they won. Here the physical rhythm appears to give a little support to the theory. But we really don't know that the 229 winning days were entirely independent cases. Some golfers had winning streaks, and the records of the women varied greatly. To exclude the nonindependence error altogether, I did another study calculating the biorhythms of 143 male golfers for the last day of the first tournament won by each of them. Now none of the figures is significantly different from what would be expected by chance. The last sports performance study looked at 95 no-hit major league baseball games won by 75 different pitchers. Since these games were widely spaced, they do represent 95 independently chosen cases. Here again there is no evidence for biorhythm theory.

Table 4
A Death Study and a Birth Study

Pitchers' Deaths	Percent of 274 Deaths on Low Biorhythm Days		
Cycle	Expected	Actual	z scores and significance
Physical	39.1%	38.0%	0.373 N.S.
Emotional	39.3%	39.4%	0.034 N.S.

Pitchers' Deaths	Percent of 274 Deaths Within One Day of the Critical Biorhythm Day		
Cycle	Expected	Actual	z scores and significance
Physical	26.1%	25.5%	0.226 N.S.
Emotional	21.4%	21.5%	0.040 N.S.

Mothers Giving Birth	Percent of 565 Births Within One Day of the Critical Biorhythm Day		
Cycle	Expected	Actual	z scores and significance
Physical	26.1%	27.6%	0.833 N.S.
Emotional	21.4%	20.4%	0.580 N.S.

Astrologers and other pseudoscientists are rightfully reluctant to predict their clients' deaths. The biorhythm books warn the reader not to use the method for this somber purpose. Nevertheless they claim that death occurs much more often at low or critical days in the emotional and physical cycles and give supportive examples of famous deaths. I calculated the rhythms of 274 major league baseball pitchers for the days of their deaths, and I found nothing. In this study and the next I gave the theory the best chance by including the days either side of the critical day itself, on the possibility that some cases might just miss biorhythm targets.

Biorhythm theory also predicts that mothers are most likely to give birth on critical days in the physical and emotional cycles. For data to test this idea I turned to the birth data published by statistician Michel Gauquelin. Some believers in the occult suspect that only data collected by a fellow believer will have a sufficiently friendly spirit to prove an occult theory. Perhaps all the golf and baseball data are too pedestrian, unsaturated by occult influences. Although this idea makes no sense to a scientist or statistician, I decided it would be only fair to use one body of data that had come from a sympathetic source. I calculated the rhythms of 565 French mothers for the date of birth of their first child, taking all the cases where the mother was born after 1900 from one of Gauquelin's data volumes. I found no evidence in favor of biorhythm theory. The 5.0 percent of the cases that fell right on emotional critical days almost achieves the 0.05 level of significance—in the wrong direction for the theory.

All six of the studies above were based on public data. Anyone can reproduce them. The three concluding studies were based on data collected from students in my large Sociology of Deviance class. Anyone with a similarly big class could replicate them. Because the two preceding studies had ignored the intellectual rhythm, .I decided to test it alone.

My first intellectual study analyzed scores my students achieved on two multiple-choice tests held 16 days apart. Because the intellectual cycle is 33 days long, most students were high on the rhythm for one test and low on the other. Therefore I was able to compare each student with himself. A majority had given me their birth dates earlier, without any knowledge of the use to which I would put them. The tests were graded by my teaching assistant, also without knowledge of my study. A total of 105 students took both tests, had given me their birth dates, had a high rhythm for one test and a low one for the other, and scored higher on one test than on the other. The chi-square of the table is 1.28, in the direction opposite to the theory, but not significant anyway.

My biorhythm mentor, Mr. Steele, pointed out to me that his theory might make a more complex prediction than the one just tested. Of course, he said, students should perform best when their intellectual rhythms are high. But they should also *learn* best when the intellectual rhythm is high. Thus they would be expected to do best near the end of the high period, when they have been in top form for several days of studying as well as for the test itself. Similarly, they should do worst near the end of the low period. I calculated the biorhythms of 150 students for the day of the midterm exam. I have divided the students into four groups, according to which quarter of the intellectual cycle each was in. As the table clearly shows, biorhythm theory flunks again.

Table 5
105 Students' Intellectual Biorhythms on Two Tests

		Test on Which the Biorhythm Was High	
		First	Second
Test on Which the Student Scored Higher	First	17	35
	Second	23	30

The last study was suggested to me by Mr. Steele. He said one of his main research projects was intended to test the hypothesis that biorhythms can be used to predict the sex of children. Biorhythm author Robert E. Smith says:

> The biorhythm theory states that when the physical biorhythm is high in the female at the time of conception, the egg cell is more likely to accept the male sperm cell and produce a boy. When the emotional biorhythm is high in the female at the time of conception the egg cell is more likely to accept the female sperm cell and produce a girl. (Smith, 1976, p. 45)

In a book on occult influences in the work of Freud and Jung, Nandor Fodor comments on the sexual aspect of Fliess' biorhythm theory:

> As to Wilhelm Fliess, he was started off on his flight of ideas by his "discovery" of a "nasal reflex neurosis," a new syndrome that he announced in

1897, claiming that dysmenorrhea was of nasal origin and that behind menstruation there was a wider process, a tendency toward periodicity in all vital activities of both sexes. The key to this periodicity was hidden in the numbers 28 and 23. The first number patently stands for the normal period of menstruation. The second may have been derived, so Jones believes, from the interval between the close of one and the onset of another. As Fliess considered all human beings bisexual, 28 was the female component and 23 the male one. These numbers were said to operate in all organic beings, and determined the biological phenomena of growth (including the sex and date of birth of the child), the date of illness, the date of death, going even beyond the human sphere to the realm of astronomy. (Fodor, 1971, 71)

Table 6
150 Students' Intellectual Biorhythms for an Exam

Quarter of the Cycle	Number of Students	Average Grade	Biorhythms Predict	Actual Result
I (high)	36	86.7%		LOWEST
II (high)	40	87.7%	HIGHEST	
III (low)	43	88.2%		
IV (low)	31	88.9%	LOWEST	HIGHEST

Mr. Steele was hoping to demonstrate the utility of biorhythms for predicting sex by collecting information on the births of a number of children. Then he would count back 280 days from the birth to estimate the date of conception, a practice endorsed by Smith, and calculate the mother's physical and emotional rhythms for that day. If one rhythm was high and the other low, the high one would determine the sex of the child. When he first told me about this project, Steele had collected information on only a half-dozen cases. He keeps his data and performs his calculations in an extremely messy fashion, so it was hard for me to follow his procedure, and I was not confident in his methodology. But I offered to provide him with all the data he needed. The students in my large class filled out a new questionnaire, giving information on their own births and on those of people close to them about whom they accurately knew the facts, a total of over 300 births. I excluded all those for which a difficult labor, induced birth, or Caesarean birth was reported. Excluding a few more at random to arrive at a nice round number, I ended up with good information about the births of 100 males and 100 females. I gave Mr. Steele the 200 birth dates to see how well he could predict sex. Despite all his boasts that he was a biorhythm engineer and researcher, he was not able to complete this simple research, at least not during the three months I was prepared to wait.

I analyzed the birth data myself. In 104 of the 200 cases, biorhythm could not make a prediction because the physical and emotional rhythms were both high, both low, or one was critical. The other 96 were good for confident predictions from biorhythm theory. I compared the predictions with reality. In 48 cases the sex prediction was correct. In 48 cases the sex prediction was not correct. This is exactly the 50 percent we would expect by chance. Steele's imaginative wife had an ingenious explanation for the cases when biorhythms fail to predict the sex of the child: Perhaps these kids grow up to be homosexuals, because their sexual identity is confused!

Conclusions

We have seen several reasons why people might believe in biorhythm theory, but scientific tests give absolutely no empirical support. Those who spread the rumor that the biorhythm theory is correct, whether through books, courses, or private conversation, are acting irresponsibly. The best evidence currently available strongly indicates that this biorhythm theory is without value.

References

Bachrach, Arthur J. and Evan G. Pattishall 1960. "An Experiment in Universal and Personal Validation." *Psychiatry*, 23, No. 3 (August): 267-270.

The Baseball Encyclopedia 1976. New York: Macmillan.

Cohen, Daniel 1976. *Biorhythms in Your Life*. Greenwich, Conn.: Fawcett.

Cowley, Susan Cheever 1975. "They've Got Rhythm." *Newsweek* (September 15): 83.

Dale, Arbie 1976. *Biorhythm. New York: Pocket Books.*

Fix, A. James 1976. "Biorhythms and Sports Performance." The Zetetic, 1, No. 1 (Fall/Winter): 53-57.

Fodor, Nandor 1971. *Freud, Jung, and Occultism*. New Hyde Park, N.Y.: University Books.

Fox, Joy and Richard Fox 1976. "Biorhythms for Computers." *Byte* (April): 20-23.

Gauquelin, Michel 1970. *Birth and Planetary Data Gathered Since 1949*, Ser. B, Vol. 1. Paris: Laboratoire d'Etudes des Relations entre Rythmes Cosmiques et Psychophysiologiques.

Gittelson, Bernard and George Thommen 1976-1977. *Biorhythm Newsletter*, 1-10. New York.

Mackenzie, Jean 1973. "How Biorhythms Affect Your Life." *Science Digest* (August): 18-22.

Melbin, Murray 1968-1969. "Behavior Rhythms in Mental Hospitals." *American Journal of Sociology*, 74: 650-665.

Nelson, Ed 1976. "New Facts on Biorhythms." *Science Digest* (May): 70-75.

O'Neil, Barbara and Richard Phillips 1975. *Biorhythms—How to Live With Your Life Cycles*. Pasadena, Cal.: Ward Ritchie Press.

Schachter, Stanley and Jerome E. Singer 1962. "Cognitive, Social and Physiological Determinants of Emotional State." *Psychological Review*, 69, No. 5 (September): 379-399.

Scharff, Robert et al. (eds.) 1972. *Golf Magazine's Encyclopedia of Golf*. New York: Harper & Row.

Smith, Robert E. 1976. *The Complete Book of Biorhythm Life Cycles*. New York: Aardvark.

Taylor, Maurice F. 1976. *Biorhythm—7 Cycles*. San Francisco: Macrovision.

———1977. *The Biorhythm 4-Cycle Bioscope and Biorhythm Meditation*. San Francisco: Macrovision.

Thommen, George S. 1973. *Is This Your Day?* New York: Avon.

Wernli, Hans J. 1976. *Bio-Rhythm*. New York: Cornerstone Library.　　●

Biorhythm Theory: A Critical Review

Terence M. Hines

The popularity of biorhythm theory has grown rapidly in the past few years. According to biorhythm theory, by determining in which phase a person is in their three separate sinusodial cycles (a 23-day physical cycle, a 28-day emotional cycle, and a 33-day intellectual cycle), predictions can be made about which days will be best (or worst) for engaging in numerous types of activities. A person's performance is claimed to be better on good, or "up," days, which make up the first half of each cycle, and worse on the "down" days of the second half of the cycle. Days on which a rhythm is shifting from the up to the down phase (or vice versa) are termed "critical days," and performance is supposed to be especially bad on these days. The worst type of day, according to biorhythm theory, is the dreaded "triple-critical day," when all three cycles are shifting phase on the same day. Although its popularity is rather recent, biorhythm theory has a history dating from the 1880s. Gardner (1966) has provided a fascinating account of that history.

Many claims have been made in the popular press and by those selling various biorhythm books and devices (slide rules, games, calculators, etc.) that biorhythm theory has been scientifically validated in studies that allegedly show that about 60 percent of accidents caused by human error occur on critical days. Since critical days make up only about 20 percent of all days, this certainly would be an impressive finding, if true. However, in the past several years much empirical work on biorhythm theory has shown that the claims made for the theory are without foundation. The present paper reviews this work.

Evidence Used to Support the Biorhythm Theory

The evidence used to support the validity of biorhythm theory can be roughly divided into two types. First, lists of disasters, such as airline crashes and deaths of famous persons, that seem to follow the pattern predicted by biorhythm theory are reported. Thus, aircrashes that took place when the pilot (or co-pilot) was on a down or critical day are said to support the theory. So are lists of famous people who died on such days. While many such cases can be found if one searches long enough, they alone provide no evidence whatsoever of the validity of the theory. Before concluding that biorhythm has anything to do with such events, one must take into account the number of times such events take place when biorhythm theory predicts they *won't*. Such cases are conveniently ignored by biorhythm proponents.

Seemingly more impressive evidence for biorhythm theory comes from studies that claim to show that an impressive number of accidents occur on critical days when compared with the number of occurrences expected by chance. Such studies are often (1) impossible to find, and thus impossible to verify; (2) contain only an assertion that a significant effect of biorhythm was found without providing sufficient detail to check the claims; or (3) contain grievous methodological and/or statistical errors, rendering the conclusions drawn from the data worthless.

Regarding the first type of study, the kind that are impossible to find, the most widely cited supposedly were carried out in Japan and found that about 60 percent of accidents fell on critical days. But Katz (1977) states that the *Toronto Star* "was unable to locate any of the original studies referred to by the biorhythm promotional literature. The Japanese Embassy in Washington, D.C., the Japanese Trade Center and the Japanese Chamber of Commerce in New York all stated that they were unaware of any Japanese firms employing biorhythm." Another example of what might be termed a "phantom study" was cited by Pittner and Owens (1975): "The Canadian Armed Forces concluded after a 7,000 case study that there is a definite relationship between accidents and critical days" (p. 44). Unfortunately, no reference for this study was given. Colonel J. Boulet, director of information services for the Canadian Department of National Defense, in a letter to me dated July 20, 1978, stated: "A search to locate the report of the biorhythm theory study allegedly carried out by the Canadian Forces proved negative."

A report by Anderson (1973) is an example of the "assertion only" type of study claiming to support biorhythm theory. Anderson devotes only one paragraph of his four-and-a-half-page paper to his claim that "we have analyzed more than 1,000 accidents during the past two years, and the amazing thing is that we have come out with more than 90 percent of the accidents occurring on critical days" (p. 19). Anderson presents no information whatsoever on the method used for picking the analyzed accidents, the method for computing critical days (a most important point, to be discussed later), or any other details of how the results were arrived at.

Finally, I will give two examples of studies claiming to support biorhythm theory that present a more detailed description of the methods of data analysis used. These studies contain errors that totally invalidate the conclusions drawn from the data. The first is by Pittner and Owens (1975). These authors analyzed accidents involving human error from the records of the Safety Office of the Army Corps of Engineers, Philadelphia District, for fiscal years 1971 through 1974. "A total of 51 accidents were computed; and of these, approximately 35 percent (18 accidents) occurred on critical days" (p. 45). If this finding were true, it would be supportive of biorhythm theory, for it is significantly different from chance. Happily, Pittner and Owens included a tabulation of the day on each of the three biorhythm cycles on which each accident occurred (their Figure A, p. 44). An examination of their raw data shows, amazingly, that only 13, not 18, accidents occurred on critical days. The number of accidents out of 51 that should fall on critical days by chance alone is 12. The chi-square for the occurrence of 13 accidents results in a figure of $\chi^2 = 0.11$ ($df = 1$, $p < .85$). This is not even close to being significantly different from chance. Thus, the actual results of the Pittner and Owens study show no support for biorhythm theory at all, in spite of what the authors, incorrectly, claimed their data showed. Unfortunately, Pittner and Owens's erroneous conclusions have been reported elsewhere (Brownley and Sandler, 1977) as supporting biorhythm theory.

Where did the extra five critical-day accidents come from in Pittner and Owens's study? Clearly it is not just a misprint, since 35 percent of 51 is approximately 18. These authors may have done what is fairly common in biorhythm studies; they must have included not only critical days but also one day before and one day after critical days (critical day plus and minus one). That is done because "an individual born shortly following midnight is actually biorhythmically closer to the preceding day than an individual born at noon. Similarly, an individual born approaching midnight is closer to the coming day" (Williamson, 1975, p. 18). As long as such a procedure is made explicit, there are no problems. However, it is easy for those not trained in statistical methodology to include accidents falling on the critical day plus and minus one and then compare the total figure to the number of accidents expected to fall only on the actual critical day by chance. Such a procedure obviously results in a serious error with a seemingly very impressive number of critical-day accidents, a number much higher than expected by "chance" when the chance figure that is used is that for a single critical day. It is interesting to note that about 60 percent of a random sample of accidents will fall on a critical day plus and minus one by chance alone. Many of the phantom studies said to support biorhythm noted above claim that about 60 percent of accidents occur on critical days.

Williamson falls directly into this trap. After stating that "critical days comprise only 20 percent of the total," he says, "Days considered to be critical are the calculated critical days plus or minus one." He then evidences surprise that out of 33 accidents, 58 percent fell on a critical day. Of 13 helicopter accidents, 61 percent occurred on the pilots' critical days. Since "critical day" here means the actual critical day plus and minus one, these results are exactly what would be expected by chance alone and show no support for biorhythm theory.

Another statistical problem in some biorhythm studies is what may be called the "shotgun" approach. In such a study, the investigator tests all possible biorhythm predictions and, when one or two turn out to be verified by statistically significant differences in the data, he claims that this supports biorhythm theory. It does not. Consider a paper by Knowles and Jones (1974). These authors examined the incidences of altercations between police and suspects for any biorhythm effects. When comparisons were made regarding the occurrence of such incidents on officers' critical days, no significant effects were found. However, the authors then went on: "All relationships of patterns of days and periods for each of the three cycles" were examined for both officer and suspect. The author can't mean this literally as there is a simply huge number of different such patterns. Even considering the simplest cases, a given rhythm can be in one of five different patterns: (1) up phase, on ascending portion of curve; (2) up phase, descending portion; (3) down phase, ascending portion; (4) down phase, descending portion; and (5) critical. Since there are three rhythms and two persons involved in each altercation, that gives 5^6 or 15,625 different possible patterns to be examined. Knowles and Jones never state exactly how many patterns they examine, but since they apparently had access to a computer it may have been a great many. Thus the fact that they found significant effects on four comparisons out of however many they ran is really not terribly supportive of biorhythm theory. It is even less supportive when one remembers that the major claims of the theory, regarding critical days, were not confirmed.

Thus it appears that the evidence claiming to support biorhythm theory does not stand up to close inspection.

Studies Failing to Support Biorhythm Theory

With the great popularity of biorhythm theory, a number of attempts have been made to assess its validity using adequate (and often quite elegant) statistical and experimental paradigms. I shall now review these studies. They uniformly offer no support whatsoever for the claims of biorhythm enthusiasts.

Accident occurrences. A fairly early study of biorhythm and accidents was carried out by Mason (1971) for the Workers' Compensation Board of British Columbia. Mason examined 13,285 accidents from the Board's files. This included all time-loss accidents occurring during the first four months of 1971. Accidents were no more likely to occur on critical days than on noncritical days. Further, various combinations of critical and noncritical days did not yield accident rates higher than chance. A possible criticism of these results is that the sample included all accidents, not just those due to human error, and that the inclusion of accidents not caused by human error reduced the effect of biorhythms. Two points answer this criticism. First, the addition of other than human-error accidents could be expected to weaken any effects of biorhythms but not to eliminate it totally. After all, human error plays some role in a large percentage of all accidents. Second, when the entire sample of accidents was divided into those more likely to be the result of human error and those less likely, no effects of biorhythm were found for either subsample.

Sacher (1974) performed a study of the effects of biorhythms on aircraft mishaps, using 4,346 accidents involving naval aircraft between 1968 and 1973. Again, no significant effects of pilots' position on the various biorhythm cycles were found.

Nett (1975) studied 400 accidents that occurred at two Army Materiel Command installations. Nett's study is interesting in that he used six different methods for calculating critical days. He then compared the number of accidents occurring at each of the two AMC installations on seven types of days (critical on any cycle; critical on a given cycle; multiple critical days) with that expected by chance. This resulted in 42 different tests of biorhythm theory for each facility. Only for one such facility did any of the 42 tests reach the 5 percent level of significance, indicating that such an event should occur by chance only once in twenty times. However, since it will be recalled that there were 42 separate tests, the occurrence of an event expected by chance one in twenty times offers no support for biorhythm theory.

Weaver (1974) studied 25 percent of the Army aviation accidents involving pilot error that occurred over a two-and-a-half-year period starting in 1971. He claimed that 49 percent of these accidents occurred on critical days. Unfortunately, this study is an example of the assertion-only type, and Weaver did not even report how many accidents were included. More adequately described studies of aircraft accident occurrences have failed to find any biorhythmic effects. Wolcott, McMeekin, Burgin, and Yanovitch (1975, 1977a, 1977b) have examined a total of 9,505 aircraft accidents, both civilian and military, and have found no biorhythm effects. Khalil and Kurucz (1977) found no biorhythm effects in their sample of 63 general aviation accidents.

Turning from aircraft to other types of accidents, Khalil and Kurucz also looked for biorhythm effects on automobile accidents. There were no such effects in a sample of 181 accidents where the driver was at fault. In perhaps the most elegant study of biorhythms to date, Shaffer, Schmidt, Zlotowitz, and Fisher (1978) examined 205 highway accidents and found no hint of any effects of biorhythm. This study is a model for the careful statistical evaluation of the biorhythm theory. Carvey and Nibler (1977) investigated 150 "work-related" vehicular accidents. No biorhythm effects were found. Brownley and Sandler (1977) examined 506 fatal driving accidents. No biorhythm effects were found.

Several studies have investigated the alleged effects of biorhythms by examining industrial accidents. Lyon, Dyer, and Gary (1978) found no such effects in a sample of 112 accidents occurring at the Oak Ridge National Laboratory. A subsample of 67 accidents where the victim was obviously at fault also revealed no biorhythm effects. Persinger, Cooke, and Janes (1978) found no biorhythm effect on 400 coal mine accidents. Carvey and Nibler likewise found no biorhythm effect in a sample of 210 on-the-job accidents for which workmen's compensation was requested. Thus, over 25,000 accidents have been examined with no sign of biorhythm effects being found.

Sports Performance. Some investigators have looked at the possibility that biorhythms have an effect on variables other than accident occurrences. One of the areas where biorhythm is used widely is sports performance. As with studies of accident occurrences, studies of biorhythm effects on sports performance have failed to demonstrate such effects. Khalil and Kurucz (1977) failed to find biorhythm effects on the performance of the University of Florida swim team or the faculty-student bowling team. Louis (1978) found no biorhythm effects on the occurrence of 100 no-hit baseball games between 1934 and 1975. Fix (1976) found no biorhythm effects on the batting performance of 70 major league baseball players during the 1975 season. There was no effect either when players were considered as a group or when each player's performance on, for example, a down day was compared to his performance on an up day. Schonholzer, Schilling, and Muller (1972) found no biorhythm effects on the occurrence at 1,051 record sports performances. Bainbridge (1978) examined the performance of golfers, both male and female, and found no biorhythm effects in the male golfers. He did find an effect of slight significance for the female golfers. However, since this is the only report of significant biorhythm effects on sports performance out of the dozens examined, it cannot be taken too seriously.

Other Variables. Turning to studies of biorhythm effects on other variables, five papers report attempts to correlate human moods with positions on the various biorhythm curves. One obvious prediction would be that mood should be most strongly under the control of the 28-day emotional cycle, with people reporting better feelings on up days than on down days. None of these five papers report any evidence to support biorhythm theory (Dorland and Brinker, 1973; Rodgers, Sprinkle, and Lindberg, 1974; Whitton, 1977; Wright, 1977; Steer, 1978). In her 1974 doctoral dissertation, Yates studied the effects of biorhythms on performance on four paper-and-pencil psychological tests measuring mood, motor performance, and intellectual performance. Ninety-six students, both male and female, served as subjects. No biorhythm effects were found on any of the measures used.

Hersey's studies (1931, 1932, 1955) of workers' emotions and their relation to on-the-job productivity are often cited as supporting biorhythm theory. In fact, this research clearly contradicts the theory's basic assumption. Hersey asked hundreds of workers to rate their moods on an 11-point scale over long periods of time. He then examined these ratings for any cyclic fluctuations. He did find that moods do show cyclic changes. However, there were large differences both within and across individuals. In one study (Hersey, 1932) the mean length of the mood cycle was 34.6 days but varied from 21 to 65 days for different individuals. A given individual's rhythm would also change greatly over time because of various environmental factors. No support for biorhythm there!

Bailey (1978) examined the deaths of presidents of the United States for biorhythm effects. Of the 35 presidents who have died, 8 have died on single critical days and 2 on double critical days. This is not significantly different from chance ($\chi^2 = 3.32$, $df = 1$, $p > .05$).

Neil and Sink (1976) had three subjects perform a choice reaction-time task daily for 70 days. Three measures were derived from a subject's performance: reaction time, movement time, and the rate at which information was transmitted by the subject. Data from each subject for each measure were Fourier analyzed to "identify periodicities in performance." Twelve significant periodicities were found, some that were close to either one of the biorhythm cycles or a multiple thereof. However, since the periods that were significant varied from subject to subject and from measure to measure, these data provide no support for biorhythm theory.

Two more-detailed analyses of reaction-time performance in relation to biorhythm theory have been carried out. Wolcott, Hanson, Foster, and Kay (1979) found no biorhythm effects on a choice reaction-time task.

Hines (1978) examined data from a study by Fozard, Thomas, and Waugh (1976). Data from this study consisted of reaction times and error rates in a choice reaction-time task from 111 of Fozard et al.'s male subjects. Scores for 112 of these subjects on the Ammons Quick Test (Ammons and Ammons, 1962), a short IQ test, were also available. The large number of subjects in this study permitted comparisons to be made between critical- and noncritical-day performance.

The data from Fozard et al. showed no differences in performance between up days and down days for any of the three measures (reaction time, error rate, Ammons IQ) used. Likewise, there were no effects of the number of rhythms in the up phase or down phase on any of the three measures. Only the analysis of critical versus noncritical days showed a single significant difference. Reaction times on emotionally critical days were slower than on emotionally noncritical days (450 msec. versus 396 msec.; $t(109) = 2.76, p < .01$). This represents the only significant difference out of a total of 21 differences examined. It thus offers no support for biorhythm theory, being due rather to the fact that if enough differences are examined, even from a set of random data, some will be "significant" by chance alone.

Hines also examined data from a study of aging by Hines and Posner (1976), which consisted of scores on three subscales of the Wechsler Adult Intelligence Scale (WAIS), a standard adult IQ test. Scores for a total of 65 subjects, 40 females and 25 males, were examined. The three subscales used were Digit Span, Block Design, and Vocabulary. Because of the relatively small number of subjects, comparisons of performance on critical and noncritical days could not be made—too few subjects were tested on critical days for meaningful results. Analysis of the data from Hines and Posner showed no differences in performance between up days and down days on any of the three subscales for any of the three biorhythms. A further comparison examined the effects of the total number of rhythms in the up or the down phase on the day the subjects were tested. On a given day a person can be in one of four conditions, depending on the phases of his or her three biorhythms. All three rhythms can be up, two can be up and one down, one can be up and two down, or all three can be down. Biorhythm theory predicts that as the number of rhythms in the down phase increase, performance should decrease. No such trend was found for any of the three WAIS subscales used.

Finally, Bainbridge (1978) has looked for biorhythm influences on several miscellaneous variables. He found none. Specifically, he found no effects of either the physical or the emotional cycle on the day of death for 274 baseball pitchers. Nor were there any such effects on the days on which 565 mothers gave birth or on the sex of their infants. The intellectual cycle had no effect on test performance of two samples of students from Bainbridge's classes, one sample of 105 students and one of 150. Khalil and Kurucz also found no biorhythm effect on a sample of 105 miscellaneous deaths.

The conclusion of the numerous well-done studies that have examined the validity of biorhythm theory is clear—there is no evidence to support the theory. This should not be surprising, since, although there are many well-documented biological rhythms (not biorhythms) in both humans and animals (Bunning, 1973; Saunders, 1977), none are of the fixed, mechanical, and unchangeable type that the three biorhythm cycles are supposed to be. Even the menstrual cycle, the best known of the human cycles, is highly variable both within a given woman and across different women (Matsumoto, Nogami, and Ohkuri, 1962). Further, this cycle has almost no effect on cognitive and motor performance (Sommer, 1973).

References

Ammons, R., and C. Ammons 1962. "The Quick Test (QT): Provisional Manual." *Psychological Reports* 11: 111-161.

Anderson, R. 1973. "Biorhythm—Man's Timing Mechanism." *American Society of Safety Engineers Journal* 18 (2): 17-21.

Bailey, K. 1978. "Biological Time and Political Behavior." Paper presented at the Annual Meeting of the American Political Science Association, New York, September.

Bainbridge, W. 1978. "Biorhythms: Evaluating a Pseudoscience." *Skeptical Inquirer* 2 (2): 40-56.

Brownley, M., and C. Sandler 1977. "Biorhythm—An Accident Prevention Aid?" In A. Neal and R. Palasek, eds., *Proceedings of the Human Factors Society 21st Annual Meeting*. Santa Monica, Calif.: Human Factors Society, pp. 188-192.

Bunning, E. 1973. *The Psychological Clock*, rev. 3rd ed. London: English Universities Press.

Carvey, D., and R. Nibler 1977. "Biorhythmic Cycles and the Incidence of Industrial Accidents." *Personnel Psychology* 30: 447-454.

Dorland, J., and N. Brinker 1973. "Fluctuations in Human Mood (A Preliminary Study)." *Journal of Interdisciplinary Cycle Research* 4: 25-29.

Fix, A. 1976. "Biorhythms and Sports Performance." *Zetetic* 1 (1): 52-57.

Fozard, J., J. Thomas, and N. Waugh 1976. "Effects of Age and Frequency of Stimulus Repetitions on Two-Choice Reaction Time." *Journal of Gerontology* 31: 556-563.

Gardner, M. 1966. "Mathematical Games." *Scientific American* 215 (1): 108-112.

Hersey, R. 1931. "Emotional Cycles in Man." *Journal of Mental Science* 77: 151-169.

—— 1932. *Workers' Emotions in Shop and Home*. Philadelphia: University of Pennsylvania Press.

—— 1955. *Zest for Work*. New York: Harper & Bros.

Hines, T. 1978. "Lack of Biorhythmic Effects on Three WAIS Subscales, the Ammons IQ Test, and Reaction Time and Error Rate in a Choice Reaction Time Task." Submitted for publication.

Hines, T. and M. Posner 1976. "Slow But Sure: A Chronometric Analysis of the Process of Aging." Paper presented at the Annual Meeting of the American Psychological Association, Washington, D.C., September.

Katz, S. 1977. "Biorhythm: Guide to Life or Fad for 'Numerical Nuts'?" *Toronto Star*, April 2, p. A3.

Khalil, T., and C. Kurucz 1977. "The Influence of Biorhythm on Accident Occurrence and Performance." *Ergonomics* 20: 389-398.

Knowles, L., and R. Jones 1974. "Police Altercations and the Ups and Downs of Life Cycles." *Police Chief* (Nov.): 51-54.

Louis, A. 1978. "Should You Buy Biorhythms?" *Psychology Today* 11 (11): 93-96.

Lyon, W., F. Dyer, and D. Gary 1978. "Biorhythm: Imitation of Science." *Chemistry* 51 (3): 5-7.

Mason, K. 1971. *An Investigation of the Biorhythm Theory*. Vancouver, British Columbia: Workers' Compensation Board.

Matsumoto, S., Y. Nogami, and S. Ohkuri 1962. "Statistical Studies on Menstruation: A Criticism of the Definition of Normal Menstruation." *Gumma Journal of Medical Science* 11: 294-318.

Neil, D., and F. Sink 1976. "Laboratory Investigation of 'Biorhythms.'" *Aviation, Space, and Environmental Medicine* 47: 425-429.

Nett, D. 1975. *A Study of the Relationship of Biorhythms to Accidents at Two AMC Installations*. Texarkana, Texas: Army Materiel Command (NTIS #AD-A009 319).

Persinger, M., W. Cooke, and J. Janes 1978. "No Evidence for Relationships Between Biorhythms and Industrial Accidents." *Perceptual and Motor Skills* 46: 423-426.

Pittner, E., and P. Owens 1975. "Chance or Destiny? A Review and Test of the Biorhythm Theory." *Professional Safety* (April): 42-46.

Rodgers, C., R. Sprinkle, and F. Lindberg 1974. "Biorhythms: Three Tests of the Predictive Validity of the 'Critical Days' Hypothesis." *International Journal of Chronology* 2: 247-252.

Sacher, D. 1974. *Influence of Biorhythmic Criticality on Aircraft Mishaps*. Monterey, Calif.: Naval Postgraduate School (NTIS #AD-783 817).

Saunders, D. 1977. *An Introduction to Biological Rhythms*. New York: Halsted Press.

Schonholzer, G., G. Schilling, and H. Muller 1972. "Biorhythmik." *Schweizerische Zeitschrift fur Sportmedizin* 20: 7-27.

Shaffer, U., C. Schmidt, H. Zlotowitz, and R. Fisher 1978. "Biorhythms and Highway Crashes." *Archives of General Psychiatry* 35: 41-46.

Sommer, B. 1973. "Effect of Menstruation on Cognitive and Perceptual-Motor Behavior: A Review." *Psychosomatic Medicine* 35: 515-534.

Steer, R. 1978. "Moods and Biorhythms of Heroin Addicts." *Psychological Reports* 43: 829-830.

Weaver, C. 1974. "Biorhythms: The Question of Ups and Downs." *U.S. Army Aviation Digest* (January): 13-17.

Whitton, J. 1977. "An Empirical Study of Biorhythm in Humans." *New Horizons* 2 (3): 32-35.

Williamson, T. 1975. "Cancel Today and Save Tomorrow!" *Airscoop* (March): 18-20.

Wolcott, J., C. Hanson, W. Foster, and T. Kay 1979. "Correlation of Choice Reaction Time Performance with Biorhythmic Criticality and Cycle Phase." *Aviation, Space, and Environmental Medicine* 50: 34-39.

Wolcott, J., R. McMeekin, R. Burgin, and R. Yanowitch 1975. "Biorhythms—Are They a Waste of Time?" *TAC Attack* (November): 4-9.

_____ 1977a. "Correlation of Aircraft Accidents with Biorhythmic Criticality and Cycle Phase in U.S. Air Force, U.S. Army, and Civil Aviation Pilots." *Aviation, Space, and Environmental Medicine* 48: 976-983.

_____ 1977b. "Correlation of General Aviation Accidents with the Biorhythm Theory." *Human Factors* 19: 283-293.

Wright, M. 1977. "Biorhythm: Fact or Fancy." In A. Neal and R. Palasek, eds., *Proceedings of the Human Factors Society, 21st Annual Meeting.* Santa Monica, Calif.: Human Factors Society, pp. 193-196.

Yates, M. 1974. *Study of Biorhythms with Students in Four Fraternities and Four Sororities at the University of Wyoming.* Unpublished doctoral dissertation, University of Wyoming (Xerox University Microfilms #75-244).   •

BOOK II: INQUIRIES INTO FRINGE-SCIENCE

Exploring Science's Fringes

Asimov's Corollary

Isaac Asimov

In my English colleague Arthur C. Clarke's book *Profiles of the Future* (Harper & Row, 1962, rev. ed., 1972), he advances what he himself calls "Clarke's Law." It goes as follows:

> *When a distinguished but elderly scientist states that something is possible, he is almost certainly right. When he states that something is impossible, he is very probably wrong.*

Arthur goes on to explain what he means by "elderly." He says: "In physics, mathematics, and astronautics it means over thirty; in the other disciplines, senile decay is sometimes postponed to the forties."

Arthur goes on to give examples of "distinguished but elderly scientists" who have pished and tut-tutted all sorts of things that have come to pass almost immediately. The distinguished Briton Ernest Rutherford pooh-poohed the possibility of nuclear power, the distinguished American Vannevar Bush bah-humbugged intercontinental ballistic missiles, and so on.

But naturally when *I* read a paragraph like that, knowing Arthur as I do, I begin to wonder if, among all the others, he is thinking of me.

After all, I'm a scientist. I am not exactly a "distinguished" one but nonscientists have gotten the notion somewhere that I am, and I am far too polite a person to subject them to the pain of disillusionment, so I don't deny it. And then, finally, I am a little over thirty and have been a little over thirty for a long time, so I qualify as "elderly" by Arthur's definition. (So does he, by the way, for he is—ha, ha—three years older than I am.)

Well, then, as a distinguished but elderly scientist, have I been going around stating that something is impossible or, in any case, that that something bears no relationship to reality? Heavens, yes! In fact, I am rarely content to say that something is "wrong" and let it go at that. I make free use of terms and phrases like "nonsense," "claptrap," "stupid folly," "sheer idiocy," and many other bits of gentle and loving language.

Among currently popular aberrations, I have belabored without stint Velikovskianism, astrology, flying saucers, and so on.

While I haven't yet had occasion to treat these matters in detail, I also consider the views of the Swiss Erich von Däniken on "ancient astronauts" to be utter hogwash; I take a similar attitude to the widely held conviction (reported, but not to my knowledge subscribed to, by Charles Berlitz in *The Bermuda Triangle*) that the "Bermuda triangle" is the hunting ground of some alien intelligence.

Doesn't Clarke's Law make me uneasy, then? Don't I feel as though I am sure to be quoted extensively, and with derision, in some book written a century hence by some successor to Arthur?

No, I don't. Although I accept Clarke's Law and think Arthur is right in his suspicion that the forward-looking pioneers of today are the backward-yearning conservatives of tomorrow,* I have no worries about myself. I am very selective about the scientific heresies I denounce, for I am guided by what I call Asimov's Corollary to Clarke's Law. Here is Asimov's Corollary:

> *When, however, the lay public rallies round an idea that is denounced by distinguished but elderly scientists and supports that idea with great fervor and emotion—the distinguished but elderly scientists are then, after all, probably right.*

---

*Heck, Einstein himself found he could not accept the uncertainty principle and, in consequence, spent the last thirty years of his life as a living monument and nothing more. Physics went on without him.

But why should this be? Why should I, who am not an elitist, but an old-fashioned liberal and an egalitarian (see "Thinking About Thinking," in *The Planet That Wasn't,* Doubleday, 1976), thus proclaim the infallibility of the majority, holding it to be infallibly wrong?

The answer is that human beings have the habit (a bad one, perhaps, but an unavoidable one) of being human; which is to say that they believe in that which comforts them.

For instance, there are a great many inconveniences and disadvantages to the universe as it exists. As examples: you cannot live forever, you can't get something for nothing, you can't play with knives without cutting yourself, you can't win every time, and so on and so on (see "Knock Plastic," in *Science, Numbers, and I,* Doubleday, 1968).

Naturally, then, anything which promises to remove these inconveniences and disadvantages will be eagerly believed. The inconveniences and disadvantages remain, of course, but what of that?

To take the greatest, most universal, and most unavoidable inconvenience, consider death. Tell people that death does not exist and they will believe you and sob with gratitude at the good news. Take a census and find out how many human beings believe in life after death, in heaven, in the doctrines of spiritualism, in the transmigration of souls. I am quite confident you will find a healthy majority, even an overwhelming one, in favor of side-stepping death by believing in its nonexistence through one strategy or another.

Yet as far as I know, there is not one piece of evidence ever advanced that would offer any hope that death is anything other than the permanent dissolution of the personality and that beyond it, as far as individual consciousness is concerned, there is nothing.

If you want to argue the point, present the evidence. I must warn you, though, that there are some arguments I won't accept.

I won't accept any argument from authority. ("The Bible says so.")

I won't accept any argument from internal conviction. ("I have faith it's so.")

I won't accept any argument from personal abuse. ("What are you, an atheist?")

I won't accept any argument from irrelevance. ("Do you think you have been put on this Earth just to exist for a moment of time?")

I won't accept any argument from anecdote. ("My cousin has a friend who went to a medium and talked to her dead husband.")

And when all that (and other varieties of nonevidence) are eliminated, there turns out to be nothing.*

Then why do people believe? Because they want to. Because the mass desire to believe creates a social pressure that is difficult (and, in most times and places, dangerous) to face down. Because few people have had the chance of being educated into the understanding of what is meant by evidence or into the techniques of arguing rationally.

But mostly because they want to. And that is why a manufacturer of toothpaste finds it insufficient to tell you that it will clean your teeth almost as well as the bare brush will. Instead he makes it clear to you, more or less by indirection, that his particular brand will get you a very desirable sex partner. People, wanting sex somewhat more intensely than they want clean teeth, will be the readier to believe.

Then, too, people generally love to believe the dramatic, and incredibility is no bar to the belief but is, rather, a positive help.

Surely we all know this in an age when whole nations can be made to believe in any particular bit of foolishness that suits their rulers and can be made willing to die for it, too. (This age differs from previous ages in this, however, only in that the improvement of communications makes it possible to spread folly with much greater speed and efficiency.)

Considering their love of the dramatic, is it any surprise that millions are willing to believe, on mere say-so and nothing more, that alien spaceships are buzzing around the earth and that there is a vast conspiracy of silence on the part of the government and scientists to hide that fact? No one has ever explained what government and scientists hope to gain by such a conspiracy or how it can be maintained, when every other secret is exposed at once in all its details—but what of that? People are always willing to believe in any conspiracy on any subject.

People are also willing and eager to believe in such dramatic matters as the supposed ability to carry on intelligent conversations with plants, the supposed mysterious force that is gobbling up ships and planes in a particular part of the ocean, the supposed penchant of Earth and Mars to play Ping-Pong with Venus and the supposed accurate description of the result in the Book of Exodus, the supposed excitement of visits from extraterrestrial astronauts in prehistoric times and their donation to us of our arts, techniques, and even some of our genes.

To make matters still more exciting, people like to *feel* themselves to

---

*Lately, there have been detailed reports about what people are supposed to have seen during "clinical death." I don't believe a word of it.

be rebels against some powerful repressive force—as long as they are sure it is quite safe. To rebel against a powerful political, economic, religious, or social establishment is very dangerous and very few people dare do it, except, sometimes, as an anonymous part of a mob. To rebel against the "scientific establishment," however, is the easiest thing in the world, and anyone can do it and feel enormously brave, without risking as much as a hangnail.*

Thus, the vast majority, who believe in astrology and think that the planets have nothing better to do than form a code that will tell them whether tomorrow is a good day to close a business deal or not, become all the more excited and enthusiastic about the bilge when a group of astronomers denounce it.

Again, when a few astronomers denounced the Russian-born American Immanuel Velikovsky, they lent the man (and, by reflection, his followers) an aura of the martyr, which he (and they) assiduously cultivate, though no martyr in the world has ever been harmed so little or helped so much by the denunciations.

I used to think, indeed, that it was entirely the scientific denunciations that had put Velikovsky over the top and that, had the American astronomer Harlow Shapley only had the *sang froid* to ignore the Velikovskian folly, it would quickly have died a natural death.

I no longer think so. I now have greater faith in the bottomless bag of credulity that human beings carry on their back. After all, consider von Däniken and his ancient astronauts. Von Däniken's books are even less sensible than Velikovsky's and are written far more poorly,† and yet he does well. What's more, no scientist, as far as I know, has deigned to take notice of von Däniken. Perhaps they felt such notice would do him too much honor and would but do for him what it had done for Velikovsky.

So von Däniken has been ignored—and, despite that, is even *more* successful than Velikovsky is, attracts more interest, and makes more

---

*A reader once wrote me to say that the scientific establishment could keep you from getting grants, promotions, and prestige, could destroy your career, and so on. That's true enough. Of course, that's not as bad as burning you at the stake or throwing you into a concentration camp, which is what a *real* establishment could and would do, but even depriving you of an appointment is rotten. However, that works only if you are a scientist. If you are a nonscientist, the scientific establishment can do nothing more than make faces at you.

†Velikovsky, to do him justice, is a fascinating writer and has an aura of scholarliness that von Däniken utterly lacks.

money.

You see, then, how I choose my "impossibles." I decide that certain heresies are ridiculous and unworthy of any credit, not so much because the world of science says, "It is not so!" but because the world of non-science says, "It is," so enthusiastically. It is not so much that I have confidence in scientists' being right, but that I have so much in nonscientists' being wrong.

I admit, by the way, that my confidence in scientists' being right is somewhat weak. Scientists have been wrong, even egregiously wrong, many times. There have been heretics who have flouted the scientific establishment and have been persecuted therefor (as far as the scientific establishment is able to persecute), and, in the end, it has been the heretic who has proved right. This has happened not only once, I repeat, but many times.

Yet that doesn't shake the confidence with which I denounce those heresies I do denounce, for in the cases in which heretics have won out, the public has, almost always, not been involved.

When something new in science is introduced, when it shakes the structure, when it must in the end be accepted, it is usually something that excites scientists, sure enough, but does not excite the general public—except perhaps to get them to yell for the blood of the heretic.

Consider Galileo, to begin with, since he is the patron saint (poor man!) of all self-pitying crackpots. To be sure, he was not persecuted primarily by scientists for his scientific "errors" but by theologians for his very real heresies (and they were real enough by seventeenth-century standards).

Well, do you suppose the general public supported Galileo? Of course not. There was no outcry in his favor. There was no great movement in favor of the earth going round the sun. There were no "sun-is-center" movements denouncing the authorities and accusing them of a conspiracy to hide the truth. If Galileo had been burned at the stake, as Giordano Bruno had been a generation earlier, the action would probably have proved popular with those parts of the public that took the pains to notice it in the first place.

Or consider the most astonishing case of scientific heresy since Galileo—the matter of the English naturalist Charles Robert Darwin. Darwin collected the evidence in favor of the evolution of species by natural selection and did it carefully and painstakingly over the decades, then published a meticulously reasoned book that established the fact of evolution to the point where no rational biologist can deny it* even though there are arguments over the details of the mechanism.

Well, then, do you suppose the general public came to the support of Darwin and his dramatic theory? They certainly knew about it. His theory made as much of a splash in his day as Velikovsky did a century later. It was certainly dramatic—imagine species developing by sheer random mutation and selection, and human beings developing from apelike creatures! Nothing any science-fiction writer ever dreamed up was as shatteringly astonishing as that to people who from earliest childhood had taken it for established and absolute truth that God had created all the species ready-made in the space of a few days and that man in particular was created in the divine image.

Do you suppose the general public supported Darwin and waxed enthusiastic about him and made him rich and renowned and denounced the scientific establishment for persecuting him? You know they didn't. What support Darwin did get was from scientists. (The support any rational scientific heretic gets is from scientists, though usually from only a minority of them at first.)

In fact, not only was the general public against Darwin then, they are against Darwin now. It is my suspicion that, if a vote were taken in the United States right now on the question of whether man was created all at once out of the dirt or through the subtle mechanisms of mutation and natural selection over millions of years, there would be a large majority who would vote for the dirt.

There are other cases, less famous, where the general public didn't join the persecutors only because they never heard there was an argument.

In the 1830s the greatest chemist alive was the Swede Jöns Jakob Berzelius. Berzelius had a theory of the structure of organic compounds which was based on the evidence available at that time. The French chemist August Laurent collected additional evidence that showed that Berzelius' theory was inadequate. He himself suggested an alternate theory of his own which was more nearly correct and which, in its essentials, is still in force now.

Berzelius, who was in his old age and very conservative, was unable

---

*Please don't write to tell me that there are creationists who call themselves biologists. Anyone can call himself a biologist.

to accept the new theory. He retaliated furiously, and none of the established chemists of the day had the nerve to stand up against the great Swede.

Laurent stuck to his guns and continued to accumulate evidence. For this he was rewarded by being barred from the more famous laboratories and being forced to remain in the provinces. He is supposed to have contracted tuberculosis as a result of working in poorly heated laboratories, and he died in 1853 at the age of forty-six.

With both Laurent and Berzelius dead, Laurent's new theory began to gain ground. In fact, one French chemist who had originally supported Laurent but had backed away in the face of Berzelius' displeasure now accepted it again and actually tried to make it appear that it was *his* theory. (Scientists are human, too.)

That's not even a record for sadness. The German physicist Julius Robert Mayer, for his championship of the law of conservation of energy in the 1840s, was driven to madness. Ludwig Boltzmann, the Austrian physicist, for his work on the kinetic theory of gases in the late nineteenth century, was driven to suicide. The work of both is now accepted and praised beyond measure.

But what did the public have to do with all these cases? Why, nothing. They never heard of them. They never cared. It didn't touch any of their great concerns. In fact, if I wanted to be completely cynical, I would say that the heretics were in this case right and that the public, somehow sensing this, yawned.

This sort of thing goes on in the twentieth century, too. In 1912 a German geologist, Alfred Lothar Wegener, presented to the world his views on continental drift. He thought the continents all formed a single lump of land to begin with and that this lump, which he called "Pangaea," had split up and that the various portions had drifted apart. He suggested that the land floated on the soft, semisolid underlying rock and that the continental pieces drifted apart as they floated.

Unfortunately, the evidence seemed to suggest that the underlying rock was far too stiff for continents to drift through, and Wegener's notions were dismissed and even hooted at. For half a century the few people who supported Wegener's notions had difficulty in getting academic appointments.

Then, after World War II, new techniques of exploration of the sea bottom uncovered the global rift, the phenomenon of sea-floor spreading, the existence of crustal plates, and it became obvious that the earth's crust *was* a group of large pieces that were constantly on the move and that the continents were carried with the pieces. Continental drift, or "plate tectonics," as it is more properly called, became the cornerstone of geology.

I personally witnessed this turnabout. In the first two editions of my *Guide to Science* (Basic Books, 1960, 1965), I mentioned continental drift but dismissed it haughtily in a paragraph. In the third edition (1972) I devoted several pages to it and admitted having been wrong to dismiss it so readily. (This is no disgrace, actually. If you follow the evidence you *must* change as additional evidence arrives and invalidates earlier conclusions. It is those who support ideas for emotional reasons only who can't change. Additional evidence has no effect on emotion.)

If Wegener had not been a true scientist, he could have made himself famous and wealthy. All he had to do was to take the concept of continental drift and bring it down to earth by having it explain the miracles of the Bible. The splitting of Pangaea might have been the cause, or the result, of Noah's Flood. The formation of the Great African Rift might have drowned Sodom. The Israelites crossed the Red Sea because it was only half a mile wide in those days. If he had said all that, the book would have been eaten up and he could have retired on his royalties.

In fact, if any reader wants to do this *now*, he can still get rich. Anyone pointing out this article as the inspirer of the book will be disregarded by the mass of "true believers," I assure you.

So here's a new version of Asimov's Corollary, which you can use as your guide in deciding what to believe and what to dismiss:

> *If a scientific heresy is ignored or denounced by the general public, there is a chance it may be right. If a scientific heresy is emotionally supported by the general public, it is almost certainly wrong.*

You'll notice that in my two versions of Asimov's Corollary I was careful to hedge a bit. In the first I say that scientists are "probably right." In the second I say that the public is "almost certainly wrong." I am not absolute. I hint at exceptions.

Alas, not only are people human; not only are scientists human; but I'm human, too. I want the universe to be as *I* want it to be and that means completely logical. I want silly, emotional judgments to be *always* wrong.

Unfortunately, I can't have the universe the way I want it, and one of the things that makes me a rational being is that I know this.

Somewhere in history there are bound to be cases in which science said "No" and the general public, for utterly emotional reasons, said "Yes" and in which it was the general public that was right. I thought about it and came up with an example in half a minute.

In 1798 the English physician Edward Jenner, guided by old wives' tales based on the kind of anecdotal evidence I despise, tested to see whether the mild disease of cowpox did indeed confer immunity upon humans from the deadly and dreaded disease of smallpox. (He wasn't content with the anecdotal evidence, you understand; he *experimented*.) Jenner found the old wives were correct and he established the technique of vaccination.

The medical establishment of the day reacted to the new technique with the greatest suspicion. Had it been left to them, the technique might well have been buried.

However, popular acceptance of vaccination was immediate and overwhelming. The technique spread to all parts of Europe. The British royal family was vaccinated; the British Parliament voted Jenner ten thousand pounds. In fact, Jenner was given semidivine status.

There's no problem in seeing why. Smallpox was an unbelievably frightening disease, for when it did not kill, it permanently disfigured. The general public therefore was almost hysterical with desire for the truth of the suggestion that the disease could be avoided by the mere prick of a needle.

And in this case, the public was right! The universe *was* as they wanted it to be. In eighteen months after the introduction of vaccination, for instance, the number of deaths from smallpox in England was reduced to one-third of what it had been.

So there are indeed exceptions. The popular fancy is sometimes right.

But not often, and I must warn you that I lose no sleep over the possibility that any of the popular enthusiasms of today are liable to turn out to be scientifically correct. Not an hour of sleep do I lose; not a minute.   •

Astrology

A Statistical Test
of Sun-Sign Astrology

John D. McGervey

It has been said that the basic premise of astrology is that the stars and planets can influence terrestrial processes. If astrology did indeed develop from such a premise by careful observations followed by testing of results against predictions in a scientific way, one could have no quarrel with it. Nobody denies that extraterrestrial influences exist.

We could even accept the fact that astrologers can identify no known physical mechanism on which to base their predictions. If the predictions of astrology come true, then the subject cannot be dismissed, even though the basis of the prediction is not understood. However, the bases of astrological predictions are so far removed from any logical cause-and-effect relationship that it becomes difficult for any logical thinker to remain open-minded. The predictions are not based on any observable or even hypothetical physical *process*; instead they are often based on superficial aspects of the appearance of celestial objects. For example, Mars is red and blood is red, so Mars has something to do with blood, and by extension, Mars governs (in some vague sense) warfare and combat.

If we try to discredit astrology simply by pointing to the stupidity of this sort of reasoning, we run the risk of being considered closed-minded. Since advances in science are often based on ideas that seem stupid when they are first proposed, we should apply unbiased tests to the *results* of a theory and not apply value judgments to the reasoning that leads to these results. Who knows? Maybe by some curious coincidence the planet Mars does have something to do with warfare.

Unfortunately it is hard to evaluate the various "one-shot" predictions that astrologers make, because nobody knows what would be a good percentage of successful predictions; there are no standards of performance, and any particular failure can be attributed to an individual astrologer's mistake rather than to the "science" of astrology. However, there are some predictions, applicable to the entire population, that result from the drawing up of horoscopes. A number of tests of planetary and solar influences in horoscopes have been reported, but all appear to suffer from either a small sample or the possibility that the cause-and-effect relation has been incorrectly diagnosed. For example, effects claimed to be associated with the rising of one of the planets could be, and probably are, the result of the fact that more people are born in the morning hours than in the evening hours (Jerome 1976).

Tests of planetary influence are difficult because of the necessity of knowing the exact time of birth as well as the date, so such tests always involve a relatively small population. It is clear that in a small number of people one can always find common traits that one can then attribute to some astrological phenomenon; even Adolf Hitler and Julie Andrews probably have some traits in common. But one element of a horoscope that can be tested with good statistics using readily available information is the effect of the "sun sign." Although "serious" astrologers say that the sun sign is simply one component of a horoscope and that the "ascendant" and planetary influences are equally or even more important, to my knowledge they have never said that the sun sign has no influence whatsoever. They may say, for example, that sun-sign astrology as given in newspapers does not *completely* determine one's destiny, but they still refer to the *influence* of the sun. Clearly, if the sun has any influence at all, it should be detectable in a large enough population.

To test the effect of the sun sign, we need a characteristic that can be determined unambiguously for each member of a large population. A person's occupation is ideal for such a study, because it can be determined unambiguously by using standard reference books. For example, Americans who have done sufficient work in science to be listed in *American Men of Science* (1965) are scientists, and others are not. Although various astrologers may disagree on the specific effects of a given sign and may even define the signs differently (some of them have now become aware of the precession of the equinoxes), virtually all of them claim some connection between one's sun sign and one's chances of success in (or aptitude for) a given occupation.[1]

In searching for such a correlation I have tabulated the birthdates of 16,634 persons listed in *American Men of Science* and of 6,475 persons

---

1. For a summary of such claims see M. Zeilik II, *American Journal of Physics* 42 (1974): 538-42, or L. E. Jerome, *Leonardo* 6 (1973): 121-30.

Table 1.
Number of births by astrological sign

Sign	Dates (inclusive)	Scientists*	Politicians°
Capricorn	Dec. 24 - Jan. 19	1241	462
Aquarius	Jan. 23 - Feb. 18	1217	445
Pisces	Feb. 21 - Mar. 19**	1193	480
Aries	Mar. 23 - Apr. 18	1158	452
Taurus	Apr. 23 - May 19	1185	471
Gemini	May 24 - Jun. 19	1153	471
Cancer	Jun. 24 - Jul. 20	1245	514
Leo	Jul. 25 - Aug. 20	1263	504
Virgo	Aug. 25 - Sept. 20	1292	507
Libra	Sept. 25 - Oct. 21	1267	513
Scorpio	Oct. 25 - Nov. 20	1246	488
Sagittarius	Nov. 24 - Dec. 20	1202	453

*Birthdays taken from consecutive pages in two different volumes listed in *American Men of Science* (1965). A small percentage of scientists (less than 1 percent, in my estimation) may choose not to be listed in this directory, but elimination of this small number from the sample can hardly have a significant effect on the overall distribution. Some of those listed may also pursue other occupations, but this does not nullify the fact that they have achieved something in science to set them apart from nonscientists.

°Virtually all of the birthdays in *Who's Who in American Politics* (1973) were used. About 1 percent of the IBM cards were punched incorrectly and not redone.

**February 29 not included.

Table 2.
Number of births on each date

Scientists	Politicians

Scientists / Politicians

```
Scientists                                    Politicians

46 40 52 50 50 51 49 31 50 39        26 14 17 18 23 19 16 08 11 15
48 48 40 28 47 45 63 45 43 36        15 20 23 18 12 16 20 14 22 24
40 55 47 53 52 47 39 47 47 36 44     20 23 15 17 15 19 11 14 16 13 15

48 59 48 41 45 43 39 34 49 35        17 16 17 12 25 20 19 14 25 14
48 46 42 49 42 48 54 35 48 52        17 19 19 19 13 09 21 14 21 23
33 47 50 43 58 41 36 55 10           11 17 18 20 21 15 18 20 07

50 47 37 26 42 38 45 45 41 38        21 13 14 15 19 20 16 21 17 19
45 44 54 48 40 44 61 46 39 42        24 14 20 20 16 19 21 12 19 20
51 52 36 45 48 43 41 43 40 42 45     19 18 16 13 11 22 19 24 12 19 12

43 38 36 46 37 45 43 41 34 49        21 13 13 13 18 16 15 18 17 20
43 36 46 54 40 51 43 50 48 58        16 24 20 14 14 18 20 14 18 16
44 51 47 56 40 43 45 40 44 31        17 18 14 16 18 15 20 11 18 16

48 43 34 44 46 47 51 50 38 36        15 17 13 16 15 16 17 20 16 20
37 57 48 48 38 42 45 35 52 53        14 21 21 22 22 22 15 18 23 13
52 48 50 46 47 44 40 54 40 50 41     17 15 10 13 19 10 11 19 10 12 16

46 32 45 33 47 47 43 44 46 45        18 18 17 23 10 12 20 20 18 18
48 37 36 34 38 40 54 36 40 49        13 31 17 16 21 26 22 17 24 13
39 52 51 44 46 45 42 46 34 52        21 19 14 15 13 17 19 20 31 19

53 49 53 52 39 43 56 37 42 39        19 12 21 24 14 21 18 19 22 26
49 45 50 44 58 53 31 57 51 35        19 17 16 21 12 24 20 21 16 18
50 42 54 38 46 42 37 52 31 45 33     10 30 20 25 23 23 25 13 19 12 17

56 52 46 45 54 50 47 44 60 47        13 16 21 33 20 16 21 18 21 22
52 52 33 46 56 43 51 46 41 56        09 17 21 18 19 18 18 12 22 17
50 58 45 55 49 53 43 42 47 43 51     24 21 13 24 16 22 24 18 17 17 28

59 56 57 39 35 56 56 57 40 40        28 24 23 18 22 17 22 14 19 17
47 39 56 55 40 44 60 40 43 45        11 23 20 18 13 21 12 12 11 20
50 49 49 47 45 57 46 38 38 34        24 27 17 21 17 16 28 16 17 18

42 57 52 55 46 57 43 57 49 52        25 22 13 17 19 14 24 15 23 28
47 51 44 46 48 52 38 54 41 45        15 21 16 09 26 17 17 19 25 18
33 61 46 37 50 43 44 69 44 48 53     18 11 15 18 16 14 19 24 23 13 12

43 42 45 34 38 40 43 47 48 42        20 21 15 15 19 18 25 13 17 20
45 43 45 49 53 58 42 39 44 55        22 17 21 21 19 16 17 15 20 16
45 49 51 51 45 53 48 33 51 48        18 17 24 17 14 23 15 17 18 20

52 30 43 33 52 48 33 40 39 48        14 17 19 17 19 13 13 08 16 20
42 47 36 45 40 43 40 57 54 51        22 20 15 12 17 20 11 21 19 16
44 42 47 53 51 39 45 46 44 51 47     22 14 22 13 06 20 20 14 20 24 18
```

listed in *Who's Who in American Politics* (1973). The results are summarized in Table 1. Because the starting and ending dates of a given sign vary from year to year, I have tabulated the totals for the central 27 dates of each sign. The dates not included in these signs show no significant deviation from the flat pattern observed in the dates that were used, as can be seen by referring to the complete tabulation in Table 2.

The number of scientists born under each sign lies between 1,153 and 1,292; the mean (m) is 1,220 and the standard deviation is 45.6. The theoretical standard deviation for a binomial distribution of this size with randomly selected signs would be 33.4. The maximum deviation observed is 2.1 times the theoretical binomial standard deviation. Corresponding numbers for the politicians are: $m = 474$, $\sigma = 26.2$, and binomial $\sigma = 20.8$. The value of the reduced chi-squared for a fit to a flat distribution is 1.70 for scientists and 1.45 for politicians. These values are slightly high, and careful study of the numbers in Table 1 shows that there is a definite trend in the dates. *Both* sets shows an excess of births in late summer and a corresponding deficiency in the spring. These deviations are somewhat too large to be random fluctuations, even though they are a small percentage (less than 5 percent) of the mean. But there is no need to invoke astrological influences for this effect; the same pattern appears in "live births by month" in the U.S. population, where an excess of about 5 percent in July, August, and September occurs (*Vital Statistics of the United States 1968-69*). Thus any effect of one's sun sign on one's choice of occupation must be considerably less than 5 percent, hardly enough to justify the vast literature on the subject.

No effect was observed in the individual dates, either; for scientists, the mean number per day was 45.6, the maximum observed was 69, and the minimum 26. One hundred twelve dates, or 30.7 percent of the total of 365 dates, had more than 52 or fewer than 39 scientists' birthdays; that is, there were 253 cases within one standard deviation of the mean—just about what one would expect for a random normal distribution. In other words, a table of birthdates serves reasonably well as a random number generator (unless a pair of twins is listed).

An astrologer might argue that the class of scientists and the class of political figures is too broad and that subsets of these groups (e.g., microbiologists, paleontologists) might favor certain signs, but that these sets would distribute themselves among the various signs so that no overall effect is seen. However, books on astrology consistently insist that "scientists" or "politicians" are favored by one sign or another. Furthermore, it is highly improbable that the various scientific disciplines could be favored by certain signs in such a way that when the groups are added together no effect of the sun sign remains. By breaking the population up into sufficiently small subsets one can undoubtedly find, in one subset or another, a surprisingly large deviation from the mean in some range of birthdates. But the significance of such a deviation must be viewed in the light of the large number of possible subsets that could be chosen, as well as the large number of ranges of dates that could be used. If an astrologer chooses the occupation and the range of dates *before* looking at the data and correctly *predicts* a large deviation on the basis of his "science," then the result might be significant. However, that has not yet happened.

In the face of this negative result some astrologers might be tempted to claim that they never attached *any* significance to sun signs. But they are then faced with the task of explaining (1) why their "science," thousands of years old, suddenly has lost one of the elements that has appeared in every book on the subject, (2) how the positions of the planets can have an influence if the sun's position does not, and (3) how the time of day when one is born can have an influence which varies with the seasons and planets if the date of the year has no influence in itself. If logic had any place in astrology, they would be faced with a hopeless task.

References

American Men of Science 1965, 11th ed. *The Physical and Biological Sciences.* New York: R. R. Bowker.

Jerome, L. E. 1976. *The Humanist,* March/April 1976, pp. 52-53.

Vital Statistics of the United States 1968-69, Public Health Service, Department of Health, Education, and Welfare; and *Vital Statistics of the United States* 1937-39, Census Bureau, U.S. Department of Commerce.

Who's Who in American Politics 1973, 4th ed. New York: R. R. Bowker.

An Empirical Test of Popular Astrology

Ralph W. Bastedo

Introduction

Astrology is everywhere. It permeates contemporary American culture. Newspapers and magazines, ranging from the conservative *National Enquirer* to the radical *Berkeley Barb*, talk about it endlessly. Horoscopes are now carried by 1,250 daily newspapers, that is, by two out of every three papers in the United States. The average citizen can't help but be inundated by the flood of articles and syndicated columns.[1]

Such a state of affairs is amply borne out by recent Gallup polls. George Gallup reports that over three-quarters of American adults know their astrological sign. Furthermore, he finds that more than one adult in four embraces astrology. This means that there are between 30 and 40 million believers nationwide. So astrology has millions more followers than most religious denominations.[2]

The San Francisco Bay area, the focus of my study, is an astrologer's paradise. Telephone directories list thirty-four professional astrologers and astrological schools in the immediate vicinity of San Francisco and Berkeley. The vicinity also features eleven astrology shops and bookstores. Glock and Wuthnow of the Survey Research Center report that nine out of ten bay area residents know their sign—a saturation level rarely encountered in polling research. Only three out of ten bay area adults are firm disbelievers.

Bay area lifestyles reflect this. An ABC affiliate, KGO-TV, sponsors astrologer Joyce Jillson on its morning news and interview program "A.M. San Francisco." The Oakland Athletics, an American League baseball team, employs astrologer Laurie Brady to consult with team players and help them throughout training. Students at U.C. Berkeley complain that bay area employers use an individual's astrological sign, deduced from one's birth date, to discriminate among job applicants. It's just too bad if you're Scorpio or Taurus—you won't get the job.

The Need for an Empirical Test

To counter astrology's evident popularity, nearly 200 scientists in 1975 signed a manifesto sponsored by *The Humanist* magazine deploring any recognition whatsoever of astrology as a science. The scientists described the astral art as characterized by "irrationalism" and "obscurantism." Included among the signers were 19 Nobel laureates.[3]

In response to the lament of the scientists, a San Francisco Bay area newspaper angrily cited arguments in defense of astrology. Its editors concluded by flinging down the gauntlet with a challenge: "With the availability of accurate measure of the universe and computers, the time to test astrology, including all its symbolism, has come."

I agree. Given the widespread popularity of astrology, a test of a few of its major tenets is certainly called for. An empirical testing of *all* facets of astrology would necessitate a lifetime of work and countless volumes of data and interpretation. That is not my goal. What I plan is considerably more modest. What I shall do is examine a few of the basic assumptions of popular astrology—that form of astrology in which over 30 million Americans believe.[4]

Popular Astrology and Judicial Prediction

I would describe popular astrology as a nontechnical and indiscriminate form of traditional, judicial astrology. It is judicial in that it attempts to foretell terrestrial life from the movements of celestial bodies. By contrast, natural astrology was used to foretell the movements of the heavenly bodies themselves. Natural astrology has quietly evolved into the science of astronomy and astrophysics.

Of the many, many types of judicial astrology that pertain to the "vitasphere," only a few directly relate to my testing. Of less concern to us are the ancient schools of (1) mundane astrology, (2) naturalist astrology, (3) agricultural astrology, and (4) astro-meteorology, or meteorological astrology. These attempt to forecast historical events and natural disasters of worldly import. They try to predict wars, earthquakes, weather changes, plagues, assassinations, droughts, conspiracies, famines, and revolutions.

Of more direct interest to us are the schools of (5) genethliacal astrology, (6) electional astrology, and (7) horary astrology. These attempt to forecast human events in more personalized terms. They try to predict a wide array of human characteristics, particularly personality traits.

The discipline of genethliacal astrology is also called genethlialogy and natal horoscopy. Genethliacal astrologers construct and interpret personalized, lifelong horoscopes known as "natal charts." Estimates of an individual's destiny are computed on the basis of longitude, latitude, and precise time of birth.

The discipline of electional astrology is the horoscopic art of choosing exactly the right moment for a person to pursue an enterprise. Astrological "election" provides advice about when an individual should hold a wedding ceremony, assume public office, begin a voyage, or lay the foundation-stone for a new building.

The discipline of horary astrology answers questions for the individual as they crop up on a daily basis. Mass-circulation newspapers have popularized this version of the astral art. As any frequent reader knows, such "horoscopes" offer day-to-day advice based solely on a person's astrological sign at the moment of birth. This sign is also sometimes referred to as a "zodiac sign" or "sun sign."[5]

Popular Astrology and the Traditional Perspective

The assumptions of popular astrology remain faithful to astrological tradition. The public is not at all troubled by the "precession of the equinoxes," that very slow shift in the night sky pattern (as observed from earth) taking 25,800 years to come full circle. But the precession of the equinoxes does bitterly divide present-day astrologers.[6]

The division arises because traditional astrology, also described as the "movable" or "tropicalist" perspective, uses as its reference point the vernal equinox. The vernal equinox moves eastward 1.4 degrees each century through the "fixed" stars of the familiar zodiac constellations. Since a shift of about 30 degrees has transpired in the past 2,000 years, traditional astrology suffers from ossification. Because zodiac signs refer to 30-degree sectors of the ecliptic and not to constellations, the traditionalist's sign of Aries now coincides roughly with the real constellation of Pisces. Traditional astrology relies on the same model of the night-sky constellations as recorded two millennia ago by ancient stargazers.

Popular Astrology vs. Technical Astrology

Yet while these issues of astrological calculation rage wildly amid astrologists, the public remains largely unaware or indifferent. The reason for this is quite understandable: popular astrology is not only both judicial and tradition-bound, it is also relatively nontechnical. Most Americans' interest in astrology focuses only on their particular constellation of the zodiac, or "sun sign." The sun sign is designated on the day of one's birth by the sun's position within a 16-degree-wide path of the ecliptic, a path otherwise known as the zodiac (hence the term "zodiac sign").

Full-time astrologers, concerned with the intricate calculations necessary for natal charts and other forecasts, go far beyond mere consideration of the sun sign. To some astrologists, it is of only secondary significance. Astrologers usually take into account the longitude, latitude, and precise moment of birth. They also calculate such relationships as the position of the sun and moon in regard to each of the eight planets and to the "ascendant"—the celestial "house" rising above the eastern horizon at the moment of birth.

The maze of calculations that professional astrologers make are far too complex to be adequately reviewed here. Suffice it to say that most folks are struck dumb by such measurements. They do not care for technical astrology with its calculations of "cusps," "medium coeli," "imum coeli," "angles," "transit," "aspects," "opposition," "conjunction," "sextile," "square," "trine," "quintile," "sesquiquadrate," "quincunx," and "orb." All they want to know is their sign.

And with good reason. Astrologists are in rare unanimous agreement that knowledge of the sun sign provides valuable, predictive information—albeit incomplete, perhaps—about human psychology and behavior. Astrologists may quibble over whether the sun sign reflects an "astral influence" or a biological process of "synchronicity," but they affirm in unison that the motions of the stars with regard to the earth correspond to an important array of deterministic influences upon human behavior.

The Birthday Sample

To find out whether people do, in fact, differ according to their sun sign, it was necessary to locate a survey sample in which respondents had been asked their birth date. Knowing both the day and month of birth for each pollee would permit accurate division of the sample among the twelve zodiac signs.[7]

The sample adopted for this study was drawn in 1971 by the Survey Research Center of the University of California at Berkeley. The center sampled the San Francisco Bay area, a five-county region of 2.5 million people and one million households. The Berkeley researchers used a cross-sectional, stratified cluster sample. The response rate was 75 percent. Interviews were completed with 1,000 adults.

Although the survey data were collected admittedly for reasons

---

List 1

The zodiac typologies and cutting points as defined by this study

Zodiac sign	Ruler	Element	Cutting points
1. Aries the ram	Mars	Fire	March 22 to April 19
2. Taurus the bull	Venus	Earth	April 21 to May 20
3. Gemini the twins	Mercury	Air	May 23 to June 20
4. Cancer the crab	Moon	Water	June 23 to July 21
5. Leo the lion	Sun	Fire	July 24 to August 22
6. Virgo the virgin	Mercury	Earth	August 24 to September 22
7. Libra the balance or scales	Venus	Air	September 24 to October 21
8. Scorpio the scorpion	Mars or Pluto	Water	October 24 to November 21
9. Sagittarius the archer or centaur	Jupiter	Fire	November 23 to December 21
10. Capricorn the goat or sea-goat or goat-fish	Saturn	Earth	December 23 to January 19
11. Aquarius the water-bearer	Saturn or Uranus	Air	January 21 to February 17
12. Pisces the fish	Jupiter or Neptune	Water	February 20 to March 19

other than my test of popular astrology, they will serve our purpose quite well. This is because the questionnaire included a large number of items on which astrologists have made explicit predictions.[8]

The Zodiac Typology as Independent Variable

From our knowledge of respondents' birth dates it was possible to allocate the respondents among the twelve sun signs. There developed one small wrinkle, however. Because, while the sun changes signs around the twentieth of each month, the exact date on which it changes for any given month is not identical year in and year out. Nor, for that matter, do all astrologers agree on what the precise changeover dates should be. The result is muddied confusion for those few respondents who have the misfortune to be born around the twentieth of each month.

To eliminate this nuisance, specific cut-off points were adopted for each of the signs so the problematic cases could be dropped from tabulation. A review of astrological literature led to the decision to exclude those 45 respondents (out of the sample of 1,000) who were born on one of 18 problematic days.[9]

The exclusion of 18 days from the entire year of birthdays is no great loss. Almost 96 percent of our sample remains intact. This solution ensures that astrologists need not worry that our zodiac categories have been polluted by misclassified members of adjacent sun signs. The twelve-category zodiac typology created for our test is free of such contamination. See list 1 for a complete listing of the zodiac signs and corresponding cut-off points.

The Content Analysis

But one last barrier faces us before we can conduct our experiment. The problem arises that not all astrologers agree on which human characteristics are to be attributed to each of the twelve signs. Astrologers constantly wrangle over such matters. To choose the sun sign interpretations of only one astrologer would inevitably prompt accusations that our analysis was parochial, selective, and inconsequential.

So a different approach was adopted. It was decided to conduct a thorough ''content analysis'' of popularly available books dealing with traditional, judicial astrology. A total of fourteen sources, representing several schools of thought, were ultimately selected for the content analysis. Most of the sources can be succinctly described as reviews by professional astrologers of prior and contemporary writings. All were obtained from public libraries.

From the content analysis of astrological literature a grand total of over 2,375 sign-specific adjectives were tabulated. From this statement it can be seen that each of the twelve zodiac signs was described roughly 200 times by assorted adjectives. About 30 adjectives were cited twice or more per sign. A little less than 100 adjectives on average were singularly mentioned. Despite this immense variety of words, the content analysis shows a remarkable degree of qualitative and substantive agreement among traditional exponents of judicial astrology. Most of the adjectives describing any one sign are either synonyms or terms sharing similar (albeit not equivalent) meanings. Antonyms are negligible.

The General and Specified Hypotheses

Two hypotheses will be tested. The first we shall label the "specified" or "specific" hypothesis. The second we shall christen the "general" or "generic" hypothesis. The specified hypothesis postulates a priori what impact each of the zodiac signs is to have upon human characteristics. This hypothesis therefore necessitates some sort of content analysis. But the generic hypothesis does not. It stipulates merely that human characteristics should differ according to sun sign. It does not postulate a priori in what direction or form these differences should be. It is non-specific.

Popular astrology will be vindicated if our data buttress the specified hypothesis. But if our evidence fortifies the generic hypothesis without lending support to the a priori specification, then popular astrology's vindication will be only partial.

Leadership Ability as the Dependent Variable

We find from the content analysis that leadership qualities are attributed to the zodiac signs of Leo, Aries, Scorpio, Capricorn, and Sagittarius. People born under these signs are predicted to be unusually strong, domineering, authoritative, masterful, tough, and kingly. They are born to be leaders.

Nonleadership traits are attributed to the signs of Taurus, Pisces, Cancer, and Virgo. People born under these signs are timid, insecure, shy, clinging, indecisive, and weak. They are born to be servants.

Examining a cross-tabulation where sign is the independent variable and "leadership ability" is the dependent variable, we find what appears to be initial confirmation of the specified hypothesis. Refer to table 1. Aries illustrates a disproportionate amount of leadership ability within the top row of the table. But, sad to say, the signs of Leo, Scorpio, Capricorn, and Sagittarius do not. Indeed, all four are actually below rather than above the mean percentage of the row. As for our negative predictions, we find that both Virgo and Pisces are indeed above the average percentage of the bottom row. But Taurus and Cancer, alas, are not. This is bad news. Our cross-tabulation provides no decisive evidence to buttress the specified hypothesis regarding leadership ability.

To test the general hypothesis we use the chi-square (χ^2) goodness-of-fit statistic. Chi-square measures the discrepancy from the null hypothesis. The null hypothesis, or H_0, stipulates that the two variables being compared are statistically independent of one another, that there is absolutely no relationship whatsoever between the two. The chi-square statistic compares the cell frequencies observed within the rows and columns of the table with those predicted by the null hypothesis.

If the chi-square is quite large compared to the degrees of freedom, then the probability value and level of significance will be less than one percent.[10] The null hypothesis is then rejected. The observed relationship is judged statistically significant.

But if the chi-square is small relative to the degrees of freedom, then the probability value and corresponding level of significance will be much greater than this percent. In that case, H_0 is accepted and the generic hypothesis is weakened. The observed relationship is deemed statistically insignificant. It is attributed to chance fluctuation.

For this particular cross-tabulation the chi-square statistic is relatively small. The relationships observed are not significant at even the .05 level. The distributions we see may be safely attributed to random chance. The data here fail to strengthen the generic hypothesis.[11]

Political Stand as the Dependent Variable

The content analysis tells us that liberalism is associated with the sun signs of Aquarius, Aries, Gemini, and Sagittarius. Folks born under these signs are forecast as distinctly unconventional, rebellious, impatient, reformist, mutable, revolutionary, radical, rash, and liberal. They

Table 1
Cross-tabulation with astrological sign as the independent variable and "leadership ability" as the dependent variable

"This question concerns various talents and characteristics. Please indicate whether you think you have a lot, a fair amount, only a little, or none at all by circling the appropriate number next to each item:

Leadership ability   (1) a lot, (2) fair amount, (3) only a little, or (4) none at all."

The a priori predictions of leadership qualities are capitalized.
The a priori predictions of servile qualities are in italics.

	ARIES	*Taurus*	Gemini	*Cancer*	LEO	*Virgo*	Libra	SCORPIO	SAGITTARIUS	CAPRICORN	Aquarius	*Pisces*	total
1)	34%	19%	13%	25%	18%	19%	26%	18%	20%	17%	16%	26%	21%
2)	39	54	60	48	51	51	40	52	48	50	53	32	48
3)	17	20	17	20	25	19	24	16	24	22	17	29	21
4)	10	8	11	7	6	11	11	14	8	11	15	13	10
	100%	100%	100%	100%	100%	100%	100%	100%	100%	100%	100%	100%	100%
N	70	65	79	81	88	90	76	77	79	88	76	84	953

degrees of freedom = 33.0   chi-square = 38.012   probability value = 0.25

are born to be leftists.

Conservative traits, on the other hand, are attributed to the signs of Taurus, Cancer, Leo, Virgo, and Capricorn. Such individuals are patient, traditional, cautious, conventional, tenacious, obstinate, persistent, and conservative. They are born to be rightists.

Looking at the cross-tabulation with zodiac sign as the independent variable and "stand politically on most issues" as the dependent variable, we find again what appears to be initial confirmation of the specific hypothesis. Refer to table 2. Aquarius, as predicted, is more liberal than average for the first and second rows. But Sagittarius, Aries, and Gemini are not. Overall, they are actually less liberal. As for the conservative predictions, we see that Virgo has easily the largest percentage for the bottom three rows. This makes it the most conservative of the signs. The figures for Capricorn are likewise greater than the typical percentages for these rows. Yet Taurus and Cancer are smaller than average. This will not do. The cross-tabulation presents no conclusive evidence to buttress the specified hypothesis regarding political stand.

To test the generic hypothesis we use the chi-square goodness-of-fit statistic. We see that for this cross-tabulation the chi-square is relatively tiny. Since the probability value is much larger than .05, the associations seen within the table are statistically insignificant. The frequencies we have observed can be attributed to chance fluctuation. The data fail to strengthen the generic hypothesis.[12]

Subjective Intelligence as the Dependent Variable

We learn from the content analysis that intellectual abilities are associated with the zodiac signs of Aquarius, Sagittarius, Aries, Gemini, and Virgo. Persons born under these signs are predicted to be exceedingly analytical, rational, logical, educated, intelligent, scientific, discriminating, ingenious, and intellectual. They are born to be academicians. Anti-intellectual qualities, on the other hand, could not be derived from the content analysis. It appears that none of the twelve signs is characterized by stupidity.

Taking a close look at the cross-tabulation with sign as the independent variable and subjective "intelligence" as the dependent variable, we discover that Aquarius and Sagittarius lend support to the specified hypothesis. As predicted, Aquarius and Sagittarius both are more intelligent than average within the uppermost row. Note table 3. But

Table 2

Cross-tabulation with astrological sign as the independent variable and "stand politically on most issues" as the dependent variable

"Suppose that the line drawn . . . shows the range of political opinion in our country. Where would you say you stand politically on most issues?

(1) radical, (2) radical/liberal, (3) liberal, (4) liberal/moderate, (5) moderate, (6) moderate/conservative, (7) conservative, (8) conservative/very conservative, or (9) very conservative."

The a priori predictions of liberal qualities are capitalized. The a priori predictions of conservative qualities are in italics.

	ARIES	*Taurus*	GEMINI	*Cancer*	*Leo*	*Virgo*	Libra	Scorpio	SAGITTARIUS	*Capricorn*	AQUARIUS	*Pisces*	total
1)	1%	3%	4%	8%	6%	5%	7%	3%	4%	6%	7%	4%	5%
2)	7	5	5	9	6	2	4	8	5	7	11	5	6
3)	12	16	24	22	23	23	18	24	22	14	18	21	20
4)	9	8	16	6	13	12	11	10	15	14	14	15	12
5)	42	43	29	41	33	25	38	26	31	37	34	29	34
6)	9	16	12	6	9	9	8	8	4	5	5	8	8
7)	20	8	9	5	11	21	11	15	13	15	10	15	13
8)	0	0	1	3	0	1	3	0	4	1	1	2	1
9)	0	2	0	1	2	2	1	7	3	2	1	2	2
	100%	100%	100%	100%	100%	100%	100%	100%	100%	100%	100%	100%	100%
N	69	63	76	79	86	87	76	74	78	87	74	83	932

degrees of freedom = 88.0   chi-square = 84.274   probability value = greater than 0.25

Table 3

Cross-tabulation with astrological sign as the independent variable and subjective "intelligence" as the dependent variable

"This question concerns various talents and characteristics. Please indicate whether you think you have a lot, a fair amount, only a little, or none at all by circling the appropriate number next to each item:

Intelligence   (1) a lot, (2) fair amount, (3) only a little, or (4) none at all."

The a priori predictions of intelligence are capitalized.

	ARIES	Taurus	GEMINI	Cancer	Leo	VIRGO	Libra	Scorpio	SAGITTARIUS	Capricorn	AQUARIUS	Pisces	total
1)	27%	22%	22%	26%	21%	23%	29%	33%	28%	25%	28%	23%	25%
2)	67	75	73	70	74	67	62	61	68	72	62	66	68
3)	6	3	5	3	6	9	9	7	3	3	11	8	6
4)	0	0	0	1	0	1	0	0	1	0	0	4	1
	100%	100%	100%	100%	100%	100%	100%	100%	100%	100%	100%	100%	100%
N	70	65	78	81	88	90	76	77	79	88	76	84	952

degrees of freedom = 33.0   chi-square = 35.371   probability value = 0.35

Gemini, Aries, and Virgo, alas, have percentages lower than average. This is quite unsatisfactory. Our cross-tabulation provides no conclusive evidence to support the specific hypothesis of subjective intelligence.[13]

To examine the generic hypothesis we apply the chi-square goodness-of-fit statistic. It appears that for this cross-tabulation also the chi-square is relatively small. The relationships we see in the table are not statistically significant at the .05 level. The cell frequencies we have observed should be attributed to random chance. The data do not strengthen the general hypothesis.[14]

Ascribed Intelligence as the Dependent Variable

Astrologists might contend that the previous test suffers from a reliance on subjective estimation by the respondent himself, that it does not rely on a more objective, outside appraisal. Luckily for our experimental design, a second item on the questionnaire asked the interviewer to judge the intelligence of the pollee as conveyed during the hour or two of questioning. Using the same derivations from the content analysis as earlier, we now have a second table against which to check our earlier predictions.

Examining the cross-tabulation with sun sign as the independent variable and "intelligence quotient" as the dependent variable, we discover that Sagittarius once again lends credence to the specified hypothesis. As predicted, Sagittarius is more intelligent than average within the top row. But our prediction this time is even further off the mark than on the last occasion. Aries, Gemini, Virgo, and Aquarius have fewer intelligent people than average. If these particular zodiac signs really do indicate intelligence, one might conclude that their members cleverly hide it from both themselves and others. They may indeed be intelligent, but neither they nor anybody else apparently realize it. Suffice it to say that such an empirical result, however explained, is unsatisfactory. Our cross-tabulation undermines the specified hypothesis regarding ascribed intelligence. Examine table 4.

Once again we apply the chi-square goodness-of-fit statistic to test the general hypothesis. But the chi-square is too little. The relationships observed within the cells of the table are not significant at even the .05 level. The frequencies we have observed can be attributed to chance fluctuation. For the fourth time the data weaken the generic hypothesis.[15]

Table 4

Cross-tabulation with astrological sign as the independent variable and ascribed "intelligence quotient" as the dependent variable

"How would you rate the respondent's I.Q.?
(1) far above average, (2) somewhat above average, (3) about average, (4) somewhat below average, (5) far below average."

The a priori predictions of intelligence are capitalized.

	ARIES	Taurus	GEMINI	Cancer	Leo	VIRGO	Libra	Scorpio	SAGITTARIUS	Capricorn	AQUARIUS	Pisces	total
1)	6%	15%	8%	8%	10%	7%	9%	7%	11%	8%	8%	12%	9%
2)	50	28	43	40	40	43	40	47	33	41	41	39	40
3)	43	52	44	45	42	49	46	38	48	38	50	44	45
4)	1	5	5	6	6	1	3	8	8	13	1	2	5
5)	0	0	0	1	1	0	3	1	0	1	0	2	1
	100%	100%	100%	100%	100%	100%	100%	100%	100%	100%	100%	100%	100%
N	70	65	77	80	87	89	76	77	79	88	76	82	946

degrees of freedom = 44.0   chi-square = 48.199   probability value = 0.31

Astrological Belief as the Dependent Variable

If people with ordinary senses are consistently unable to detect their own true traits—as some astrologists might indeed surmise from our evidence—then perhaps at least the sensually gifted will differ. We can hope that people with occultist and similar extrasensory abilities will be able to divine the truth on such matters.

From the content analysis we find that occultist qualities are attributed to the sun signs of Pisces, Scorpio, and Aries. People born under these signs are predicted as unusually clairvoyant, shrewd, mystical, magical, spiritual, psychic, and occultist. They are born to be astrologers. Anti-occultist properties could not be derived from the content analysis. It would appear that the occult has only friends, and no enemies, among the twelve zodiac signs.

Examining the cross-tabulation with sign as the independent variable and "belief in astrology" as the dependent variable, we find what looks like initial confirmation of the specified hypothesis. See table 5. Pisces and Aries display higher percentages of astrology-prone belief than the overall mean of the top and second rows. But Scorpio, alas, displays a lower percentage. This is tragic. It sure beats me why so, so many people with occultist insight should disbelieve in astrology. Once again, it must be noted, the inconsistency within the cross-tabulation undermines the specific hypothesis.

To test the generic hypothesis we use the chi-square goodness-of-fit statistic. Our chi-square here is relatively small. The relationships seen in the table are not at the .05 level of significance. The distributions within the cross-tabulation should be attributed to random choice. The data once again, for the fifth time, undermine the generic hypothesis.[16]

The Twenty-eight Additional Variables

To supplement the five dependent variables already cited, twenty-eight additional items were run and then examined for statistically significant results. Nine of these additional items tap other personality traits. The nine are subjective measurements of one's (1) music ability, (2) artistic ability, (3) ability to make friends, (4) ability to organize well, (5) ability to feel deeply, (6) ability to make things, (7) self-confidence, (8) creativity, and (9) gift of gab.

Six additional questionnaire items tap feelings toward the occult.

Table 5

Cross-tabulation with astrological sign as the independent variable and "belief in astrology" as the dependent variable

"People who believe in astrology claim that the stars, the planets, and our birthdays have a lot to do with our destiny in life. What do you think about this:

Are you (1) a firm believer in astrology, are you (2) somewhat doubtful, are you (3) very doubtful, or are you (4) a firm disbeliever?"

The a priori predictions of occultist abilities are capitalized.

	ARIES	Taurus	Gemini	Cancer	Leo	Virgo	Libra	SCORPIO	Sagittarius	Capricorn	Aquarius	PISCES	total
1)	9%	5%	6%	14%	9%	9%	13%	9%	6%	8%	8%	10%	9%
2)	46	45	44	30	47	37	39	39	36	33	45	51	41
3)	28	19	22	20	25	18	20	26	24	26	23	16	22
4)	18	32	28	37	19	37	29	26	33	33	24	24	28
	100%	100%	100%	100%	100%	100%	100%	100%	100%	100%	100%	100%	100%
N	68	65	79	81	88	90	77	77	78	88	74	83	948

degrees of freedom = 33.0   chi-square = 33.043   probability value = 0.47

The six are subjective measurements and evaluations of one's (10) knowledge about astrology, (11) knowledge about horoscopes, (12) knowledge about one's own astrological sign, (13) interest in one's horoscope, (14) experience with extrasensory perception, and (15) type of ESP experience.

Four items tap the respondent's social philosophy and outlook on life. Three measured the degree to which the pollee felt that (16) most of life is decided for us rather than by us, that (17) some people are born lucky and others are born unlucky, and that (18) people suffer because they haven't learned to find inner peace. The fourth (19) recorded the respondent's view of the purpose of life.

Three items tap sociological characteristics. The three are categorical variables describing the subject's (20) occupation, (21) religion, and, if Protestant, (22) denomination.

Lastly, five items describe the respondent's physical appearance. Four were judgments by the interviewer of the respondent's (23) attractiveness, (24) physical disabilities, (25) height, and (26) weight. The fifth (27) was a subjective evaluation of one's good looks.

None of these twenty-seven supplemental variables yielded results that were statistically or substantively noteworthy.[17] No support was found for either (a) the general hypothesis or (b) the specified hypothesis within any of the twenty-seven supplemental cross-tabulations.

But since astrological prediction depends so much on the date of birth, it was suggested that age in years be used as a control variable. Although neither technical astrology nor popular astrology presupposes any zodiac irregularities according to age, this seemed a reasonable request given the time-dependent nature of the astral art. Running chi-square for each of the five original and twenty-seven supplemental variables while controlling for age again produced statistically uninteresting results. The sixty-two partial tables created in this endeavor yielded no surprises.

Conclusions

Popular astrology is clearly rooted in traditional, judicial astrology. It should not be confused, however, with the technical astrology of some proponents of the astral art.

Our empirical test of popular astrology confirms neither (a) the general hypothesis nor (b) the specified hypothesis of astrological prediction. The test sustains the null hypothesis.

The impotence of the general hypothesis was repeatedly documented by the chi-square goodness-of-fit statistic. An examination of thirty-three variables, measuring everything from physical attractiveness to belief in astrology, found no variable that was statistically significant at the .01 level when cross-tabulated with the astrological sign of the respondent. Indeed, thirty-two of the thirty-three were not significant at even the .05 level, an outcome fully in accord with probability theory. All of our results can be attributed to random chance.

An intensive look at five dependent variables for which content analysis permitted detailed a priori prediction found the specified hypothesis to be lacking in every instance. The null hypothesis was sustained in each of the cases. The five measured (1) leadership ability, (2) political stand, (3) subjective intelligence, (4) ascribed intelligence, and (5) astrological belief.

From such results one can only conclude that quite a number of the tenets of popular astrology are untenable.

To the good-humored scientist and layman alike, I wish to end with a note that my test has been thoroughly scientific. My dispassionate approach has been broad-minded, detached, unprejudiced, and unbiased. The methods I have used have been logical, rational, and trust-inspiring. I have incorporated my natural intelligence, genius, brilliance, and inventiveness in this truth-thirsting endeavor. Above all else, I have remained sincere, honest, idealistic, and serious in my never-ending quest for the truth.

It could not be otherwise. Because I am an Aquarian. And Aquarians are like that.

Acknowledgment: My thanks for their assistance goes to Ann Stannard, Dennis Rawlins, Persi Diaconis, and especially Charles Y. Glock.

Notes

1. The best-selling astrological periodical in the nation is *Horoscope*, published by Dell. *Horoscope* has a monthly circulation of a quarter-million, a readership many times larger than that of most astronomy and general science publications. Even so prestigious a group as the 6,000-member Association for Humanistic Psychology, publisher of the *Journal of Humanistic Psychology*, has flirted with the subject. Instruction in "Experiential Astrology" was sponsored by the AHP at its latest national convention.

2. According to Gallup, believers in astrology tend to be much less educated, without a higher-status job, nonwhite, poor, under 25 years old, single, and female. Studies of public opinion in foreign lands yield similar findings. They confirm the existence of millions of

believers in Australia, Brazil, France, Great Britain, India, Japan, and West Germany.

3. The manifesto, its signers, and their criticisms are cited by Bart J. Bok and Lawrence E. Jerome (eds.) in *Objections to Astrology* (Buffalo, New York: Prometheus Books, 1975).

4. I am treading in the footsteps of P. R. Farnsworth, J. A. Hynek, A. Müller, B. I. Silverman, M. Whitmar, and A. Standen. I shall try to avoid the pitfalls into which P. Choisnard, K. E. Krafft, C. G. Jung, R. J. Pellegrini, and M. Gauquelin have fallen.

5. A good review of these and related subjects is given by Louis MacNeice in *Astrology* (Garden City, N.Y.: Doubleday, 1964), pp. 8-35.

6. The point of their disagreement is over how to handle the "synetic vernal point," that is, the degree and constellation in which the rising sun is located on the first day of spring (the vernal equinox, about March 21). Sidereal astrology, also known as the "fixed" perspective, argues that by using zero degrees Aries the traditionalists are two millennia out of date. The siderealists claim that the traditionalists have neglected to compensate for the movement of the earth's axis relative to the stars.

7. This method is much better than asking subjects what their signs are, since all too often they give wrong answers. For instance, within this particular sample, more than one person in twenty gave an incorrect sign. Moreover, one person in ten could not give any answer. Experimenters far too often neglect this problem.

8. As an aside, it should be noted that the researchers purposely overselected for young adults 30 years old and below. Respondents born between 1942 and 1957 numbered 565. Those born between 1880 and 1942 numbered just 435. A proportionately representative sample would have had 328 respondents 30 years old and under and 672 respondents over age 30. We shall examine the sample in its unweighted form to avoid any charge of unfair manipulation of the data. Although neither popular astrology nor technical astrology presuppose any zodiac irregularities according to age groups, we shall adopt age as a control variable later in our analysis to allay fears of a hidden distorter effect.

9. The excluded birthdates are April 20, May 21 and 22, June 21 and 22, July 22 and 23, August 23, September 23, October 22 and 23, November 22, December 22, January 20, February 18 and 19, and March 20 and 21.

10. The degrees of freedom is the product of (a) the number of rows minus one and (b) the number of columns minus one. It is usually expressed as $(r-1)(c-1)$.

11. This finding parallels one test of significance for the specified hypothesis. The test was run by collapsing the cross-tabulation into two rows and two columns, eliminating those three zodiac signs for which predictions were lacking. With 1.0 degrees of freedom and a chi-square of 0.135, the probability value was .72.

12. This finding parallels a test of significance for the specified hypothesis. The test was done by collapsing the table into two rows and two columns, eliminating those three signs for which predictions were absent. With 1.0 degrees of freedom and a chi-square of 0.400, the probability value was .53.

13. The cross-tabulation, however, does suggest why unintelligent qualities could not be derived from the content analysis. It would appear that stupidity is an amazingly rare trait—at least according to subjective estimates.

14. This finding parallels a test of significance for the specified hypothesis. The test was run by collapsing the cross-tabulation into two rows and two columns. With 1.0 degrees of freedom and a chi-square of 0.017, the probability value was .89.

15. This finding also parallels a test of significance for the specific hypothesis. The test was run by collapsing the table into two rows and two columns. With 1.0 degrees of freedom and a chi-square of 0.017, the probability was .89.

16. This finding, too, parallels a test of significance for the specified hypothesis. The test was run by collapsing the cross-tabulation into two rows and two columns. With 1.0 degrees of freedom and a chi-square of 2.903, the probability value is .09 (the best probvalue we have yet had, but still lacking statistical significance).

17. None of the 27 supplemental variables were significant at the .01 level. Indeed, 26 of the 27 were not significant at even the .05 level. Such results are in accord with probability theory. Needless to say, none of the 27 tables bore a relationship in harmony with the

predictions of astrologers.

References

1. Astrological Experiments and Tests

Choisnard, Paul 1921. *Preuves et bases de l'astrologie scientifique*. Paris: n.p.

Farnsworth, P. R. 1937. "Aesthetic Behavior and Astrology." *Character and Personality* 6: 335-340.

Gauquelin, Michel 1969a. *The Cosmic Clocks*. New York: Avon.

——— 1969b. *The Scientific Basis of Astrology: Myth or Reality*. Translated from the French by James Hughes. New York: Stein and Day.

Jung, Carl Gustav 1955. "An Astrological Experiment." *The Interpretation of Nature and the Psyche*. Ed. by C. G. Jung and W. Pauli. New York: Pantheon Books.

Krafft, Karl Ernst 1939. *Traité d'astrobiologie*. Paris: Legrand.

Müller, Arno 1958. "Eine statistische Untersuchung astrologischer Faktoren bei dauerhaften und geschiedenen Ehen." *Zeitschrift für Parapsychologie* 1: 93-101.

Pellegrini, Robert J. 1973. "The Astrological 'Theory' of Personality: An Unbiased Test by a Biased Observer." *Journal of Psychology* 89, Part 1 (September): 21-28.

——— 1975. "Birthdate Psychology: A New Look at Some Old Data." *Journal of Psychology* 89, Part 2 (March): 261-265.

Silverman, Bernie I. 1971. "Studies of Astrology." *Journal of Psychology* 77, Part 2 (March): 141-149.

Silverman, Bernie I. and Marvin Whitmar 1974. "Astrological Indicators of Personality." *Journal of Psychology* 87, Part 1 (May): 89-96.

Standen, Anthony 1975. "Is There an Astrological Effect on Personality?" *Journal of Psychology* 89, Part 2 (March): 259-260.

2. Surveys of Belief in Astrology

American Institute of Public Opinion. 1976. "32 Million Look to Stars for Help in Conducting Daily Affairs." *Gallup Opinion Index* No. 132 (July): 25-27.

Defrance, Philippee, et al. 1971. *Le retour des astrologues*. Paris: Les Cahiers du Club du Nouvel Observateur.

Delaney, James G. and Howard D. Woodyard 1974. "Effects of Reading an Astrological Description on Responding to a Personality Inventory." *Psychological Reports* 34, No. 3 (June): 1214.

Gallup, George H., Jr. 1975. "32 Million Americans Express Belief in Astrology." *Gallup Poll*, weekly news release published by the American Institute of Public Opinion (October 19): 1-3.

——— 1978. "Surprising Number of Americans Believe in Paranormal Phenomena." *Gallup Poll*, weekly news release published by the American Institute of Public Opinion (June 15): 1-3.

Glock, Charles Y., and Robert J. Wuthnow. "The Religious Dimension: A Report on Its Status in a Cosmopolitan American Community." *Emerging Dimensions of Religious Consciousness.* Ed. by Rocco Caporale and Antonio Grumelli (in press).

Gorer, Geoffrey 1955. *Exploring English Character.* London: Cresset.

Indian Institute of Public Opinion 1973. "What Kind of Mood the Australians Are In." *Monthly Public Opinion Surveys* 19, No. 2, issue 218 (November): 11-14.

Jahoda, Gustav 1969. *The Psychology of Superstition.* Baltimore: Penguin.

Levitt, E. E. 1952. "Superstitions: Twenty-five Years Ago and Today." *American Journal of Psychology* 65: 443-449.

Maitre, Jacques 1966. "The Consumption of Astrology in Contemporary Society." *Diogenes* 53 (Spring): 82-98.

Moore, Marcia 1960. *Astrology Today: A Socio-Psychological Survey.* New York: Lucis.

Nixon, H. K. 1925. "Popular Answers to Some Psychological Questions." *American Journal of Psychology* 36: 418-423.

Ralya, L. R. 1945. "Some Surprising Beliefs Concerning Human Nature Among Pre-medical Psychology Students." *British Journal of Educational Psychology* 15: 70-75.

Roper Public Opinion Research Center 1974. "British Shun Mysticism." *Current Opinion* 2, No. 3 (March): 36.

Schmidtchen, Gerhard 1957. "Soziologisches über die Astrologie. Ergebnisse einer reprasentativ-Befragung." *Zeitschrift für Parapsychologie und Grenzgebiete der Psychologie* 1: 47-72.

Snyder, C. R. 1974. "Why Horoscopes Are True—The Effects of Specificity on Acceptance of Astrological Interpretations." *Journal of Clinical Psychology* 30, No. 4 (October): 577-580.

Snyder, C. R., Daniel L. Larsen and Larry J. Bloom 1976. "Acceptance of General Personality Interpretations Prior to and After Receipt of Diagnostic Feedback Supposedly Based on Psychological, Graphological, and Astrological Assessment Procedures." *Journal of Clinical Psychology* 32, No. 2: 258-265.

Truzzi, Marcello 1975. "Astrology as Popular Culture." *Journal of Popular Culture* 8, No. 4 (Spring): 906-911.

Warburton, F. W. 1956. "Beliefs Concerning Human Nature in a University Department of Education." *British Journal of Educational Psychology* 26: 156-162.

Wuthnow, Robert J. 1976. "Astrology and Marginality." *Journal for the Scientific Study of Religion* 15, No. 2 (June): 157-168.

―――― 1976. *The Consciousness Revolution.* Berkeley: University of California Press.

Wuthnow, Robert J. and Charles Y. Glock 1974. "The Shifting Focus of Faith: A Survey Report." *Psychology Today* 8, No. 6 (November): 131-136.

3. Critiques of Astrology

Barth, James R. and James T. Bennett 1974. "Astrology and Modern Science Revisited." *Leonardo* 7, No. 3: 235-237.

Bok, Bart J., Lawrence E. Jerome and Paul Kurtz 1975. "Objections to Astrology—A Statement by 186 Leading Scientists." *The Humanist* 35, No. 5 (September/October): 4-6. (Reprinted by Bart J. Bok and Lawrence E. Jerome in *Objections to Astrology*. Buffalo, N.Y.: Prometheus, 1975.)

Bok, Bart J. 1975. "A Critical Look at Astrology." *The Humanist* 35, No. 5 (September/October): 6-9.

Bok, Bart J. and Margaret W. Mayall 1941. "Scientists Look at Astrology." *Scientific Monthly* 52 (March): 233-244.

Erikson, W. Keith 1976. "Inaccuracy of Astrological Research." *The Humanist* 36, No. 6 (November/December): 43-44.

Jerome, Lawrence E. 1976. "Planetary 'Influence' Versus Mathematical Realities." *The Humanist* 36, No. 2 (March/April): 52-53.

——— 1975. "Astrology—Magic or Science." *The Humanist* 35, No. 5 (September/October) 10-16.

——— 1973. "Astrology and Modern Science: A Critical Analysis." *Leonardo* 6: 121 ff.

Ratzan, Lee 1975. "The Astrology of the Delivery Room." *The Humanist* 35, No. 6 (November/December): 27.   ●

The Moon and the Maternity Ward

George O. Abell
and Bennett Greenspan

The role of the moon as the principal raiser of tides on the earth has been known for many hundreds of years and was first explained in terms of gravitational theory by Isaac Newton in the seventeenth century. Actually the gravitational force of the sun on the earth is more than 100 times stronger than that of the moon; but because the moon is nearer, the difference between its gravitational pull on the side of the earth nearest it and the pull on the other side farthest from it is greater than the corresponding difference in the sun's pull by a factor of about 2.5. Consequently, while the sun also plays a role in raising tides, the moon's role is dominant.

The importance of the moon in producing solar eclipses has been known for thousands of years; and similarly, eclipses of the moon, which occur on those infrequent occasions when the full moon enters the shadow of the earth, have been known since antiquity. Indeed the correct explanation of eclipses is found in the writings of Aristotle and must have been known much earlier.

Sunlight reflected to the earth from the moon, especially when the moon is near full, has been an important influence on man's affairs since antiquity. The harvest moon, for example, which occurs in autumn, when the angle that the moon's rising path makes with the horizon is such that there can be bright moonlight in the early evenings for several days in a row, has been a longtime boon to farmers. Obviously, the bright light of full moon is an aid to certain nighttime activities, just as the time near new moon, when the sky is darkest at night, is optimum for other activities, such as astronomical observations of faint objects. Song and poetry are full of allusions to the moon as an aid to romance. Perhaps there are psychological effects of the moon upon man as well.

The few obvious effects, such as tides and eclipses, and other spectacular phenomena associated with the moon give rise to a vast amount of lunar folklore and superstition. It has been claimed, for example, that crops fare best when planted just after full moon. The full moon has been associated with fertility, with women's menstrual cycles, with violence, especially murder and suicide, with incidences of epilepsy, with evil spirits, with madness, and even with lycanthropy (the curse of the werewolf). For a sensational, if rather inaccurate, account, see *Moon Madness* (Abel, 1976).

Many of the alleged influences of the moon are often cited as evidence that celestial bodies (the moon, at least) can affect humans and their affairs, and thus as support for astrology.*

Full Moon and the Time of Birth

One of the most widely held beliefs about lunar effects is that there are many more human births at the time of full moon than at other times of the lunar cycle. The belief is even widespread among nurses in maternity wards and among some gynecologists as well. One of our colleagues recalls that his first child was born during full moon and that when he arrived at the hospital there were expectant women waiting in the halls for available rooms; the nurses all explained that "it always happens this way" at the time of full moon. Astrologer Sydney Omarr frequently refers to the incidence of crime and violence, as well as the higher birthrate, at the time of full moon and recommends that the moon be saved for romance.

There are some published studies that seem to support the idea that the full moon favors a higher than average rate of births. E. J. Andrews (1960), for example, reports that in the Tallahassee Memorial Hospital during the period 1956 to 1958 there were 401 babies born within two days of full moon, 375 within two days of new moon, and 320 within two days of first quarter. In a study of more than 510,000 births in New York City during the ten-year period beginning in 1948, Menaker and Menaker

---

*It must be noted, however, that these alleged lunar influences on humans are not, in general, the ones that would have been predicted from the appearance of the moon in the natal horoscope—the chart showing the positions in the heavens of the planets at the time and place of the birth of an individual.

(1959) claim that the birthrate was about one percent higher during the two weeks following full moon than before. However, one of those authors (Menaker, 1967) later studied another half-million births in New York during 37 lunar months from 1961 to 1963 and reported a one percent excess in the birthrate during the two-week period *centered* on full moon. Subsequently, Osley, Summerville, and Borst (1973) reported on a study of yet another half-million births in New York, during a later unspecified three-year period, showing a one percent excess in the birthrate during the two weeks *preceding* full moon. In contrast, Rippmann (1957) analyzed 9,551 natural births over a ten-year period in Danville, Pennsylvania, and found no correlation at all with the phase of the moon.

Again astrologers invoke the alleged influence of the full moon on the birthrate as evidence that celestial bodies can affect human affairs. It is not clear, however, how such a correlation would have anything to do with the interpretation of natal horoscopes according to the rules of Ptolemy (handed down from the second century) by which astrologers interpret the personalities and predict the futures for those individuals.

Still, if true, any correlation between the birthrate and the phase of the moon would be a very surprising result. Moreover, if properly interpreted, such a correlation would tell us something of the greatest importance to our understanding of the human reproductive cycle, to say nothing of influences over a distance of a quarter of a million miles that evidently have nothing to do with any of the known forces of nature. Thus we felt that the matter was worth a very careful check. Our analysis has been reported in the *New England Journal of Medicine* (Abell and Greenspan, 1979).

Data

We have tallied all births, live and dead, from the records of the UCLA Hospital maternity ward during the period March 17, 1974, to April 30, 1978. In some instances a birth would occur at home before the mother left for the hospital or en route to the hospital. To be consistent, we counted only births that actually took place in the hospital. During that interval there were 11,691 live births at the hospital. Of these, 3,549 were induced (by drugs or Caesarean section), leaving a total of 8,142 births that occurred naturally. To take account of the possibility that inducing a birth might have caused it to occur on a date earlier than normal, we treated total live and natural births separately. During the same interval there were 141 instances of multiple births, all of which were twins except for four sets of triplets and one set of quadruplets. In some of these multiple births, one of the infants was born before the mother reached the hospital. Finally, there were 168 stillbirths—amounting to 1.4 percent of the total.

The Lunar Phase

The average time that the moon requires to pass through a complete cycle of phases—that is, from new moon to new moon, or full moon to full moon, an interval known as the *synodic month*—is 29.530588 days. Because our birth data are recorded by calendar days (midnight to midnight) we had to divide our period of 1,506 days into synodic months of either 29 or 30 days. Our procedure was as follows.

The local date of each full moon during the interval was obtained

TABLE 1
Live Births at UCLA Hospital, March 17, 1974, to April 30, 1978

Phase Day	Total Live Births	Natural Live Births
1	406	273
2	390	273
3	425	291
4	412	287
5	391	270
6	396	282
7	399	293
8	398	276
9	383	260
10	380	271
11	364	243
12	398	292
13	372	261
14	411	279
15 (Full)	385	268
16	397	270
17	397	276
18	401	290
19	394	278
20	416	303
21	371	241
22	383	252
23	397	268
24	411	294
25	385	265
26	423	305
27	418	280
28	405	285
29	372	264
30	211	152
Totals	11,691	8,142

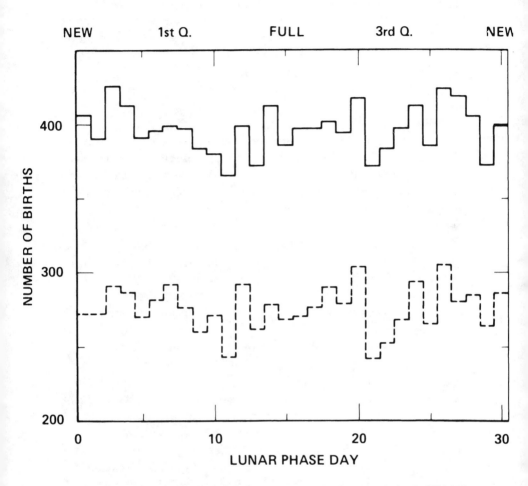

FIGURE 1. The numbers of total live births (solid line) during a four-year period at the UCLA Hospital plotted against lunar phase, and the numbers of natural live births (dashed line) during the same period.

from the U.S. Naval Observatory's annual publication *The American Ephemeris and Nautical Almanac*. In each case we took the local date in Los Angeles (between midnight and midnight, either Pacific Standard Time or Pacific Daylight Time, as appropriate). We defined the date of full moon as the fifteenth day in the lunar cycle. The previous day would then be lunar day number 14, and the day before that lunar day number 13, and so on. Thus by counting back 14 days from the date of full moon, we obtain day 1, which we define as the beginning of the lunar cycle; it is approximately the date of new moon. By counting forward from full moon, we define day 16, 17, and so on, until we reach the beginning of the next lunar cycle. In this way we determine whether the particular cycle has 29 or 30 calendar days. In the 51 synodic months over the period we surveyed, there were 27 30-day months and 24 29-day months. Thus the mean length of the synodic month covered in our sample was 29.529412 days—very close to the known mean length of the synodic month.

Results

We next totaled all of the births that occurred on a first day of a lunar cycle, on a second day of a lunar cycle, and so on. These figures for the total number of live births, and also the noninduced, or natural, live births, are displayed in columns 2 and 3 of Table 1, and are displayed in Figure 1. The much smaller totals corresponding to the thirtieth day in the lunar cycle are not due to an avoidance of the day before new moon but to the fact that there were only 27 30-day months among the 51 synodic months surveyed. (In Figure 1, the totals for day 30 have been increased by the factor 51/27 so that they could be compared with the data for other dates.)

There are no obvious peaks in the numbers of births either at full moon or at any other time in the lunar cycle. There are, of course, small fluctuations in the actual total numbers of births among the successive days of the lunar cycle, but such random fluctuations are expected.

We can, in fact, use a standard statistical test to check the hypothesis that day-to-day fluctuations as large as those observed would be obtained in a random sampling from a hypothetical population in which births occur uniformly throughout the lunar cycle. If, in fact, the phase of the moon has no effect whatsoever on numbers of births, we would expect a total number of live births on each day during the period covered that is close to the average number of 395.91, except for day 30 when the expected number would be only 209.60 (which is 27/51 of 395.91). The corresponding expected numbers of natural, or noninduced, births are 275.73 and 145.97, respectively. The standard χ^2 test shows that, for the total numbers of all live births, fluctuations from day to day in the lunar cycle at least as large as those obtained here would be found about 95 percent of the time. Fluctuations in the day-to-day totals of natural births as large as those found here would occur 65 percent of the time. In other words, fluctuations such as those we have obtained are entirely to be expected, and there is no reason whatsoever to ascribe any influence to the moon on the numbers of births during the lunar cycle.

Multiple Births

The numbers of multiple births occurring during the period March 17,

TABLE 2
Multiple Births and Stillbirths at UCLA Hospital,
May 17, 1974, to April 30, 1978

Phase Day Range	Number of Multiple Births	Number of Stillbirths
3-7	27	26
8-12	27	33
13-17	15	23
18-22	24	26
23-27	31	30
28-2	17	28
Totals	141	166

1974, to April 30, 1978, are given in column 2 of Table 2. The numbers in Table 2 refer to the number of instances of multiple births (that is, sets of twins), whether or not one of the twins may have been born before the mother reached the hospital; the numbers do not refer to individual births. We have grouped the numbers of multiple births in 5-day intervals in the lunar cycle to obtain numbers of births large enough to make meaningful comparisons. Note that the interval containing full moon (day 15) contains the lowest number of incidences of multiple births of any of the intervals. However, this minimum is not significantly different from fluctuations expected by chance.

If the incidence of multiple births is completely unaffected by the phase of the moon, we would expect the 141 sets of multiple births to be distributed evenly among the intervals with a mean number per interval of 23.87, except for the interval containing day 30, for which the expected number of multiple births is 21.63. Again we used the χ^2 test to compare the actual incidence of multiple births with the expected ones and found that fluctuations as large as those observed here would occur 20 percent of the time. In other words in one-fifth of all such experiments as this we would find fluctuations from the expected numbers at least as large as those we have obtained. There is, therefore, no reason to ascribe to the moon any effect on multiple births.

Stillbirths

The distribution of 168 stillbirths among the same six 5-day intervals in the lunar cycle is exhibited in column 3 of Table 2. Again, the number of stillbirths obtained during the interval containing full moon is actually the lowest in the list. But again, there is no significance to these small fluctuations. The expected number of stillbirths, if the moon has no effect, is 28.108 for all of the intervals except the one containing day 30, when the expected number should be 25.462. A χ^2 test shows that fluctuations of the size obtained here would occur 78 percent of the time.

Conclusions

Our analysis of the nearly 12,000 live and dead births occurring at the UCLA Hospital maternity ward in an interval of 51 lunar months from 1974 to 1978 reveals no correlation between the numbers of births and full moon or any other phase of the moon. This negative result occurs whether we consider all live births, just those that are completely uninduced (that is, natural), incidence of twinning or other multiplicity, or even stillbirths.

These results were a considerable surprise to several of the nurses at the UCLA Hospital maternity ward: for they, like so many others, fully expected to find a strong correlation of birthrate with full moon. To be sure, if one combs through all of the data, an occasional lunar month can be found in which there is a greater than average number of births at or near full moon, but these are only random fluctuations; there are just as many months in which there are fewer than the expected number of births near full moon. Probably the nurses simply remember those months in which they noticed there was a full moon during a particularly busy night. Perhaps it is similar to the tendency to remember those dreams that seem to come true and forget the vast majority that do not.

Our results appear to be at odds with some of the published surveys referred to above. At least among those mothers in the Los Angeles area who are cared for at the UCLA Hospital there appears to be no influence of the moon whatsoever on the times of their deliveries. We strongly suspect that the moon has no influence on birthrates anywhere in our society and are perplexed at the seemingly discordant results, especially those obtained in the New York samples. Because three of those studies purport to cover approximately a half-million births each, a very substantial amount of effort would have been required to carry out the tests properly. Especially in view of their discordant results, perhaps a new look at those data would be appropriate.

We know of other cases where claims of effects of the full moon have not been verified by careful statistical studies. Among these are Pokorny's (1964) study of 2,497 suicides and 2,017 homicides in Texas between 1959 and 1961; Pokorny and Jachimczyk's (1974) analysis of 2,494 homicides in Texas between 1957 and 1970; Lester, Brockopp, and Priebe's (1969) analysis of 339 suicides in Erie County, New York; and Pokorny's (1968) analysis of 4,937 mental hospital admissions—none of which shows any correlation either with the phases or the distance of the moon.

The moon has unquestioned influence on the tides and on certain other phenomena, but several pieces of evidence at hand suggest that many of the "incredible facts" about the influence of the moon on man are simply not facts at all.

References

Abel, E. C. 1976. *Moon Madness*. Greenwich, Conn.: Fawcett.

Abell, G. O., and B. Greenspan 1979. "Human Births and the Phase of the Moon." *New England Journal of Medicine* 300: 96.

Andrews, E. J. 1960. "The Cyclic Periodicity of Postoperative Hemorrhage." *Journal of the Florida Medical Association* 45: 1362-1366.

Lester, D., G. W. Brockopp, and K. Priebe 1969. "Association Between a Full Moon and Completed Suicide." *Psychological Reports* 25: 598.

Menaker, W. D. 1967. "Lunar Periodicity with Reference to Live Births." *American Journal of Obstetrics and Gynecology* 98: 1002-1004.

Menaker, W., and A. Menaker 1959. "Lunar Periodicity in Human Reproduction: A Likely Unit of Biological Time." *American Journal of Obstetrics and Gynecology* 77: 905-914.

Osley, M., D. Summerville, and L. B. Borst 1973. "Natality and the Moon." *American Journal of Obstetrics and Gynecology* 117: 413-415.

Pokorny, A. D. 1964. "Moon Phases, Suicide, and Homicide." *American Journal of Psychiatry* 121: 66-67.

———— 1968. "Moon Phases and Mental Hospital Admissions." *Journal of Psychiatric Nursing* 6: 325-327.

Pokorny, A. D., and J. Jachimczyk 1974. "The Questionable Relationship Between Homicides and the Lunar Cycle." *American Journal of Psychiatry* 131: 827-829.

Rippmann, E. T. 1957. "The Moon and the Birth Rate." *American Journal of Obstetrics and Gynecology* 74: 148-150.

Results of U.S. Test of the 'Mars Effect'

Paul Kurtz, Marvin Zelen, and George Abell

Introduction

The research of Michel and Françoise Gauquelin has attracted considerable attention on both sides of the Atlantic. Although they are critics of traditional astrology, they nonetheless claim to have found a statistical correlation between the professions and personalities of certain individuals and the positions of certain planets in the sky at the time and places of their births. The alleged correlation between the position of the planet Mars in the heavens at the time of birth and the incidence of being a sports champion was taken as a test case of the Gauquelins' hypothesis.

According to the Gauquelins, if the sky is divided into 12 sectors (similar to the Placidean "houses"), Mars appears in the sectors just above

The authors would like to thank Frank Dolce, Germain Harnden, and Neal Radice for their assistance in this study.

the eastern horizon and just west of the meridian (the first and fourth sectors) at the time of birth of sports champions more often than can be expected by chance. The Gauquelins have compiled data for a total of 2,088 European sports champions. If there is an equal probability of an athlete being born in any of the 12 sectors, then we would expect that 1/6 (approximately 16.7 percent)[1] of the athletes would be born in the key sectors. However, the Gauquelins show that in their sample this proportion is approximately 22 percent. For many years the Comité Para of Belguim investigated the Gauquelins' research but was unable to confirm it. The Gauquelins have held that the Comité Para replicated their results; the Comité, however, questioned the Gauquelins' procedure for calculating the theoretical frequency with which Mars appears in the "key" sectors at the birth times of the general population, in comparison with which the alleged deviations were found.

In 1977, Professor Marvin Zelen, a Fellow of the Committee for the Scientific Investigation of Claims of the Paranormal, proposed a new test of this Mars effect, which is now called the *Zelen test*. In Zelen's proposal the Gauquelins would select at random a subset of sports champions from their original sample and then for each champion obtain birth data for all other people born in the same areas on the same day. The plan was later modifed to select individuals born within ± 3 days of the champions. If the champions were born in the first and fourth sectors significantly more often than were the people in this new control sample, this would tend to support the Mars effect hypothesis. The Gauquelins selected names from the *chief-lieux* of the departments and provinces of France and Belgium, areas where 303 champions from the original sample were born. According to the Gauquelins, the Zelen test confirmed the Mars effect, resulting in a statistically significant difference ($P = .03$). (Cf. *Humanist*, Nov./Dec. 1977.)

The authors of this paper are not convinced by the Gauquelins' interpretation of the results of the Zelen test. The Mars effect was significant only in the Paris part of the sample, and was relatively weak or nonexistent for the rest of France and for Belgium. If it was a real effect, one would expect to verify the phenomenon with data from diverse geographical areas. Moreover, the Gauquelins did not employ a random selection as originally planned: They used the entire list of Paris champions but in other areas used only champions born in the chief towns. Since our 1977 report, we further learned that in the city of Paris they did not even compare champions born in each of the 20 arrondissements but only in the fourteenth arrondissement. (This was not reported in their paper and we only learned of it from the Gauquelins recently, after questioning them about it.) The reason for comparing champions and nonchampions from the same village, town, or arrondissement was to equalize any environmental factors that might influence active participation and achievement in sports.

Our independent calculation, based only on male sports champions, resulted in a *P*-value of .04. The Gauquelins have objected to our omitting women from the analysis of sports champions. But surely women have not had the same opportunities men have had to pursue sports. It is obvious that sexual discrimination and sex roles are important factors. Accordingly, it seemed unjustifiable to compare the sample of 303 sports champions, which contained 294 males and 9 females, with the total larger population of nonchampions, which included both males and females on roughly a 50-50 basis. In our analysis using the Zelen test, therefore, we dropped the females in that sample and compared the male champions with male nonchampions only.[2]

In any case, the central issues are whether the original sample of 2,088 champions is representative of outstanding athletes in general, whether the Mars effect shown by that sample is real or a statistical fluctuation, and whether the Mars effect is an artifact of the selection process or the application of the test. Consequently, Kurtz, Zelen, and Abell, in cooperation with the Gauquelins, agreed to conduct an entirely independent study of U.S. sports champions to see if the Mars effect could be replicated.

The U.S. Test

At a meeting in July 1977, Kurtz, Zelen, Abell, and M. Gauquelin outlined plans for the U.S. test. A representative sample of U.S. sports champions was to be selected from directories of sports champions.

At the outset we did not know how much information we would obtain, e.g., whether the hour of birth was generally recorded, or whether the data would be released to us even if it were. Gauquelin himself has encountered similar difficulties in obtaining complete data from some countries. We selected all 340 American champions listed in the *Lincoln Library of Sports Champions* (Frontier Press, 1974).[3] We also selected 218 names from *Who's Who in Football* (Arlington House, 1974), primarily, but not exclusively, the players who made the All-Star or All-Pro teams, and 47 All-Star players from *Who's Who in Basketball* (Arlington House, 1973). Thus our first proposed sample was composed of a total of 605 sports champions. This total was reduced by 41 names for which birth data were not available in the directories, leaving a total of 564 names. To avoid any bias by Kurtz, Zelen, and Abell, the actual selection of the champions was made by two neutral researchers, Frank Dolce and Germain Harnden.

Once the list was compiled, we wrote to the state offices of birth registry, pointing out the scientific character of our inquiry. Birth data were not available for many of the champions, primarily because of the newly enacted U.S. Privacy Act, which prohibits states from providing such information without the consent of the individuals themselves. Twenty-one states refused to send us information, on the basis of the Privacy Act, and five states, plus the District of Columbia and Puerto Rico, did not reply. Five states had no champions in them, and one state did not respond to our original request until six months later (those data were included in our second sample). Nevertheless, 18 states waived the rule and sent information, from which we compiled data on 128 champions. Of these 128 champions, 19.5 percent were born with Mars in key sectors. Although statistically this proportion was not significantly different from the expected 16.7 percent, we felt that the sample was too small and M. Gauquelin agreed with this judgment.

Accordingly, we decided to expand the sample by requesting information on more champions from those states that responded to our first canvass. We selected the remaining champions who resided in those states and who are listed in *Who's Who in Football* (330 names) and *Who's Who in Basketball* (145 names). This list included many All-Star and All-Pro players in basketball and football not included in the first sample. Since we had no idea initially of how much information would be forthcoming, not all the All-Star and All-Pro players were selected in the first sample. We also added the names of champions listed in *Who's Who in Track and Field* (Arlington, 1973), which contains 111 names of American champions from those states, and in *Who's Who in Boxing* (Arlington, 1974), which contains 92 such champions. We sent requests for data on these champions to the 18 cooperative states, but received replies and birth information from only 14 of them. Even those 14 states did not have records of 186 of the champions. In all, our second canvass yielded data on 197 additional champions.

In a third canvass we wrote to those states initially refusing information, and from all states we requested data on athletes listed in the directories but whose names had been omitted in the first and second canvasses. Three additional states subsequently responded to our second and third inquiries. The final number of states providing information was 22. From those cooperating states, we requested information on 682. (See Table 1.) Thus the total sample that resulted from all three canvasses contains 408 names, a number we deemed large enough for a preliminary study.

Our view has been that there should be no bias expressed by our selection process. Accordingly, requests for birth information for *all* of the sports champions listed in the various directories who were born in the states offering information were sent by Germain Harnden. According

TABLE 1

States That Sent Information on Birth Data

State	Number Requested	Number Received
Alabama	39	20
Arizona	3	2
California	175	60
Colorado	12	7
Delaware	1	1
Hawaii	4	2
Kansas	31	25
Kentucky	52	35
Nevada	2	1
New Hampshire	3	1
New Jersey	69	38
North Carolina	31	25
North Dakota	3	3
Massachusetts	20	9
Minnesota	19	17
Montana	2	2
Ohio	93	73
Oregon	25	17
South Carolina	22	14
Utah	15	15
Virginia	23	16
Wisconsin	38	25
Totals	682	408

to our understanding, they are famous sports champions because they are listed in *Who's Who*s of sports figures who have distinguished themselves in their respective sports by outstanding achievements.

Results of Statistical Analysis of American Sports Champions

Dennis Rawlins calculated the "Mars sector" of the sky corresponding to the birth time and place of each athlete for whom we received birth data. Rawlins's calculations have been spot-checked by one of us (Abell) and found to be accurate. The Gauquelins themselves have had Rawlins's calculations checked and found them accurate. The number of American sports champions born with Mars in each of the various sky sectors is shown in Table 2.

According to the Gauquelins' hypothesis, the key sectors are 1 and 4 and one would expect to find a significantly larger number of champions born with Mars in these sectors than would be expected in a random distribution. The observed proportion of births falling in the key sectors is $(30 + 25)/408 = 0.135$. In a population of which our sample of 408 is representative, the 95 percent confidence interval is 13.5 ± 3.3 percent. In other words, one would expect that there was only a 5 percent chance that the true percentage of births in the key sectors would lie outside the range of 10.2 to 16.8 percent.

If there were an equal chance of a birth time falling within each sector, one would expect the true proportion in sector 1 and in sector 4 to be $2/12 = 0.167$. This value is barely within the 95 percent confidence interval calculated from our data. On the other hand, in Gauquelin's sample of 2,088 champions, 21.6 percent are in the key sectors. This value is clearly outside the 95 percent confidence interval of the present sample of American athletes. (The 95 percent confidence interval in Gauquelin's sample is 21.6 ± 1.8 percent.)

Since the observed proportion (13.5 percent) is less than the theoretical result (16.7 percent) if there was no "Mars effect," it is not necessary to carry out a statistical test of significance. Nevertheless, we

---

TABLE 2

Distribution of American Sports Champions Among 12 Celestial Sectors
Relative to the Position of Mars at Birth

Sectors	Number of Champions Born in Sector
1*	30
2	39
3	35
4*	25
5	33
6	22
7	39
8	37
9	29
10	47
11	36
12	36
Total	408

* Sectors 1 and 4 are "key" sectors.

TABLE 3

Summary of Test of Significance Comparing Events in Key
Sectors with Theoretical Probability of $P = 1/6$

Sectors	Observed Number in Sector	Theoretical Expectation	Observed Minus Expectation
Key (1 and 4)	55	68.0	-13.0
Other	353	340.0	13.0
Total	408		

$$\text{Chi-square} = \frac{(13)^2}{68} + \frac{(-13)^2}{340} = 2.98$$

Two-sided test: $P = .090$
One-Sided test: $P = .945$

have done so for completeness. The calculations are summarized in Table 3. If we wished to test the hypothesis that the proportion, p, is $1/6$ (no relationship) against the alternative that p should be greater than $1/6$, a one-sided significance test would result in a $P = .95$. That is, the data are well within what would be expected if one-sixth of all births fall in each sector. Alternatively, if we compare the hypothesis $p = 1/6$ with the alternative that p can be either greater than or less than $1/6$, a two-sided test shows that the P value is .09.

Another way of examining the data is to compare the number of births in each sector with its theoretical expectation based on the condition that there was an equal chance of birth times falling within each of the 12 sectors. The theoretical expectation would then be $408/12 = 34$ births in each sector. The standard chi-square test shows that in random samples from such a theoretical population we would obtain larger deviations from a uniform distribution of births among the sectors 18 percent of the time (chi-square = 14.98; 11 degrees of freedom). Hence the data are consistent with an equal chance of birth times falling within each of the 12 sectors.

We conclude that the analysis of American sports champions shows no evidence for the Mars effect.

Notes

1. Dennis Rawlins has calculated the theoretical expectations for sectors 1 and 4 to be 17.17 percent (See his article, p. 26.)

2. We have not deleted females from the U.S. test because they are not being compared with nonchampions. In the U.S. study there are 9 female sports champions, with one born in a key sector. If the women were deleted, it would not affect the results significantly.

3. One champion, Louis Groza, was inadvertently omitted from the first sample because of confusion with his brother Alex Groza; but he was added later, making a total of 341 from the *Lincoln Library*.

Land and Sea

Critical Reading, Careful Writing, and the Bermuda Triangle

Larry Kusche

In August 1977, the College Entrance Examination Board issued a report concerning the causes of the continuing decline of the skills and capacities of high school students going into college. Although some of the conclusions concerning the effect of television and the breakdown of the family were necessarily subjective because of inconclusive evidence, several firm conclusions were reached. Among them is that the decline is partly because "less thoughtful and critical reading is now being demanded and done, and that careful writing has apparently about gone out of style."

Although the reference to the lack of careful writing is apparently directed toward the students themselves, it should also be taken to include the writings that the students read, virtually all of which is done by writers a generation or two older, who do not have the same excuse for sloppy research and slovenly logic and writing techniques.

Much of the reading that students now do is called "high interest reading." It has to be high interest in order to grab their attention, to compete with the likes of the Six Million Dollar Man, Woman, and Dog, the Fonz, and Darth Vader. There does not seem to be much reading being assigned, at least in the field of the "paranormal," that is highly logical or accurate. It almost seems as if there is a belief among publishers that interest and logic are inversely proportional to each other. They are not, of course, but that seems to be the prevailing belief.

A typical example of high interest reading, taken from one of the subjects of the season, the Bermuda Triangle, follows.

> The *Sandra* was a square-cut tramp steamer, decorated here and there with rust spots along her 350-foot length. Radio-equipped and loaded with 300 tons of insecticide, she leisurely thumped her way south in the heavily traveled coastal shipping lanes of Florida in June 1950.
> The crewmen who had finished mess drifted to the aft deck to smoke

and to reflect upon the setting sun and what the morrow might bring. Through the tropical dusk that shrouded the peaceful Florida coastline they watched the friendly blinking beacon at St. Augustine. The next morning all were gone. Neither the ship nor the crew were ever seen again. They had silently vanished during the night under the starlit sky. No clue to help solve this baffling mystery has been found to this very day.

Mysterious wasn't it? A tranquil sea. Quiet circles of smoke slowly drifting from the deck. Twilight. A clear sky. Ah, peace. The fate of the *Sandra* has been a matter of curiosity for millions of readers in the past few years, but I wonder how many of the readers have thought about it long enough to have noticed the glaring flaws in the story. I wonder how many readers have a high enough *CQ* (Curiosity Quotient) to take just a few seconds to analyze the case.

Those with a low *CQ* ask questions like, "What strange force could possibly have caused this inexplicable loss? Why has nothing from the *Sandra* been found even to this very day? What is wrong with the area out there?" (Note that the low *CQ* questions are the same as those asked in the currently popular pro-mystery books on the subject.)

The reader with the high *CQ* would have seen warning flags all over the story of the *Sandra*. Alarm bells should have rung. Yes, there is something wrong, not so much with the *Sandra* or "out there," but with the *telling of the story itself.*

If the *Sandra* disappeared that very night, how could anyone have known and reported what the crewmen were doing as the sun set? Did the men saunter over to the rail to smoke and chat about the sunset? How could the writer have known that? Did they really see the lights of St. Augustine? Was the sea really tranquil? All these points are crucial to the loss because they indirectly set the scene—a quiet, peaceful evening. That is, after all, why the loss of the *Sandra* is considered strange. If conditions had been stormy, the loss would not be considered unusual.

Even before taking the time to check into the weather (why bother doing that—it's all documented, isn't it?), the curious, intelligent reader should already be questioning the account of the *Sandra*. How was the writer able to know what the sailors saw, thought, or said that night? Was the writer perhaps on the ship himself, luckily lifted off by helicopter or a small boat in time to miss the disappearance? Unlikely. If I had been that writer I'd have plainly stated that that was what happened.

Did the radioman send this crucial information about the scene to shore? Again, quite unlikely. One doesn't usually paint pastel pictures over the radio. How then, could the writer know that much about what happened on the ship? How could he know if the men saw the light of St. Augustine? Did he know where the ship was at all?

The answer to the thinking person with the faintest shred of curiosity and intelligence is that the writer could *not* have known any of the "facts" he "reported." He had to have assumed them or have lifted them from someone else.

Is it nit-picking to observe that the "facts" could not logically have been known? Are these "facts" important? Obviously, they are crucially important. The writer was using a common, blatant writing technique that I call "setting the scene." The writer indirectly informs the reader that all was calm, all was right as the steamer chugged along. The crewmen obviously were not worried about any impending danger. There were no storms. It makes the "disappearance" all the more mysterious. The ship was "known" to have been off St. Augustine, practically pinpointing its area of disappearance, and making the lack of debris even more mysterious.

But, was the ship really near St. Augustine? Based on the writer's information, we cannot really know that. He says that *the crewmen saw the light,* not that anyone ever saw the ship. But he can't know what the crewmen saw. Did the ship "silently vanish"? If no one knows what happened, if no reporter was nearby taking notes, how do we know it was nice and quiet? Maybe they were fighting for their lives, but the "silent vanishment" treatment is far more mysterious. It certainly has a higher interest, as compared to just an ordinary old sinking.

All this the intelligent reader might have deduced for himself, without doing any outside research. There is not much that the writer gave us that appears to be solid. Perhaps he's right, perhaps not. Give an infinite number of monkeys an infinite number of typewriters. . . .

The writer has used another technique which I call "undue familiarity." He mentions the "rust spots along her 350-foot length," implying that he, personally, knows about the old *Sandra*. After all, if he can describe it in that precise detail, he must have some firsthand knowledge. Perhaps he was the one who spotted it off St. Augustine. He really does his research, doesn't he?

There is just one problem here. Upon checking with Lloyd's of London I learned that the *Sandra* was only 185 feet long, just about half what the writer said it was. Now about those rust spots and the writer's apparent familiarity with the ship. . . .

Neither the length of the ship nor the spots are crucially important, of course; but they do point out, once again, that the writer's credibility is very low. Almost everything he has said is blatantly in error, or is speculation. The true length of the ship is of some importance, however, since we can probably assume that a 350-foot ship would handle weather, if it were bad, better than a 185-footer.

But the incident is still unexplained despite all the obvious erroneous assumptions and errors. The (rare) diligent reader who is interested in following up on the case might contact the weather bureau's record center in Asheville, North Carolina, and ask for the records for June. The result is that he would find that the weather *was* excellent, just like the mystery purveyors said. Now we're really stumped. Perhaps there is something "out there" after all. A ship simply cannot disappear without a trace in perfect weather.

So our diligent researcher keeps trying. Any research on a missing ship would be incomplete without contacting Lloyd's of London. Lo! What do we find there? The mystery monger made another error! The *Sandra* did not sail in June, it sailed on April 5. The weather records are now checked for the proper month, and this time we find that beginning the day the *Sandra* sailed from Savannah, and for the next few days, the Atlantic shipping lanes off the southeast United States were buffeted by winds up to seventy-three miles an hour, only two miles an hour under hurricane strength.

All the basic "facts" as presented in the mystery of the *Sandra* are now shown to be wrong. Read again the "mystery" and compare what the writer said to what really occurred. Crewmen drifting to the deck to smoke? Watching the peaceful Florida coastline? Seeing the friendly little old beacon at St. Augustine? Silently vanishing? The near hurricane does change the situation just a bit.

Yet, a number of writers have used the *Sandra* as further "proof" that something strange is going on "out there." They failed to prove their theory, but they have helped confirm one of mine, that the less a writer knows about his subject, the better equipped he is to write a mystery about it. Ignorance of the subject is, in fact, a major technique in writing about the mystery of the Bermuda Triangle and other subjects in the so-called paranormal as well. Some critics of the Bermuda Triangle refer to it as science fiction, but that is an unfair description. Unfair, that is, to science fiction. The Bermuda Triangle, as well as many other "paranormal" topics, might more properly be called "fictional science."

Many people find the "mystery," full of illogic and errors that they are unaware of, to be of a higher interest than the correct answer, of the detective work necessary to track it down. Those people who revel in the uncritical claims of pseudoscience, of the "paranormal," might properly be called the "pseudocurious" or the "paracurious." They claim to want the truth, but they really don't want it. They watch Alan Landsburg's "In Search Of" television program and believe that, because *TV Guide* and Leonard Nimoy say so, that it is a documentary. They read the tomes of Berlitz, Winer, Spencer, Jeffrey, Godwin, and Sanderson and boggle their minds, as they say. Yet all these books are chock full of examples such as the *Sandra*, confirming the complaints of the College Entrance Examination Board. Careful writing, at least in the area of the pro-paranormal, has gone out of style, and the readers *are* less critical. The readers claim to be seeking illumination but are, in fact, only seeking light entertainment. There will always be plenty of Barnum, Bailey, and Berlitz writers and publishers around to satiate their hunger and further erode their logic. •

Cattle Mutilations: An Episode of Collective Delusion

James R. Stewart

During the late summer and early fall of 1974 the areas of northeastern Nebraska and eastern South Dakota experienced a rash of "cattle mutilations." In most instances dead cattle were discovered with parts of their anatomy missing. The parts of the body most frequently mutilated were the sex organs, ears, and the mouth. The episode reached its zenith during early October when discoveries of mutilated cattle were being reported on a daily basis to law-enforcement officials throughout the area. It subsided almost as quickly as it had begun, and although other areas of the country have subsequently reported the same phenomenon, there has been no media coverage of any further mutilations in this area.

The cause of the mutilations was, and in fact still is, controversial. Some persons believed that it was the work of members of a Satan-worshipping cult whose ceremonies called for the blood and parts of animals. Others believed that the mutilations were the work of extraterrestrial beings whose purpose was to examine the physiological makeup of cattle, or simply to terrorize human beings. Still others felt that the mutilations were the work of small predators who, after having discovered the carcasses of already dead animals, proceeded to eat the accessible parts. What follows is a detailed account of the outbreak, culmination, and precipitous decline of the cattle-mutilation episode.

The episode

The mutilation episode was apparently triggered by the more-or-less simultaneous reports of a cattle mutilation and the alleged sightings of UFOs and a "monster-thing" in the area. The initial discoveries of mutilations occurred in northeastern Nebraska during the latter part of August. The "thing" and UFO sightings were reported on the front pages of area newspapers during the first week in September. The close occurrence of

288

these events provided the perceptual framework for a new view of the cause of the cattle mutilations.

The authorities had been at a loss to explain the first mutilations and had offered little speculation about their cause. Area residents quickly assimilated the new information, and the bizarre explanations became increasingly popular. The authorities played an important role throughout the entire episode. In explaining the first mutilations they had been extremely cautious and conservative; but after the mysterious sightings they came to the conclusion that not only couldn't the bizarre explanations be discounted, but they might, in fact, be correct.

Newspaper accounts after the first week of September indicate that most local authorities had become convinced that the mutilations could not have been the work of normal human beings or the postmortem activity of small predators. Either blood-thirsty cultists or extraterrestial beings were considered by an increasing number of persons to be responsible for the mutilations. The director of male admissions at the South Dakota State Mental Hospital offered yet another explanation. He reasoned that the mutilations were the work of a deranged, psychotic personality. He also warned residents that such individuals often graduated to humans as their next victims. He even went so far as to offer a hypothetical description of the person (or persons)—a young male from a farm background with high levels of hostility toward his parents and other authority figures.

The inability of the social-control agencies to satisfactorily explain the early reports of cattle mutilation was probably one of the major factors contributing to the significant increase in both the number of reported mutilations and the area over which they occurred. Within a short time reports of mutilations had spread from northeastern Nebraska throughout eastern South Dakota. In fact, the episode can be more adequately analyzed by breaking it down into two rather distinct phases—the Nebraska phase, lasting roughly from August 15 until September 30, the South Dakota phase, lasting approximately from September 15 until October 31. This distinctness is best illustrated by contrasting the number of newspaper column inches in the major area newspapers devoted to each state's reported mutilations during these time periods.[1]

It is apparent that as the Nebraska episode waned in interest the South Dakota episode gained momentum. However, at the time these

---

1. The newspapers used in this study were the *Sioux City Journal*, the *Yankton Press and Dakotan*, the *Sioux Falls Argus-Leader*, and the *Aberdeen American-News*.

events were not analyzed as separate because of their geographical proximity and because the mass media in the area generally report the news of both states.

Column inches of news concerning cattle mutilation in four daily area newspapers		
	Nebraska mutilations	South Dakota mutilations
August 11-20	0	0
August 21-31	34	0
September 1-10	51	0
September 11-20	78	10
September 21-30	5	52
October 1-10	0	123
October 11-20	0	82
October 21-31	0	13
November 1-10	0	0

The precipitating factors created a growing anxiety and gave legitimacy to the UFO or blood-cult explanations of the mutilations. During the peak of the episode, radio and television broadcasts and newspapers contained daily accounts of any newly discovered mutilations. In addition media coverage generally contained interviews with law-enforcement officials, veterinarians, or other knowledgeable persons. These interviews often confounded rather than elucidated the search for the cause of the mutilations and frequently contained unsupported personal opinions. The following headlines are typical of the confusion which prevailed during the episode:

"Stories of a 'Thing' Told by Others Here" (*Sioux City Journal*, 1 September 1974).

"Veterinarian Says Flying Objects, Cattle Mutilations May Be Related" (*Sioux City Journal*, 6 September 1974).

"Farmers Check Herds in Wake of Mutilations" (*Sioux City Journal*, 20 September 1974).

"Mutilators Psychotic, Says Yankton Mental Unit Officer" (*Sioux City Journal*, 3 October 1974).

"Doctor Says Cattle Mutilations May Switch to Human Victims" (*Sioux Falls Argus-Leader*, 3 October 1974).

"Cattle Mutilations Have Farmers Jittery; Officials Blame Predators" (*Sioux Falls Argus-Leader*, 8 October 1974).

"Mutilations Spook Area" (*Aberdeen American-News*, 17 October 1974).

Law-enforcement officials warned people to be on the lookout for strange incidents in their area. They also encouraged local residents to band together and form patrols to survey the farm lands during the night hours when most of the mutilations were thought to have occurred. Many groups followed this suggestion and some groups even went so far as to arm themselves. Despite the increased vigilance, the night patrols never observed anything suspicious or unusual. A few groups also offered rewards for information leading to the arrest and conviction of the person or persons responsible for the mutilations. However, no local residents were ever questioned or arrested.

There were 75 to 100 reports of mutilated cattle before the episode ended during the latter part of October. The decline in the number of reported mutilations seems to have begun in the same manner in both Nebraska and South Dakota. Meetings were held which were attended by law-enforcement officials, state veterinarians, and interested farmers. The general conclusion reached at these meetings was that in the vast majority of instances the cattle had simply died of natural causes and the teeth marks and tearing actions of small predators had caused what was termed the "mutilations." While not everyone agreed with the findings, it was obvious that after sifting the evidence, officials were convinced that the mutilations were, in fact, a natural phenomenon. The reports of the two state veterinary-diagnostic laboratories stated that every animal brought to them had died of natural causes and that predators, by tearing away the soft parts of the carcass, had been responsible for the apparent "mutilations." When the reports of these two meetings were broadcast by the area mass media the episode quickly subsided.

That the episode occurred at all shows that there was a partial breakdown of the normal social-control forces. The failures of the social-control agents stemmed from their being suddenly confronted with an unusual situation which was not adequately covered by the accepted norms of explanation. Local law-enforcement personnel have little, if any, experience in determining causes of cattle deaths. Consequently, they were inclined to adopt the farmer's explanations in the absence of any solid refuting evidence of their own. The same was true of some local veterinarians. Rarely do they examine dead cattle; instead that are usually asked to treat living animals.

It should be pointed out, however, that many officials remained skeptical throughout the entire episode. These persons preferred to wait until the evidence was examined by knowledgeable experts before coming to any conclusions regarding the cause of the mutilations. Their cautious statements reflected the doubts they harbored, but their disclaimers of cult or UFO involvement were usually overshadowed by the more exciting accounts offered by persons with questionable credentials.

As the episode progressed there continued to be a lack of evidence to support the claims of the "believers," while at the same time the prestigious reports of the state laboratories supported the skeptics' version of the mutilations. The death knell of the episode was sounded when the South Dakota Crime Bureau issued a statement which said that, in their estimation, the deaths of the cattle were natural, and they could find no evidence that would support a more detailed investigation. Thereafter no more reports of mutilations appeared in the newspapers in the area.

Analysis

This episode appears to be a classic case of mild mass hysteria.[2] Accounts of similar hysteria have been previously reported in Seattle, Washington (Medalia and Larsen 1958), Mattoon, Illinois (Johnson 1945), and southern Louisiana (Schuler and Parenton 1943). Those instances and the cattle-mutilations hysteria share a common theme—for inexplicable reasons people suddenly perceive the mundane in a new, bizarre fashion. Everyday occurrences (i.e., nicks in windshields, not feeling "up to par," or dead cattle) are given a new, exciting, anxiety-producing definition. The extraordinary interpretation defies logical explanation. No refuting evidence can initially be mustered by authorities, and further accounts of the behavior are increasingly reported by the affected group of persons. Skeptics, however, persist in their naturalistic explanations and, inevitably, after a brief period of time their interpretations prevail. Lack of scientific proof or verified observations eventually result in the termination of the episode, and it usually dies away rather quietly and inconspicuously.

Usually there are a number of pre-existing structural elements in a

---

2. The analysis of this episode used the value-added theory of collective behavior of Neil Smelser.

society which make it possible for an episode of this nature to occur. The two conducive features in the cattle-mutilation episode appear to have been (1) the prevailing method of raising cattle and (2) the high potential for communication that existed in the area.

Most of the mutilated cattle were not discovered until two or three days after they had died. This situation greatly complicated the accuracy of autopsies subsequently performed by authorities. During the time of the year that the mutilations occurred most farmers and ranchers in the area allow their cattle to forage on recently harvested fields and pastures. The grazing cattle are generally some distance from the farmstead and are not observed on a daily basis, unlike a feeder-lot operation, where the deaths would immediately be discovered and would therefore probably have been attributed to natural causes. As it was, the dead cattle weren't discovered until decomposition had already started. The attribution of cause of death by veterinarians and animal scientists is made more difficult or, in some instances, impossible if decomposition of the carcass is in an advanced state. While this type of cattle raising certainly didn't cause the subsequent episode of mutilations, it did provide a setting which allowed the episode to develop.

The most salient conducive feature in this situation was the high potential for rapid blanket communication of the generalized anxiety of the actors. Virtually every household subscribes to at least one of the daily or weekly newspapers published in the area. (The four daily newspapers used in this study as barometers of the growth and spread of the episode are dominant, but local weekly newspapers are also an important source of news to residents.) Equally if not more important is the presence in each home of either a television or radio set, and most residents have both. Thus there is great accessibility to the media, and reports of the "mutilations" were both immediate and extensive.

In addition to the mass media, informal groups, common gathering places, and a general "small-town" atmosphere greatly contribute to the dissemination of newsworthy topics. Friends and neighbors are prime sources of information in a *gemeinschaft* communication network and were frequently mentioned as important sources of knowledge by persons interviewed in conjunction with this study.

These two features established broad parameters for collective behavior, and only when combined with straining features did they become important factors in the determination of the episode.

The strain in the cattle-mutilations episode involved the ambiguities surrounding two circumstances—the plight of the cattle market and lack of information about the real cause of the mutilations.

The summer and fall of 1974 brought increased anxiety to farmers who raised livestock as a major source of income. While the price of livestock at the various markets remained relatively high, the cost of feeding cattle to a marketable weight reached an all-time high. The farmers were caught in a vicious cost-price squeeze. The situation became so serious economically that many farmers resorted to the deliberate slaughtering of their calves because they could no longer afford to keep them until they were ready for market. The feelings of anxiety, tension, and uncertainty generated by this situation were extremely widespread among cattle raisers both before and after the outbreak of the cattle-mutilations episode. The condition of the livestock market certainly cannot be considered a direct cause of the mutilations, but it did contribute to the formation of a susceptible state of mind. The helplessness before market conditions (i.e., the high cost of grain and the relatively lower prices received for marketed cattle) created and nurtured a growing sense of frustration, anger, and despair.

The initial reports of mutilated cattle not only acted as the precipitating factor but also contributed to the growth of strain in the episode because of the absence of a precise, well-defined explanation of the cause of the mutilations. The naturalistic explanations could not be proven beyond a reasonable doubt. The "sudden" appearance of the mutilations lessened the credibility of the natural-death-coupled-with-predators account of the mutilations. Many questioned why, if the naturalistic explanation were correct, mutilations hadn't been reported before and why the deaths weren't generally recognized by the authorities as natural. There seemed to be no routine answer adequate to satisfy the lingering doubts that existed. Most of the social-control agents (i.e., local sheriffs and veterinarians) were puzzled by the findings and conceded that they were at a loss to explain all of the mutilations. In some instances the cause of death could be established; however, most of the remains were too badly decomposed to allow adequate examinations. In the absence of definite proof to the contrary, the social-control agents were placed in the position of being unwitting accomplices to those persons who formulated a new, extraordinary interpretation of the mutilations.

Other, somewhat less important, factors are associated with the personality variables of individuals and focus on explaining the differential susceptibility to hysteria. These personality variables contribute to the likelihood of panicking or, as in this episode, adopting the bizarre rather than the mundane explanation of the cause of the mutilations. Religious fundamentalism, lower levels of education, and lower socio-economic class status have all been demonstrated to contribute to this susceptibility (Cantril 1947, pp. 113, 131, 157, 197). All of these factors are generally overrepresented in rural areas, where religious beliefs tend to be more traditional (Slocum 1962, pp. 451-55), education levels lower than in urban areas (U.S. Census of Population 1970, Table 88), and a greater proportion of the population is blue-collar or lower-middle class (U.S. Census of Population 1970, Part 29, Tables 51, 55; and Part 43, Tables 51, 55).

These factors combine to produce an orientation which is less scientific and more likely to perceive the world as mysterious and somewhat incomprehensible. The role of these characteristics is difficult to assess in this episode of collective delusion, but previous studies have demonstrated that persons with these characteristics are more willing participants in incidents of mild or severe hysteria.

There develops in episodes like that of the cattle mutilations a pattern of beliefs that gives a more precise definition to the generalized state of anxiety created by the social strain. Uncertainty is the major ingredient in the participant's state of mind. The ambiguity was created by the unusual appearance of mutilated cattle and the lack of any satisfactory explanation regarding the cause. The elimination of the ambiguity may be accomplished by the provision of believable and noncontradictory information from the related agencies of social control. However, in this episode the agents of social control actually fostered the ambiguity and uncertainty not only by not denying but, in many instances, by directing, aiding, and abetting the creation of the bizarre explanation. Most local law-enforcement officials who investigated the mutilations concurred with the judgment of farmers who believed that the mutilations represented the work of cult members or alien life-forms from UFOs. Even local veterinarians often confirmed the suspicions of the farmers and reported that the mutilations had not been the work of predators.

In contrast, the reports of university-based veterinarians in Nebraska and South Dakota indicated that every animal upon which they had performed autopsies had died of natural causes, and the subsequent "mutilations" were believed to have been the work of small predatory animals. This was the first denial of the claims of the "believers" and probably was the major factor responsible for the eventual decline of the episode. Nevertheless, "believers" correctly pointed out that the state diagnostic laboratories had examined only a small number of the actual reported cases and that the vast majority of mutilations remained unexplained. For every cattle death that could be documented as having resulted from disease, there were many that, at least in the minds of some residents, still defied explanation.

The inability of social-control agencies to eliminate the uncertainties surrounding the episode resulted in a failure to decrease the anxiety levels of the "believers." A frequently used method of allaying the increased anxiety is to restructure the situation in believable terms. The form and shape of the restructured interpretation is very much a function of cultural legends, myths, and folklore. While the appearance of mutilated cattle appears to be without precedent, there have been similar inexplicable events in the past. The continuing traditions of sea monsters, ghosts, goblins, and other mysterious phenomena represent a vast reservoir of pre-existing cultural anxiety. When ordinary explanations prove to be insufficient, the extraordinary pool of mystical lore is frequently tapped for an explanation. Usually the adoption of the bizarre explanation (e.g., blood cultists or UFOs) does not significantly reduce anxiety, but at least the cause of the ambiguity has now been identified.

In addition to pre-existing cultural lore, a series of recent dramatic events may have also played an important part in the formation of the bizarre explanation. Certainly the recent resurgence of occultism, as represented by the great popularity of books and motion pictures such as *Rosemary's Baby* and *The Exorcist*, plus the infamy and notoriety of the Manson Family, contributed in some manner to the origination of the novel interpretation of the mutilations. Moreover, UFOs are frequently identified as being responsible for various unexplained occurrences. The periodicity with which that explanation is employed might possibly allow it to be considered a routine explanation and, therefore, reduce the anxiety level of residents because it is perceived as being little cause for alarm.

Another element in the formation of the extraordinary explanation was the sighting of "The Thing." This unknown "man-animal" was described by the few persons who allegedly observed it as a hairy creature that walked on all fours and quickly vanished when seen by humans. All sightings took place at night and no one actually got a good look at the "animal." Accordingly, most authorities quickly denied the existence of this unusual creature because of the lack of hard evidence. Zoology students from the University of Nebraska also attempted to discover its identity. However, when they examined the areas where the creature was reportedly seen they found no traces of it. The report of the sighting appeared on the front pages of the two largest area newspapers only two days before the number of reported mutilations increased greatly.

Conclusions

Given the pre-existence of certain conducive and straining features in the local area, the episode developed in the aforementioned fashion. However, the evidence presented by various authorities leads one to conclude that the episode was in fact the result of collective delusion. The most convincing explanation of the episode is as follows: For reasons associated with strain and anxiety people started to interpret an everyday occurrence (the deaths of cattle) in a new, bizarre manner. The process described is virtually identical to an episode of windshield pitting that occurred in Seattle, Washington, in the mid-1950s. In a study of that episode the investigators concluded that it was caused by the fact that people suddenly started looking *at* their windshields rather than *through* them. The delusion resulted from attaching new significance to something which was commonplace, i.e., pits in windshields. In a similar fashion, some people in the two-state area of Nebraska and South Dakota came to interpret the natural deaths of cattle as something strange and unusual. Cattle deaths in this area probably occur at about the same rate over the years, but the widespread reporting of these incidents gave the appearance that there was a sudden inexplicable increase in the deaths of cattle.

The mutilation of dead cattle is also something which is perfectly natural and was unquestionably done by small predators. These animals are seldom observed in the daylight hours, and their nocturnal habits prompted the assumption that they didn't exist in numbers large enough to perform the mutilations. The apparent surgical cuts on the dead animals can likewise be explained by the extremely sharp side teeth of the small predators. According to animal scientists, the shearing of the soft meaty parts of the carcass by a predator gives it the appearance of having been cut by a knife. The similar pattern in the missing parts (i.e., ears, lips, tongues, or genitals) in almost all of the reported mutilations can also be simply explained. The hides of the cattle are too tough to be easily penetrated by the small predators, who naturally gravitate to the most exposed and softest parts of the carcass.

Another factor which seemed to give credibility to the UFO

explanation was the virtual absence of any tracks or markings at the scenes of the mutilations. This, too, can be explained naturalistically. Small animals leave few, if any, tracks, and tracks that remained at the scene were quickly obliterated by the persons who examined it.

The absence of blood in some of the animals was a factor that gave credence to the blood-cult explanations. People assumed that the animal was drained of blood so that the blood could be used in some sanguineous ritual of a sect. However, veterinarians pointed out that after a few days the blood of dead animals coagulates and gives the impression that the body has been drained.

The general public, with little knowledge of, or experience with, cattle deaths, could easily fall prey to the bizarre interpretation, but why would farmers who undoubtedly had experience with dead cattle prior to the episode also be caught up in the collective delusion?

There are two possible answers to this question. First, perhaps, given the conditions of strain and anxiety, farmers with first-hand experience and knowledge were simply caught up in the delusionary spiral. Second, there exists the possibility that some farmers reported mutilations because their insurance policies would reimburse them for acts of vandalism, but not for deaths resulting from natural causes. I have no evidence that claims were paid for mutilations, but this possibility might motivate some to report mutilated, and not simply dead, cattle.

Other mundane explanations include the possibility that in one or two instances the cattle were actually killed by people or even a pack of wild dogs. Not infrequently cattle are slaughtered in the field by persons who are stealing beef because of the high prices. This explanation is even more likely when one considers that the over-the-counter meat prices reached unprecedented highs during the period of the mutilations. There have also been isolated reports at various times of packs of dogs running wild, and it's conceivable that they might possibly have been responsible for a small number of the deaths.

In conclusion, it should be pointed out that while reports of mutilations have ended in the area discussed herein, other parts of the country have experienced similar episodes. Texas, Colorado, Oklahoma, and Kansas are among states that have reported similar outbreaks. It seems likely that one episode acts as a triggering mechanism for future episodes in neighboring areas. If this is what happens, one might expect reports of cattle mutilations to continue for some time, perhaps until most cattle-raising areas have experienced a similar episode.

References

Cantril, Hadley 1947. *The Invasion from Mars.* Princeton, N.J.: Princeton University Press.

Johnson, Donald 1945. "The 'Phantom Anesthetist' of Matoon: A Field Study of Mass Hysteria." *Journal of Abnormal and Social Psychology* 40: 175-86.

Medalia, Nahum, and Otto L. Larsen 1958. "Diffusion and Belief in a Collective Delusion: The Seattle Windshield Pitting Epidemic." *American Sociological Review* 23: 221-32.

Schuler, Edgar, and Vernon J. Parenthon 1943. "A Recent Epidemic of Hysteria in a Louisiana High School." *The Journal of Social Psychology* 17: 221-35.

Slocum, Walter 1962. *Agricultural Sociology: A Study of Sociological Aspects of American Farm Life.* New York: Harper & Row.

Smelser, Neil 1962. *Theory of Collective Behavior.* New York: The Free Press.

U.S. Bureau of the Census 1970. *U.S. Census of Population, 1970: United States Summary, Part 1.* Washington: Government Printing Office.

U.S. Bureau of the Census 1970. *U.S. Census of Population, 1970: Nebraska, Part 29.* Washington: Government Printing Office.

U.S. Bureau of the Census 1970. *U.S. Census of Population, 1970: South Dakota, Part 43.* Washington: Government Printing Office.

A Controlled Test
of Dowsing Abilities

James Randi

When a series of five one-hour television documentaries, *Indagine sulla Parapsicologia* (Inquiry into Parapsychology), was released in Italy in 1978, there was a storm of protest. Singled out for large amounts of crank mail and press censure was Piero Angela, Italy's best-known TV journalist, who had hosted, and indeed conceived, the program series. Angela had begun his investigation with an open mind, but he was soon soured by the discovery that most of the parapsychologists he interviewed throughout Europe, the United States, and South America either had no good evidence to show him or would not answer direct questions about their work. The TV series concluded that there was no basis for a belief in the paranormal.

Italy is quite dedicated to belief in such matters. Numerous organizations rose to criticize Angela and RAI-TV, the producers of the documentary, and the fire was further fanned by publication of Angela's book *Viaggio nel Mondo del Paranormale* (Journey in the World of the Paranormal). My personal offer of $10,000 to anyone performing a paranormal feat under proper observing and control conditions was outlined both in the TV program and in the book. Immediately, applicants began to register their desire to claim the reward.

I delegated Piero Angela to handle the preliminaries for me by asking that all applicants sign an agreement saying (a) that they could perform in the presence of a skeptic, and (b) that they would pay their own travel costs to visit Rome to be tested. By this means, most of them were eliminated. But some 40 remained who were willing to be tested; and RAI-TV arranged, in return for the filming rights of the event, to pay my travel and living expenses during the tests, which were held from March 22 to March 31, 1979. I visited there in the company of a colleague, William Rodriguez, who assisted with the supervision of the controls.

The list had dwindled to 11 by the time we arrived. Of these, only 9 eventually showed up, and of that small number, four were dowsers, claiming the ability to find water using various devices. This article will deal exclusively with those individuals, leaving the table-tippers, ESP artists, and others to a future writing.

The TV people had arranged for a location some 30 miles outside of Rome, near the town of Formello. I had prepared a guarded plan (see Figure 1) of a set of three pipe-patterns that were to be laid out and then covered with 50 centimeters (20 inches) of earth, within a plot measuring 10 meters square (33' × 33'). A set of carefully explicit rules was written up that had to be agreed to by the dowsers in advance of their participation. There were three days of haggling over conditions, each performer having his own variations, but eventually everything was resolved.

It was required that each dowser agree that (1) he felt able to perform that day under the stipulated conditions; (2) he was able to find water flowing at a minimum rate of 5 liters a second in a pipe 8 cm in diameter buried under 50 cm of earth; (3) the presence of skeptical persons and TV cameras with other electrical equipment was not a negative factor; (4) he would scan the area for any natural streams running underground before beginning the test, thus eliminating any possible interference with the test itself; (5) he would demonstrate the detection of running water in the exposed section of the pipe, while water flowed in it, as a preliminary; (6) detection of running water in the correct path was to be the goal of the tests, and rationalizations or excuses would not be acceptable; (7) the performer would place between 10 and 100 pegs along the path he found, each peg to be placed within an area 20 cm (8 inches) wide, centered on the actual pipe; the contestant must have at least two-thirds of the pegs on any path placed within these limits, and must succeed in doing this in two out of three tests.

I agreed to give the check for $10,000—which was placed in the hands of the presiding lawyer—to any contestant who met these conditions.

After numerous delays (hailstorms and such) the tests finally got under way. The first testee, Giorgio Fontana, had shown us the night before his uncanny ability to dowse atlas maps, and we were treated to a confidential demonstration that revealed a vast river of crude oil that flowed underground from Greenland, past England, across France and Italy to Tunisia. There, Fontana told us, we were being robbed by the Tunisians who tapped off all that oil.

Fontana's dowsing tool was a pendulum. Popular as a "psychic tool," in France in particular, such a device is simply a string or chain with a weight on it. Fontana rushed about waggling his pendulum over the ground, pointing where he wanted pegs placed by the surveyor. (See Figure 2.) As determined by random selection, path C had been used first. Fontana pegged out a path that ran almost straight from the inlet valves to the reservoir, which closely approximated actual pipe B! I had used this one optimally simple and direct route on a whim, to show that the shortest distance between the two points could not be found. I had not counted on

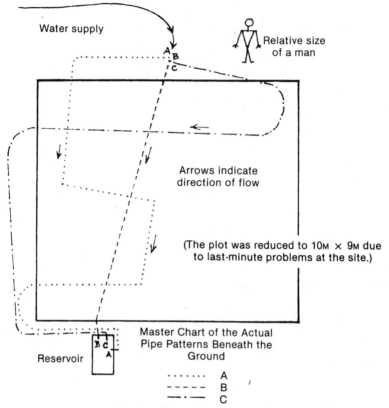

Water supply

Relative size of a man

A
B
C

Arrows indicate direction of flow

(The plot was reduced to 10M × 9M due to last-minute problems at the site.)

Master Chart of the Actual Pipe Patterns Beneath the Ground

Reservoir

B C
A

.... A
- - - - - B
— · — C

FIGURE 1. Plan of the test plot.

such a possibility as being outguessed, and though I believe Fontana's almost-success here was the result of naiveté rather than planning, it was a strong lesson for me. Regardless, the guess was totally wrong. Only one peg out of 30 was within the limits.

Next, he traced another zigzag path that was supposed to be—again— path C. This time, 2 out of 32 pegs fell within limits, at right angles to the true flow. Finally, when path B was chosen at random, Fontana showed signs of giving up; but when reminded that he had to do three trials, he merely indicated we could use his first try as a repeat. And that was nearly correct, though he did not actually dowse it out! Even then, only 6 of 30 pegs would have been correct, and Fontana was a loser.

Then Professor Lino Borga stepped up. He was enthusiastic, effervescent, and loquacious in the extreme. He declared that the prize was in the bag, and he was about to show us a thing or two. He apologized in advance for having to take my money. He was using a device I'd not seen before— two rigid sticks hinged together near one end, forming a "V" that spun about between his hands. His first path was in response to pipe B, and was

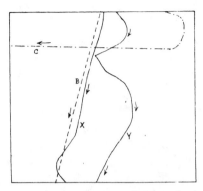

FIGURE 2. Fontana traced the vertical line (X) as path "C" (horizontal path at top), then the other (Y) as another attempt at "C." Finally, he did not trace his third try, but said we could use his first (X) as his third guess. It was actually "B"—the direct line between the valve and the reservoir!

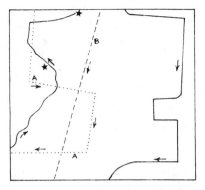

FIGURE 3. Borga's three attempts were far off the mark. He traced the pattern at the right as path "B" (dashed line), then traced it again as path "A" (dotted line), and then made the line at upper left to represent "A" again. The section between the stars indicates where water had ceased flowing, yet the dowser continued, not knowing this.

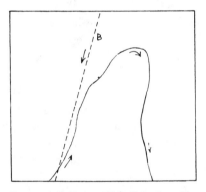

FIGURE 4. Path traced by Senatore. He was only able to make one attempt, due to lack of time. He crossed the actual water path only once, going the wrong way.

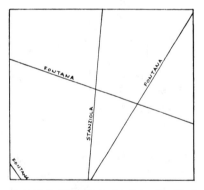

FIGURE 5. Only two dowsers decided there was any "natural" water on the site. Thus they disagreed with the other two, who said there was no water there. As seen in this chart, there was no agreement between any of them.

never even near it. When A was chosen, he decided it was again the same path he'd just traced; but he traced it again to be sure, repositioning several pegs by as much as a centimeter for greater accuracy. But again, this path *did not even intersect* the correct path. Finally, Borga gave us a wiggly response to A once again, which was not only incorrect but went in the wrong direction. Only 2 out of the 27 points he chose were within limits. All in all, he'd needed 58 pegs properly placed. He got 2, entirely by accident.

But the most interesting point about Borga's demonstration was this:

For the last few minutes of his last trace, we noted that the water had run out of the supply truck! Consult the map (Figure 3). Between the two spots marked by stars *there was no water flowing in the pipe at all.* Yet Borga continued his frenzied plotting with no water flow, and certainly not in the right place or direction. As we called his attention to the cessation of the water, saying that it was "running down," though it had actually *stopped* some time before, Borga exclaimed that it was evident from his motions. Sure enough, as he reached the end of the trace he was making, his rods slowed down, as if in response to the water stopping. This leads one to the suspicion that dowsing is the result of imagination and involuntary actions, and not a response to water or any other substance.

Then we turned to Mr. Stanziola, a pupil of Professor Borga. He probably suspected chicanery, for when he was asked to show that his dowsing-rod (a straight stick held in a bowed position) reacted to the water in the exposed part of the pipe as it flowed, he said he got no reaction. We terminated the test there, since even the fundamental dip of the rod was absent. I think Stanziola believed we had no water flowing at that point, but it was flowing copiously, as witnessed by the engineers present, at more than twice the 5-liters-a-second rate they had demanded. Exit Stanziola.

Vittorio Senatore was next, an intense young man using a piece of canelike stick, broken in the center so as to be flexible. He looped this into a script L shape and walked about entranced, eyes closed, though peeking a bit. Several times the stick flipped out of his hands, though under what mysterious power we could not tell, since he missed the chosen path (B) grandly—wrong path, wrong direction (Figure 4). Since time was running out, as well as water, he had the opportunity for only one trace, but agreed that his reputation would rest on that one attempt. In fact, Fontana, Borga, and Senatore all declared that they had been 100 percent successful, and were confident they had won the prize.

Since they had not been allowed to consult with one another after their trials, they had no idea that not only had they all failed, but two of them had declared there was no natural water flowing at the site, and the other two had plotted "rivers" and disagreed with each other. See Figure 5.

I will not detail the confrontation that took place after the tests when we retired to a local restaurant amid many bottles of spirits to discuss the results. Suffice it to say that our conclusion was that dowsing was very surely not demonstrated; yet the dowsers themselves, after due consideration of the day's events, thoroughly convinced themselves that some strange influence was afoot, since none of them had ever failed before!

I await other claimants to the prize offered.   •

Extraterrestrial Visitors

Von Däniken's Chariots: A Primer in the Art of Cooked Science

John T. Omohundro

"I am not a scientific man, and if I had written a scientific book, it would have been calm and sober and nobody would talk about it."
—Erich von Däniken

Playboy: Are you, as one writer suggested, "the most brilliant satirist in German literature for a century"?
Von Däniken: The answer is yes and no. . . . In some part, I mean what I say seriously. In other ways, I mean to make people laugh.

Were it not for the fact that Erich von Däniken has millions of otherwise intelligent people discussing his book and theories seriously, I would prefer to write a parody of his style. But I fear his readership might believe me too. I ignored his books for four years, but now I cannot teach my students or talk to my academic colleagues without his name souring my day. It is out of his hands, now, this chariot thing. It has reached the people, and for reasons that are their own they have made von Däniken a prophet (profit?) and me a defender of the Establishment.

Why is this book so popular? Von Däniken, it seems, has written one of the scriptures of a new cult. What he says, people obviously want to hear.

Throughout history, cultures subjected to stressful situations have responded with cataclysmic religious reformations, often as a substitute for or supplement to political rebellion. The Zulu Uprising in Africa, the Sepoy Rebellion in India, the New Guinea Cargo Cults, the Ghost Dance of the Plains Indians, the Taiping Rebellion in China, and the Luddites and Anabaptists in Europe are some of the famous examples. Anthropologists call them revitalization movements, messianic cults, and so forth, and take them quite seriously. Though they vary greatly, they have certain characteristics in common: a humorless fanaticism, prophets, a new world view, and a stiff distaste for the Establishment. Most of these movements are rooted in obvious and serious crises, and frequently are part of a religious and political change in the culture.

The entire von Däniken affair, even much of the UFO interest associated with it, is, I think, very much like these movements. Only hindsight will give a good perspective on this point in American history, but the "we are not alone" attitude has become an important element of our culture's religious cosmology. A frustration with science's not having delivered all that it promised, a distaste for the specialization of scientific research, and a continuing need to believe in an intelligence beyond our own are the main characteristics of this antiscience mysticism. It does not take much imagination to see that science has been for many in our culture the New Religion, with its white-frocked priests talking in strange tongues about a universe we couldn't even understand. (Try to grasp the idea of a boundless universe doubling back on itself, a la Einstein.) The priests' accomplishments in a few areas like technology and medicine were enough to satisfy the faithful. But as a religion science didn't stand the test of time. The contrast between what we could do in space with what we could do for ourselves on earth was like watching a priest celebrate mass with his zipper down. Science is rather stale as a religion, and it cannot substitute for one. The man-in-the-street prefers a richer religion than that.

If von Däniken's thesis is part of your religious cosmology, so be it. I don't argue religion; I try to study it and see how it relates to human life. But if von Däniken seems like science to you, shame on you.

What follows is an attempt to lay open von Däniken's approach as a warped parody of reasoning, argumentation, as well as a vigorous exercise in selective quotation, misrepresentation, and error based on ignorance (presumably, if it is not intentional fibbing). For students his work does serve two valuable purposes: first, it raises their interest in the cultures and myths which he so badly mishandles; and second, like Lewis Carroll's *Alice in Wonderland, Chariots of the Gods?* is a challenge to study and determine all that is wrong with it. So it is by no means a complete waste of one's time (either his, yours, or mine).

Briefly stated, *Chariots of the Gods?* proposes that scientists have overlooked or refused to inform the world of the many pieces of evidence which suggest that we have been visited, probably several times, by intelligences from other planets. Von Däniken argues that an open-minded approach to the ruins of past cultures and their art and myths raises many unanswered questions which can best be answered by accepting the hypothesis of extraterrestrial visitors. Data from Incan, Mayan, Sumerian, Egyptian, and many other cultures which suggest the hypothesis include cave painting, architectural and technological accomplishments, and mythological events of great similarity around the world. Von Däniken says that the explanations given by scientists of these data are too smug, and that now that space travel is possible for us, we must at least admit that his hypothesis is as viable as anyone else's.

Some of my professors used to tell me that hypotheses are a dime a dozen; people make them up all the time. Making an hypothesis is not science; it's what you *do* with an hypothesis that more or less is science and is to be judged by others. Von Däniken is entitled to his hypothesis. But what does he *do* with it?

The straw horse, red herring, and other ruses

Argumentation is an art which can easily be perverted. One technique to make yourself sound good is the straw horse: misrepresent the thing you wish to argue against. Von Däniken's characterizations of archaeology and anthropology—fields which focus on precisely the kind of data he studies—are abysmal.

> . . . in the future, archaeology can no longer be simply a matter of excavation. The mere collection and classification of finds is no longer adequate. Other branches of science will have to be consulted and made use of if a reliable picture of our past is to be drawn. (p. 14. This and future references from von Däniken are to *Chariots of the Gods?*)

> Let us say that someone decides to become an anthropologist and he reads and learns a lot about anthropology, about bones and apes and all those details. (Ferris, p. 58)

By denying the breadth of these fields and the wealth of data in them he has left somewhat of a vacuum into which to float his own ideas, which I hope to show are clearly not based on any background in archeology or anthropology. These are not the only disciplines he chops. His critics are nearly unanimous in accusing him of misrepresenting or failing to understand even the rudiments of geology, mythology, psychology, chemistry, astronomy, and physics (Ostriker 1973, p. 239).

His technique is successful in part because there are many presumably educated people who don't understand these fields, or even the ways of scientists in general. He has played to the prejudices and stereotypes of those who are not "scientists" (priests of the old religion). The tone of "you and I, dear reader" places him and his readership in an underdog position against the monolithic Establishment of picky pedants who represent the scholars.

Another technique that works well for misleading the mind is the red herring. The object is to confuse the reader by introducing an extraneous issue so that he will not catch you on your main point. Politicians might introduce Motherhood and Apple Pie, but von Däniken has his reflections on truth, atomic war, and propagandizing for space research. His comments in these red herrings seem startlingly in contrast to his arguments.

> We owe it to our self-respect to be rational and objective . . . (p. 5)

> We may be as religious as our fathers, but we are certainly less credulous. (p. 37)

> Anyone who really seeks the truth cannot and ought not to seek it under the aegis and within the confines of his own religion. (p. 53)

> It is unworthy of a scientific investigator to deny something when it upsets his working hypothesis and accept it when it supports his theory. (p. 66)

> It is depressing what many people—and sometimes whole occult societies—make out of their ostensible observations. They only blur our view of reality and deter serious scholars from dealing with verified phenomena . . . (p. 120)

These comments are quite irrelevant to his arguments and serve only to glaze the reader's critical judgment.

One final technique that is useful in argumentation is to warn the reader in advance about the criticisms which will be leveled by one's opponents. This is not the same thing as dealing with those criticisms, but neatly puts the critic on the defense when listeners say, "Aha! Von Däniken said you would say that!" thus somehow scoring a point for the home team. For example: "Impossible? Ridiculous? It is mostly those people who feel that they are absolutely bound by laws of nature who make the most stupid objections" (p. 84).

Our ancestors, the dummies

What most depresses my fellow anthropologists and me is the way people accept von Däniken's unnecessarily anthropocentric and ethnocentric views of other people in the world and in history. Anthropocentricism is the assumption that other living, sentient, or intelligent creatures must feel and think or evolve as humans do. Ethnocentricism is the even more narrow assumption that other people must think, behave, or evolve as we do. Further, there is usually a heavy flavor of cultural superiority in such assumptions.

Chariots of the Gods? plays upon most people's inability to break out of these assumptions. It implies that up until the last thousand years or so the world was filled with primitives, heathens, savages, dummies. Their intelligence matched their simple technologies; their languages were simple, their cultures were primitive, they were brutes. If they seem to have come up with something quite fantastic by our standards, someone smarter than them must have given it to them. They then proceeded to garble it up in their ingenuousness; they certainly didn't do those things for the same reason that we would have.

(Caucasoid-like figurines of Summer): . . . very difficult to fit into the schematic system of thought and its concept of primitive peoples. (p. 27) [*The reasoning: primitives can't be Caucasians.*]

Since we are not prepared to admit or accept that there was a higher culture or an equally perfect technology before our own, all that is left is the hypothesis of a visit from space! (p. 28) [*A sillygysm! See the next section.*]

(The Yahweh of the Semites): It is . . . difficult for enlightened children of this age to think of an infinitively good Father who gives preference to "favorite children" such as Lot's family. (p. 37) [*The reasoning: You modern Christians aren't going to believe all this Biblical stuff about a harsh God.*]

. . . descriptions of extraordinary things that could not have been made up by any intelligence living at the time the tablets were written, any more than they could have been devised by the translators and copyists who manhandled the epic over the centuries. (p. 49) [*The reasoning: Only moderns have enough intelligence to be imaginative.*]

Since the question of space travel did not arise 100 years ago, our fathers and godfathers could not reasonably have had thoughts about whether our ancestors had visits from the universe. (p. 4) [*Just plain wrong.*]

The Mayans were intelligent; they had a highly developed culture. . . . it is difficult to believe that it originated from a jungle people. (p. 55) [*The reasoning: Jungle people are somehow dumber than most. Just look at Africa.*]

(The Egyptian, Chinese, and Incan civilizations): Who put the idea of rebirth into the heads of these heathen peoples? (p. 63) [*The reasoning: Heathen heads are empty.*]

(Egyptians): How did such a highly developed civilization arise at such an early date? . . . Who gave them their incredible knowledge of math and a readymade writing? (p. 65) [*Anyone that was civilized before us cheated. What ever happened to the Honor Code?*]

(Ancient storytellers had strong imaginations): So it must be that the ancient storytellers had a store of things already seen, known, and experienced at hand to spark their [*otherwise dull?*] imaginations. (p. 65)

How on earth could people in the dim past arrive at different perceptions of one and the same thing, when the horizon was very limited? (p. 66)

(Egyptian mummification): Who put the idea of corporeal rebirth into the heads of the heathen? (p. 81)

If the stone age cavemen were primitive and savage, they could not have produced the astounding paintings on the cave walls. (p. 87)

(The Mexican flying serpent Quetzalcoatl): How could anyone worship this repulsive creature as a god, and why could it fly as well? (p. 104) [*Answer: government subsidy.*]

These are really just a handful of examples which reveal ethnocentricism. Von Däniken's reasoning, conservatively stated, is: there are some real mysteries in the past because it is obvious that people who lived then are not solely responsible for those remarkable things. There are indeed real mysteries in the past, but they are usually not the ones von Däniken sees. When one consciously puts aside the prejudices of his own culture and examines the cultures of the peoples mentioned in *Chariots of the Gods?* one begins to see the way myth, art, architecture, politics, kinship, and technology relate to one another, reflect and react to one another. The "fit" of many of these seemingly bizarre practices in the rest of their culture is often in itself a wonder to behold.

The sillygism and cooked science

Von Däniken's book is a virtual goldmine of logical fallacies, implications by innuendo and rhetorical questions, and failures to apply "Occum's Razor." Alicia Ostriker, who interviewed von Däniken for *Esquire*, wrote, "So what if the fallacies fly by in flocks like mallards heading south?" She was captivated by the man's enthusiasm and chose to overlook his "gee-whiz style fit only for kiddies." She chose to overlook his flaws—but many other people don't see them.

A non sequitur, or logical fallacy, makes a conclusion which does not follow from the premise. The book starts out with a few non sequiturs. On page vii von Däniken argues that if you ignore his book, then you are a layman who refuses to face the adventurous and mysterious past. On page 2 he says that if one accepts the possibility of developed life elsewhere in the universe, then it must have been a civilization. Here is an example, phrasing the main thesis of the book: "Since we are not prepared to admit or accept that there was a higher culture or an equally perfect technology before our own, all that is left is the hypothesis of a visit from space!" (p. 28).

A rhetorical question places the entire burden of proof on the reader, who either acquiesces because of the generally bewildering style of the argument or passes the burden of proof on to the "scholars." When contemplating the ruins of Tiahaunaco, in Bolivia, von Däniken writes: "Had our forefathers nothing better to do than spend years—without tools—fashioning water conduits of such precision?" (p. 21)

Applying Occum's Razor means that when two explanations for one set of facts are possible. one adopts the simplest explanation, that is, the one that assumes the least number of "ifs." Von Däniken has argued (Ferris 1974) that space travel is a simple explanation, since it is now possible by us. However, it is not the possibility of space travel or of extraterrestrial intelligence that is questionable. The thesis of *Chariots of the Gods?* fails by Occum's Razor because it constructs a gigantic house of cards, each card requiring a new "if." The "ifs" are held together by faith alone and patently contradict most of the principles which "science" had begun to see as a rather unified system. Look, for example, at von Däniken's thesis that modern humans are the act of deliberate breeding by extraterrestrial intelligences. The fossil record of humanlike creatures and the culture they possessed stretches back more than a million years.

Through the millennia, by rather gradual steps, we see the body approaching modern shape and the brain approaching modern size. Cultural developments like fire, sophisticated stone tools, burials, tailored clothing, and so forth appear long before modern Homo sapiens. To see ourselves as a continual development of those trends, moving and adapting to the changing climates, creatures, and contours of the land, is much tidier than introducing some undefined, undated appearance of superior "breeders."

Von Däniken plays heavily on the reader's readiness to conclude that a long string of random possibilities equals a certainty. By the same reasoning, it is a virtual certainty that you will get six heads in six coin tosses, since there is a real possibility (50 percent, to be exact) that one toss will come up heads.

Last, and perhaps most disturbing, is von Däniken's misrepresentation of the very process of "doing science." He does not exhibit, nor does he anticipate in the reader, any real facility in the nature of a "fact," an hypothesis, developing a theory, and proof (or more accurately, demonstration). At one point von Däniken disclaims that he is compiling a sequence of proofs of prehistoric space travelers: "that is not what I am doing. I am simply referring to passages in very ancient texts that have no place in the working hypothesis in use up to the present" (p. 66).

The author doesn't know what a working hypothesis is, nor is he embarrassed to stamp "Q.E.D." on an enormous gaggle of tautologies (assume something, create an hypothesis, test, claim to have proved your assumption). He avoids ever stating anyone else's explanation in reasonable terms. He is loose with his concept of proof, with which he bludgeons unidentified others for not producing. More than any other characteristic, it is this blithely ignorant toying with the method of scientific reasoning which marks the book's shabbiness.

Just plain wrong

A review of *Chariots of the Gods?* in *Book World* says, "To check his 'facts' would take months of research, since he never cites his authorities." His highly selective choice of what to introduce as data follows absolutely no discernable criteria. His translations make critics howl (with glee if they have a sense of humor, with rage if they do not). Many of the "facts" which von Däniken presents have been checked out. A few of these are presented below.

The Piri Reis maps (p. 14). (Amazing maps, but far from accurate.)

The Tiahuanaco culture of Bolivia (p. 20 ff). (Cf. Lanning to remove a few of the mysteries von Däniken sees here.)

The Sumerians (p. 24). (Braidwood and Adams among others have quite fine ideas about where the Sumerians came from.)

"Isn't there something rather absurd about worshipping a 'god' whom one also slaughters and eats?" (p. 33). (No. The world has a number of people who do so: Australian Aboriginies, Mesopotamians, Ainu, and others.)

The copper furnaces at Ezion Geber (p. 44). (The dating given is wrong; also the source has withdrawn his speculation: the rooms are storage rooms.)

The breeding experiments of space travelers on prehumans (p. 52). (The *Esquire* interviewer points out that von Däniken doesn't even believe this stuff himself. Then why say it? It defies all the principles of genetics and evolution. Ironically, he calmed down in *Gods from Outer Space*, and the book didn't sell.)

The "suddenness of Egypt" according to Egyptologists (p. 74). (This leads you to suspect someone put it there—bingo. In fact it developed out of a Neolithic farmer culture a thousand years after the civilizations began in Mesopotamia.)

The Cheops pyramid: the height formula, *pi* formula, and wood hypothesis. (Cf. Wilson. Even von Daniken's math is bad.)

"Did the Egyptians learn the possibility of mummification from nature? If that were the case, there ought to have been a cult of butterflies or cock-chafers . . . there is nothing of the kind" (p. 84). (Worship of the scarab beetle was widespread.)

The Chinese tomb with 41 dead without violence (p. 86). (How about disease or starvation, a common threat in Chinese civilization?)

Terra cotta heads in Jericho ten thousand years ago: "That, too, is astonishing, for ostensibly this people did not know techniques of pottery making" (p. 87). (Wrong on several counts, one being that terra cotta had been made into statues for over ten thousand years before this.)

"I would suggest, on tolerably good grounds, placing the incident I am concerned with in the Early Paleolithic Age—between 10,000 and 40,000 (p. 88). (The early paleolithic ended about 200,000 years before this. What he is describing is called the Upper Paleolithic and Mesolithic.)

The "Chinese" jade necklace in Guatemala (p. 93). (This one threw me for a minute, but Wilson says jade is indigenous to Central America.)

". . . in Christ's day the concept of a heaven with fixed stars taking into account the rotation of the earth did not exist" (p. 105). (Let me quote Ostriker again: "What the average reader of von Däniken probably doesn't know is that the idea of life on other worlds is not exactly a new one." She further points out that before Ptolemy's geocentric world view came along, a number of cultures were not far off the view we hold now.)

Outline drawings of animals which simply did not exist in South America ten thousand years ago, namely camels and lions (p. 106). (Perhaps they are llamas and pumas, native to the area.)

"There are artificially produced markings, as yet unexplained, on extremely inaccessible rock faces in Australia, Peru, and Upper Italy" (p. 106). (Speaking just for Australia, the aborigines have been seen to make the same markings in their totemic rituals).

Engravings of cylindrical rocketlike machines in Kunming, China (p. 107). (They are! The Chinese invented gunpowder and shot rockets.)

These are some of the items I caught. Others are pointed out in the articles mentioned in the bibliography. Rustless iron columns in India, the Easter Island stones, and so forth are not quite the mystery von Däniken claims.

This review has been aimed at those readers of von Däniken who feel that in the interests of science and reasonableness we should consider his argument. I have sketched some of the reasons why, when one considers his argument, one discovers no science or reasonableness in it. The mass popularity of *Chariots of the Gods?* doesn't derive ultimately from any interest in science or reasonableness but, as I have suggested, stems from a reaction against it. There is some justification for such a reaction; I even advocate a dose of insanity in everyone's life. Von Däniken's book is a good read if you need a dose of enthusiastic delirium. But I do not mix my insanity and my science.

REFERENCES

Ferris, Timothy 1974 "Playboy Interview: Erich von Däniken." *Playboy.* August. Pp. 51 ff.

Ostriker, Alicia 1973. "What If We're *Still* Scared, Bored and Broke?" *Esquire,* December. Pp. 238 ff.

BIBLIOGRAPHY

If your interest has been stimulated by the controversy surrounding *Chariots of the Gods?*, I recommend the following works as just as interesting but more sound.

Braidwood, Robert. *Prehistoric Men.* 7th ed. Glenview, Ill.: Scott, Foresman, 1967. Through the stone age and into the civilizations of Mesopotamia.

Deetz, James. *An Invitation to Archaeology.* Garden City, N.Y.: Doubleday, Natural History Press, 1967. Scientific archaeology.

Eiseley, Loren. *The Immense Journey.* New York: Random House, 1957. Human evolution.

Frankfort, Henri, et al. *Before Philosophy.* Baltimore: Penguin, 1966. An excellent exposition on an old myth.

Lanning, Edward. *Peru Before the Incas.* Englewood Cliffs, N.J.: Prentice-Hall, 1967. One of the few general books.

Sanders, William T., and Barbara J. Price. *Mesoamerica: The Evolution of a Civilization.* New York, Random House, 1968.

Shklovskii, I. S., and Carl Sagan. *Intelligent Life in the Universe.* New York: Dell, 1966. Exobiology by experts.

Wallace, Anthony. *Religion: An Anthropological View.* New York: Random House, 1966. A way of looking at non-Western religions.

Wilson, Clifford. *Crash Go the Chariots.* New York: Lancer Books, 1972.

Von Däniken's Golden Gods

Ronald D. Story

Gods will doubtless survive, sometimes under the protection of vested interests, or in the shelter of lazy minds, or as puppets used by politicians, or as refuges for unhappy and ignorant souls.
—Julian Huxley, *Religion without Revelation* (1927)

Stan Lee, publisher of Marvel Comics, once said: "Nowhere else but in comic books can you still recapture the fairy tale fun of finding characters bigger than life, plots wilder than any movie, and good guys battling bad guys with the fate of entire galaxies hanging on the outcome." Lee overlooked, however, one of the hottest sensations in publishing history, the phenomenally popular, ex-hotel-manager from Davos, Switzerland —Erich von Däniken—the author who has been translated into thirty-five different languages around the world, totaling up book sales approaching thirty-four million copies.

Von Däniken's theory—that superbeings from outer space came to earth in the distant past to bestow upon us intelligence and all the rudiments of early civilization—is a view some have termed *fiction science*. The reason for this label should become clear after reading any one of von Däniken's books. There is a dearth of supporting data, an endless stream of false and misleading information, a sprinkling of truth, and some of the most illogical reasoning ever to appear in print. To compose a work of fiction-science is not difficult. As one reviewer put it: "[It is] an old cosmological recipe: simply *ad astra,* mix feverishly and half bake."

Let me say that I have no objection to honest speculation. Nor do I find anything wrong with the idea that intelligent beings from another planet *could have* visited earth in ancient times. But I have found that if

318

you take von Däniken's "indications," as he calls them, check them out, and subject them to the normal rules of evidence, they fall apart. For one thing, the level of technology required for the construction of the various artifacts and monuments in question never exceeds the capacities of earthmen working on their own in the normal context of their own cultures. The archaeological "wonders" that are alleged to prove, or at least "indicate," ancient astronauts is largely a collection of interesting finds, superficially described and taken out of context. But perhaps the most serious deficiency of von Däniken's whole reasoning process is the omission of highly relevant, key information that, if known, would cast an entirely different light on the subject at hand. What follows are some examples from von Däniken's showcase of "proofs" to illustrate what I mean.

The Palenque Astronaut

In the ancient Maya city of Palenque (on the Yucatan Peninsula, in the state of Chiapas, Mexico) stands a seventy-foot-high limestone pyramid called the Temple of the Inscriptions. Until 1949, the interior of the structure had remained unexplored. But when the Mexican archaeologist Alberto Ruz Lhuillier noticed finger-holes in one of the large floor-slabs, he raised the stone and discovered a hidden stairway that had been deliberately filled in, centuries ago, with stone rubble and clay. After four years of clearing away the blockage, Ruz and his workers had descended sixty-five feet into the pyramid, where he came upon a secret tomb. Little did Ruz know that twenty years later this discovery would be used as one of the "proofs" of the existence of ancient astronauts. What has attracted the attention of ancient-astronaut fans everywhere is the stone carving that decorates the tomb lid. Von Däniken describes it this way: "On the slab [covering the tomb is] a wonderful chiseled relief. In my eyes, you can see a kind of frame. In the center of that frame is a man sitting, bending forward. He has a mask on his nose, he uses his two hands to manipulate some controls, and the heel of his left foot is on a kind of pedal with different adjustments. The rear portion is separated from him; he is sitting on a complicated chair, and outside of this whole frame you see a little flame like an exhaust."[1]

Could it be that the Palenque tomb lid actually depicts a man piloting

1. Quoted from a transcript of *The Lou Gordon Program* (WKBD-TV), video-taped in Detroit, Michigan on February 27, 1976, on which the writer appeared with von Däniken in a televised debate.

a rocket? The notion becomes less plausible once the various elements that make up the overall design are examined separately, in detail. Notice first (in the figure below) that the "astronaut" is not wearing a space suit, but is practically naked. The man in this scene is barefoot, does not wear gloves (both fingernails and toenails are illustrated), and is outfitted in nothing more than a decorative loin cloth and jewelry. In other words, he is dressed in typical style, characteristic of the Maya nobility as to be expected at around A.D. 700. Actually, this is the tomb of the Maya king Lord-Shield Pacal, who died in A.D. 683.

The details of his royal history are well established. The glyphs carved on the frame of the sarcophagus lid, as well as other glyphic evidence found in other temples at the Palenque site, trace his ancestry and give the exact dates of when he was born, when he ruled, and when he died. When the illustration on Pacal's tomb lid is oriented correctly (vertically instead of horizontally), we can see that the "rocket" is actually a composite art form, incorporating the design of a cross, a two-headed serpent, and some large corn leaves. The "oxygen mask" is an ornament that does not connect with the nostrils, but rather seems to touch the tip of Pacal's nose; the "controls" are not really associated with the hands, but are elements from a profile view of the Maya Sun God in the background; the "pedal" operated by the "astronaut's" foot is a sea shell (a Maya symbol associated with death); and the "rocket's exhaust" is very likely the roots of the sacred maize tree (the cross), which is symbolic of the life-sustaining corn

The Palenque "astronaut."

plant. The whole scene is a religious illustration, not a technological one, and is well understood within the proper context of Maya art.

The Nazca Spaceport

Another ancient astronaut idea concerns the now famous desert markings on Peru's Nazca plain. About 250 miles southeast of Lima, between the towns of Nazca and Palpa, lies a barren plateau covering 200 square miles, that served as a gigantic drawing board for its ancient inhabitants. There, discernible only from the air, are over 13,000 lines, more than 100 spirals, trapezoids and triangles, and nearly 800 huge animal drawings, all etched into the desert floor by the removal of the dark surface stones exposing the lighter-colored soil underneath. Most of the lines radiate from several star-like centers and extend for miles. These markings were probably constructed (over a period of several hundred years) sometime between 400 B.C. and A.D. 900.

In his book *Gods from Outer Space,* von Däniken tells us his theory: "At some time in the past, unknown intelligences landed on the uninhabited plain near the present-day town of Nazca and built an improvised airfield for their spacecraft which were to operate in the vicinity of the earth" (Bantam paperback edition, p. 105).

There are several good reasons why the lines probably were not ancient landing strips: (1) there simply would be no need for a runway, several miles long, to accommodate a space vehicle that should be capable of a vertical landing; (2) many of the lines run right into hills, ridges, and the sides of mountains; (3) the soft, sandy soil would not be a suitable surface for any kind of heavy vehicle to land on. As Maria Reiche, probably the world's leading authority on Nazca, has said, "I'm afraid the spacemen would have gotten stuck."

Another version of the spaceport theory maintains that the exhaust from hovering spacecraft was responsible for blowing away the sand and thus creating the lines. Again, a nice try, but this idea would seem to prove just the opposite of what its proponents intend. It is not the light-weight soil that was removed to create the lines, but rather the heavier rocks that are actually stacked in linear piles all along the sides of the lines.

What then, could have been the purpose of such an enormous array of lines, shapes and animal figures created more than a thousand years ago.

The first systematic study of the Nazca markings came in 1939 (twelve years after their actual discovery in modern times), when Professor Paul Kosok of Long Island University first mapped and photographed them from the air. His most significant finding came, however, while he was standing on the ground gazing down one of the lines toward the setting sun on June 22, 1941. This happened to be the day of the winter solstice in the Southern Hemisphere; and the apparent alignment gave Kosok the startling idea: perhaps the lines represented "the largest astronomy book in the world." He later confirmed more than a dozen such alignments, some for the soltices and others for the equinoxes, indicating that the Nazca "landing field" very likely comprised a gigantic astronomical calendar and observatory.

It has also been found that several of the large animal drawings have solstice lines associated with them. After all, how would the Nazcans be able to recognize which lines were which, if they had not devised some reference system by which to find them later?

But the question is still asked: why would the ancient Nazcans go to such trouble to construct these markings (and especially the drawings of animals) that are recognizable only from the air? In fact, the most ideal vantage point for viewing them is *not* from hundreds of miles up, where one might expect to find an orbiting satellite or a spaceship, but rather, at a point in mid-air, about 600 feet above the plain. How then, could the

Courtesy of International Explorers Society. Spider depicted on Nazca plain.

early Peruvians have seen and appreciated their work without the advantage of something like an early-model helicopter?

According to a theory recently tested by the Florida-based International Explorers Society, the "chariots of the gods" that sailed over Nazca might well have been early-model smoke balloons, piloted not by alien beings but by ancient man himself. In a new book entitled *NAZCA: Journey to the Sun* (Simon & Schuster, 1977), IES member Jim Woodman presents an impressive array of evidence to support his contention that the early Peruvians knew the secret of lighter-than-air flight long before the first hot-air balloons had ever been flown in Europe.

For example, all along the Nazca plains are thousands of ancient grave-sites containing finely woven textiles (perfectly suited for a balloon envelope), braided rope (another item useful in balloon-making), and ceramic pottery. On one of the clay pots is a picture that resembles a hot-air bag complete with tie ropes. There is also the little known fact that even in modern times, the Europeans were not the first to make manned balloon flights. In the city plaza of Santos, Brazil, stands a monument to "the flying man" Bartolomeu de Gusmao, who made his first flight on August 8, 1709.

On November 28, 1975, author Jim Woodman and the noted balloonist Julian Nott actually flew a crude hot-air balloon, appropriately named *Condor I,* over the Nazca plain to prove their point. The ten-story mountain of fabric and smoke was a reconstruction based on what resources the ancient Nazcans actually had. The envelope was made of a cotton fabric similar to the cotton pieces uncovered at the grave-sites; the gondola basket (for the pilot and co-pilot) was woven from totora reeds grown in the vicinity of Lake Titicaca nearby, and was held by lines and fastenings also made from native fibers.

It was a remarkable example of archaeology-by-experiment, in which an ancient possibility had been demonstrated without resorting to the space-god theory.

The Piri Re'is Map

But then what about the ancient map that we have all heard about, that could only have been made from an aerial photograph taken from a great height (not from just a few hundred feet up, but from an altitude of several hundred miles above the earth)? According to von Däniken, the Piri Re'is map is "absolutely accurate—and not only as regards the Mediterranean and the Dead Sea. The coasts of North and South America and even the contours of the Antarctic were also precisely delineated on Piri Re'is's maps. The maps not only reproduced the outlines of the continents but also showed the topography of the interiors! Mountain ranges, mountain peaks, islands, rivers, and plateaus were drawn in with extreme accuracy." And furthermore, "Comparison with modern photographs of our globe taken from satellites showed that the originals of Piri Re'is's maps must have been aerial photographs taken from a very great height" (From *Chariots of the Gods?*, Bantam paperback edition, pp. 14-15).

The fallacy here is simply that the map in question is not at all as accurate as von Däniken claims. The original, dated 1513, was found in 1929 in the old palace of Topkapi, as it was being converted into a

Courtesy of International Explorers Society. Condor I balloon in flight.

museum. What's more, all of the marginal notes that appear on the map have been translated; and the complete story of its origin is well established thereon.

Piri Re'is, a Turkish Admiral and noted cartographer of his day, drew the map from about twenty other charts which he reduced to one scale. And in so doing, certain errors appeared, which testify clearly to his limited knowledge of the geography of the world. These errors include: the omission of about 900 miles of South American coastline, a duplication of the Amazon River, the omission of the Drake Passage between Cape Horn and the Antarctic Peninsula (representing nine degrees on the map), and a nonexistent landmass (presumed by von Däniken and his disciples to be Antarctica) that is drawn about 4,000 miles north of where Antarctica should be. This says nothing of the fact that none of the mountain ranges, islands, rivers, and the coastlines are drawn true to form (especially when it comes to matching up the southernmost portion of the map with Antarctica). In other words, the map fits in perfectly well with other sixteenth century cartography and in no way can be reasonably regarded as the product of space-beings engaged in prehistoric aerial reconnaissance.

The Stone Giants of Easter Island

Next, we travel to one of the loneliest places on earth. The natives call it *Te Pito o te Henua* or The Navel of the World; but because the Dutch Admiral Jacob Roggeveen discovered this tiny dot of land in the South Pacific on Easter Sunday, 1722, it has since been known to the rest of the world as Easter Island. Of all the inhabited places on earth, Easter Island is one of the most isolated by the sea. Its location is 2,300 miles due east of Chile and 1,300 miles west of Pitcairn Island. Partly because of such extreme isolation, its prehistoric inhabitants have been regarded as one of the most curious cultures of the ancient world. They had their own system of picture-writing (unlike any other language in the world), a fairly sophisticated political and religious structure, and a tradition of stone-working that resulted in over 600 colossal heads (up to forty feet in height and weighing eighty tons) carved from volcanic stone.

The statues have inspired a long tradition of pseudoscientific speculation, the most recent "theory" being that the heads must be modeled after space-beings since (it is claimed) they do not resemble any people on earth. Actually, the main stylistic features of the stone faces do resemble the predominant facial features of the Easter Island natives, despite false statements to the contrary.[2]

But what the ancient astronaut believers really get excited about is the "mysterious" fact that the statues were ever carved out of the "steel hard" volcanic rock in the first place. Surely, they reason, this could not have been accomplished with the primitive stone picks that have been found in the local quarry.

In 1955-56, explorer-anthropologist Thor Heyerdahl led an expedition of archaeologists to Easter Island for the purpose of finding the answers to this and other puzzles. And on this expedition, which lasted six months, not only was the carving process demonstrated by the islanders themselves, but the transporting and raising of one of the statues onto its *ahu* platform was demonstrated as well.

First, the carving: the initial step was to soften the surface of the volcanic tuff with an application of water. Then, a special flaking motion was used by the carver, with the basalt tool, to cut a pair of grooves in the rock, leaving a keel in the middle, which he later knocked out. The result was an extremely efficient process by which six men actually carved the entire outline around a small-sized statue in just three days. The transporting of a twelve-ton statue was accomplished by 180 men pulling ropes attached to the stone giant's head; the body rested on a wooden sled. According to archaeologist Edwin Ferdon, who witnessed the demonstration: "To begin with, they had to pull it out of deep sand, but once it got up onto hard soil, we could have cut that crew down by at least one-half. Once they got out of the sand, they really started tearing with this thing, and we had to stop them, or they would have pulled it away from the actual site. They must have pulled it a hundred yards before we stopped them." Another statue, this one weighing about twenty or thirty tons, was raised onto an elevated masonry platform. This was done through the use of levers (three large wooden poles) and an ingenious under-building of stones. It took twelve men eighteen days to complete the job—the point, of course, being that outer-space technology was not required.

Those Gold-Filled Caves

The most controversial of von Däniken's books was *The Gold of the Gods* (1972), in which he claimed to have seen the "Golden Zoo" (a fantastic

2. A very nice comparison is shown in an illustrated article entitled "Easter Island and Its Mysterious Monuments" by H. La Fay and T.J. Abercrombie in the *National Geographic*, Vol. 121, No. 1, January, 1962, p. 99.

collection of animal statues made of solid gold) and "Metal Library" (two or three thousand gold-leaf plaques embossed with an unknown script) in a subterranean tunnel system 800 feet beneath Ecuador and Peru. In this, von Däniken's third book, we find a collection of photographs, purportedly of gold treasures, which had been entrusted to one Padre Crespi. Von Däniken writes: "Today I know the biggest treasure from the dark tunnels is not on show in South American museums. It lies in the back patio of the Church of Maria Auxiliadora at Cuenca in Ecuador. . . ." (from *The Gold of the Gods*, Bantam paperback edition, p. 21). What is the significance of this find? Quoting a South American Professor, Miloslav Stingl, von Däniken publishes this statement: "If these pictures are genuine, and everything indicates that they are, because no one makes forgeries in gold, at any rate not on such a large scale, this is the biggest archaeological sensation since the discovery of Troy" (*Ibid.*, p. 46). More about Father Crespi and the "artifacts" later.

In March, 1972, von Däniken met an Argentine adventurer-explorer Juan Moricz, who, according to the account given in *The Gold of the Gods*, was the discoverer of the caves. It is also stated by von Däniken that Moricz took him on a personal tour through the mysterious underworld. In fact, von Däniken begins his book by saying: "To me this is the most incredible, fantastic story of the century. It could easily have come straight

Courtesy of Thor Heyerdahl. Raising the statue on Easter Island.

from the realms of science fiction if I had not seen and photographed the incredible truth in person" (*Ibid.*, p. 1). But when the German news magazine *Der Spiegel* dispatched a reporter to interview Juan Moricz in Ecuador, Moricz said: "Däniken has never been in the caves—unless it was in a flying saucer. If he claims to have seen the library and other things himself then that's a lie." According to Moricz, during the first week in March, 1972, von Däniken invited him over for a meal at the Atahualpa Hotel in Guayaquil, Ecuador. Here they began discussing the cave story, one detail leading to another. Moricz said: "I told him everything. For hours, for days, he squeezed it out of me." This information was later passed on to von Däniken's readers as his own experiences. What, apparently, von Däniken did not know was that the cave story was not even original with Moricz. It is said that the legend can be traced back more than thirty years to a deranged army captain named Jaramillo.

There was a cave expedition, led by Moricz, in 1969. He was accompanied by fourteen persons including a local Indian chief, Nayambi, of the Coangos tribe. Although they did find an artificially carved stone archway and some walls of carved, granite blocks, there were no gold treasures nor any evidence of our alleged astral ancestors.

Concerning Father Crespi, it is reported by *Der Spiegel* that others besides von Däniken have seen the Crespi treasure, and although there are

Archaeologist Pino Turolla in Father Crespi's "collection."

some excellent pieces (pre-eminently the stone ones), the German source says, "most were found to be imitations, made of tin and brass, like one can buy by the dozens in souvenir shops in Cuenca."

Archaeologist Pino Turolla of Miami, Florida, confirms this report and accuses von Däniken of writing a fraudulent book (*The Gold of the Gods*) with phony pictures. Turolla has, in fact, taken his own photos of the "priceless artifacts" in Crespi's collection (see photo on pg. 32) and revealed their origin. Turolla has seen the little factory clearing where, he says, the stuff is actually made by local Ecuadorian Indians. The natives then trade what is mostly junk to Father Crespi for some clothes or small sums of money. Crespi, it turns out, is a much-loved but eccentric old man who collects this stuff, which, according to one reporter, is "closer to copper plumbing than God Gold." Among the treasure, Turolla said he even saw a copper toilet-bowl float.

The von Däniken Mystique

It is time to attempt an answer to the inevitable two-part question: Why are von Däniken's theories so popular; and, Why do so many people take him seriously? I have a list of answers, none of which, in my opinion, *completely* explains the ancient astronaut craze; but perhaps, someone, someday, will probe the depths of the human brain and look more deeply into the human psyche than has been possible up to now. In the meantime, however, here is my list:

(1) A large segment of the population, a majority perhaps, does not take kindly to the concept of human evolution that modern science has developed. It is simply unflattering to them to believe that our ultimate ancestors were a form of prehistoric ape or "monkey," so to speak. A more pleasing notion, and a more traditional one at that, is that man has a supernatural origin as, for instance, is told in the Bible. Von Däniken is abiding in this respect in that he saves the supernatural part, and even makes it *appear* compatible with modern-day science.

(2) The very concept of God that most of us have been taught consists in this familiar mental image: God is a super-being from a super-world from somewhere "out there" (i.e., among the stars). This whole frame of reference fits perfectly the theme that God was an astronaut.

(3) Another feat accomplished by von Däniken was to (seemingly) reconcile modern science with a literal interpretation of the Bible. Speculations are generally more popular anyway, if they are overly simple.

Abstruse theology is just as forbidding to the mass public as is academic science. The real answer to a difficult problem oftentimes requires more mental effort than many are willing or able to muster.

(4) The theme of salvation is no doubt central to the ancient astronaut myth. What could be more appealing than beings who are godlike in their technical knowledge (which is threatening but still means so much to us) and in their wisdom (so we assume), and who could direct us in the use of advanced technology for the ultimate good of mankind? Since the gods may be our salvation, we *want* to believe in them, whether we realize it or not.

(5) A characteristic of many ancient myths is the twin godhead: the benefactor for Good and the scapegoat for Evil. In the new mythology, the astronaut-gods are primarily benefactors for good. However, they also may personify evil. A recent study undertaken by the Center for Policy Research under the direction of Dr. Clyde Z. Nunn showed that from 1964 to 1973, belief in the existence of the devil increased from 37 percent to 48 percent. A news article entitled "Demonic Believers Increase" says that Dr. Nunn "attributes the growing popular belief in the devil to a mood of 'uncertainty and stress, when things seem to be falling apart and resources seem limited for coping with it.' Nunn suggested that in a fearful world people tend to look for 'scapegoats' such as the devil." (In the light of this study, it is interesting to note some of the titles published about ancient astronauts, such as Eric Norman's *Gods, Demons and Space Chariots* and *Gods and Devils from Outer Space.*)

(6) I think it is fairly certain that two other widespread beliefs have contributed greatly to von Däniken's success: One is that something can't be printed unless it's true. The other is that if von Däniken is wrong, then scientists and theologians will come out and refute him. While corresponding with a number of scientists and editors of scientific publications, I found that most academicians, of those that were certainly capable of refuting von Däniken, simply thought it beneath their dignity to do so. The idea, to them, was not to dignify such a subject as ancient astronauts by discussing it. This attitude accomplished only one thing. It helped to enhance von Däniken's popularity.

As I mentioned at the beginning of this article, the von Däniken books have sold an astonishing thirty-four million copies. This total exceeds the twenty-five million figure of Dr. Benjamin Spock's *The Common Sense Book of Baby and Child Care*, the twelve million total of Grace Metalious's *Peyton Place*, and the twelve million copies sold of

Valley of the Dolls by Jacqueline Susann. So, whether you consider von Däniken "nonfiction" (as his publishers declare) or "fiction science," he is, as a matter of historical fact, one of the most successful authors of all time. And, although he may not have told us very much that is true about our past; he has done something that is, perhaps, more important than that. The whole ancient astronaut controversy should leave us with new insights about ourselves. If we do *not* learn anything from this experience—it could be that the world is in even more trouble than anyone thought.   •

Chariots of the Gullible

William Sims Bainbridge

Erich von Däniken's theory that human civilization is the legacy of ancient visitations by extraterrestrial beings is a leading example of a host of similar notions that have become surprisingly popular in recent years. This article will not add to the chorus of denunciation that answered von Däniken's irresponsible claims (Story 1977; Omohundro 1976; Wilson 1970, 1975; Lunan 1974). Rather, it will begin to answer the important question of why so many people react favorably to the myth of ancient astronauts. The data come from a pair of questionnaire studies completed with the help of 235 university students.

Questionnaire A (N = 114 students): On the first day of the study, the students were shown the film *In Search of Ancient Astronauts*, which presents von Däniken's basic ideas in a clear and vivid manner. The following day, I gave a short lecture presenting the main tenets of *biorhythm theory*, another pseudoscientific doctrine. Then students filled out Questionnaire A, which focused on biorhythms but also included items measuring students' acceptance or rejection of von Däniken's theory.

The findings reported in this article are expressed by three standard measures of statistical association: Pearson's *r*, Kendall's tau, and gamma. Although they are computed with different formulas, they can be interpreted in the same way. The coefficients range from -1.00 through zero to $+1.00$. A strong *positive* coefficient linking questionnaire statements X and Y indicates that people who agree with X tend to agree with Y also. A strong *negative* coefficient indicates that people who agree with X tend to disagree with Y. Coefficients close to zero indicate that there is no relationship between X and Y.

Questionnaire B (N = 121 students): Several months later, a similar two-day sequence was carried out. First, students were shown *In Search of Ancient Astronauts* so they would be familiar with von Däniken's ideas. On the second day, students were given Questionnaire B, which is rather long and contains items designed to test a number of theories. After each questionnaire had been completed, I gave a lecture debunking von Däniken, so that students would be informed rather than deceived by the material presented in the studies.

The 235 student volunteers were recruited from two sections of a low-level university course in sociology. Students majoring in the social sciences are overrepresented (about 25 percent of each class); but the course is frequently used as a distribution credit by students who are not even interested in sociology, and a majority of those taking it have not yet declared a college major. Analysis of the data from both questionnaires indicated that students' academic interests and experience did not seem to influence their opinions on von Däniken. For example, an item in Questionnaire A revealed that the theory of ancient astronauts was equally well accepted by those who preferred either the social sciences or the physical sciences (tau = −0.01). While not perfectly representative of our student body, this large group shows great diversity in interests and opinions. A random sample of the general population would have been ideal, of course; but college students are more than merely expedient and inexpensive research subjects. They are also one of the most interesting segments of the American public. A recent national poll (Gallup 1976) found that persons in the 18–24 age group were twice as likely as older citizens to believe in astrology. Once we learn why many students accept occult and pseudoscientific beliefs, we can design efficient research to determine whether the same factors explain belief among nonstudents.

The questionnaires were designed to test the comparative success of four types of theory in explaining acceptance of von Däniken's space-age mythology. Frequently used to explain many different kinds of deviant behavior, these four types are: strain theory, control theory, cultural-deviance theory, and trait theory (Stark 1975). Travis Hirschi has described the first three concisely:

> According to *strain* or motivational theories, legitimate desires that conformity cannot satisfy force a person into deviance. According to *control* or bond theories, a person is free to commit delinquent acts because his ties

to the conventional order have somehow been broken. According to *cultural-deviance* theories, the deviant conforms to a set of standards not accepted by a larger or more powerful society. (Hirschi 1969:3)

Trait theory holds that persons deviate from conventional standards because of individual characteristics, whether innate or acquired. These traits may be either handicaps that prevent them from behaving like other people or unusual desires and perceptions that compel them to violate norms. There are several variants of each type of theory, and here we will restrict ourselves to a few of those with the greatest relevance and plausibility. It is convenient to discuss the major types in pairs, beginning with *strain* and *control*.

Strain Theory and Control Theory

Strain theory holds that a person will deviate from standards of conventional or correct belief when he suffers from personal unhappiness caused by a failure to satisfy normal desires within the context of an ordinary life. Unhappiness and psychological tension may drive people to seek new alternatives, to accept exotic beliefs, as well as to experiment with novel lines of action (Smelser 1963; Lofland and Stark 1965).

In Table 1 we can see that strain theory apparently fails to explain belief in von Däniken among our students. In this table and in the others based on Questionnaire B, we show the associations (expressed by Kendall's tau) between several questionnaire items and agreement with two logically opposite statements: "Von Däniken's theory of ancient astronauts is probably true" and "Von Däniken's theory of ancient astronauts is probably false." Thirty-four (28 percent) of the 121 students had agreed with the first statement, while 37 (33 percent) agreed with the second. Naturally, people who agree with one of these statements tend to disagree with the other (tau = -0.75, gamma = -0.90). By including both statements in our tables, we provide important confirmation of apparently significant findings. To be at all convincing, a relationship between any questionnaire item and acceptance of von Däniken's theory has to show two statistically significant coefficients of opposite sign, for example, a strong positive correlation with the opinion that von Däniken's theory is true and an approximately equal negative correlation with the opinion that it is false.

The first item in Table 1 is a question frequently used by the Gallup Poll: "On the whole, would you say you are satisfied or dissatisfied with

---

TABLE 1

Correlations (tau) Between Opinions About
von Däniken and Five Strain Variables

(N = 121 students)

Strain Variable	Von Däniken's Theory Is True	Von Däniken's Theory Is False
Student is "satisfied with the future facing you and your family."	−0.10	0.11
"I am often bothered by the feeling of loneliness."	0.10	−0.07
"I am fairly satisfied with the progress I am making at college."	−0.03	0.05
"I often wonder about the meaning and purpose of life."	0.06	0.00
"I am a basically happy person."	−0.05	0.02

(In most cases, the student was asked to respond to the statement on a five-point agree-disagree scale.)

---

the future facing you and your family?" There is a slight tendency for satisfied students to reject the idea that von Däniken's theory is true (tau = −0.10) and to accept the idea that it is false (tau = 0.11). However, these coefficients are extremely small. Statistical analysis shows that we really cannot have any confidence in figures this low, and we would have to see numbers near 0.20 before we could be reasonably sure that dissatisfaction was in fact a cause of belief in ancient astronauts. The four other measures of strain or unhappiness show even smaller coefficients.

Although strain theory has proved useful in explaining many kinds of deviant behavior, control theory is often more successful. Control theory holds that persons who are strongly tied to the conventional intellectual establishment will be prevented from accepting unconventional speculations. Conversely, people who are less strongly attached will be free to adopt deviant beliefs. A student who is committed to college, who is deeply involved in his studies, and who has received much instruction notions of von Däniken. Enthusiasm for conventional science and a positive attitude toward technological progress should indicate that a person is committed to the standard intellectual establishment against which von Däniken rails with such vehemence and therefore that he will reject such a deviant belief.

Unfortunately for control theory, Tables 2 and 3 show that these apparently reasonable propositions have no power to explain acceptance of von Däniken's theory. Students who were seniors and about to graduate from college were *not* more likely than freshman students to reject the theory. It did not matter whether students had taken courses in astronomy, anthropology, ancient history, social science, or physical science.

TABLE 2

Correlations (tau) Between Opinions About
von Däniken and Eight College (Control) Variables

(N = 121 students)

College Variable	Von Däniken's Theory Is True	Von Däniken's Theory Is False
Year in College (Freshman–Senior)	0.02	0.01
"I have not yet taken any college courses in astronomy."	−0.03	0.09
"I have not yet taken any college courses in anthropology."	−0.04	0.09
"I have not yet taken any college courses in ancient history."	−0.06	0.04
"I have already taken several courses in the social sciences."	−0.06	0.14
"I have already taken several courses in the physical sciences."	0.08	0.03
Number of correct answers on the astronomy quiz.	0.04	0.02
Student agrees "a college education is pretty much a waste of time."	0.02	0.02

Questionnaire B included a simple eight-item true-false astronomy quiz, designed to measure the student's basic knowledge of this subject. Students ignorant of astronomy were not more likely to accept the theory than those who were somewhat knowledgeable. The final item in Table 2 measures the student's personal commitment to college; those who felt college was a waste of time were not the slightest bit more likely to accept the theory.

Similarly, positive attitudes toward science and technology seemed to have no power to inhibit belief. If anything, Table 3 shows a slight tendency for students who hold favorable attitudes toward science and technology to accept the idea of ancient astronauts. But the coefficients are so small that we must conclude there is no significant relationship.

TABLE 3

Correlations (tau) Between Opinions About
von Däniken and Attitudes Toward Science and Technology

(N = 121 students)

Attitude	Von Däniken's Theory Is True	Von Däniken's Theory Is False
"Science has done a lot more good than harm for the world."	0.11	−0.09
"The potential dangers of nuclear energy are outweighed by its potential benefits."	−0.03	0.07
"Machines have thrown too many people out of work."	0.02	0.07
"Technology does more good than harm."	0.07	−0.14
"It would be nice if we would stop building so many factories and go back to nature."	0.04	0.03
"Technology has made life too complicated."	−0.12	0.11

The failure of strain and control theories is not only sociologically interesting but also surprising and shocking. Both theories have shown their worth in explaining other kinds of deviance, such as crime, drug abuse, and even mental illness (Stark 1975; Faris and Dunham 1939). The failure of control theory is probably most remarkable. Apparently our university does not give students the knowledge to protect them from intellectual fraud. Of course it is rare for a college professor to mention von Däniken. Teachers of anthropology and ancient history never bother to refute the theory that human culture was received from ancient astronauts. Astronomers seldom debunk astrology or discuss UFOs. Psychology textbooks do not contain chapters on ESP or on such exotic spiritual practices as Yoga, Zen, or Transcendental Meditation. Thus, the failure of control theory may simply reflect a failure of higher education.

The failure of control theory may explain the failure of strain theory. If the intellectual establishment fails to define truth and to enforce conformity, then no special motivation is required to explain deviance. If there is no penalty for believing in unsubstantiated speculations, then anyone might do so. Strain theory explains why some people break free from social control. If there is no control, then there is no need to break free.

Cultural-Deviance Theory and Trait Theory

According to cultural-deviance theories, a person will believe von Däniken because he belongs to a subculture, to a group or social network of people who share beliefs and values that favorably dispose them to believe. There are at least four subcultures that might be important influences in causing belief in ancient astronauts: antiscience, the occult, traditional religion, and the youth counterculture. We used Table 3 to show the inadequacy of control theory, but it also shows that belief in von Däniken is not simply a reflection of the antiscience and antitechnology cultural trends that have emerged recently in the United States and Europe.

Table 4 presents data from Questionnaire A showing the associations between belief in von Däniken and belief in five other theories. Students were asked to indicate their opinion of the chance that each theory was true. It is not surprising that believers in ancient astronauts also think "there is intelligent life on other planets." Much more interesting is the fact that they also tend to accept the pseudoscientific biorhythm theory and tend to believe in ESP. These two strong associations support the impression that von Däniken's theory is part of a generalized occult subculture.

TABLE 4

Correlations (Pearson's *r*) Between Acceptance of
von Däniken's Theory and Five Other Theories

(N = 114 students)

Acceptance of Other Theories	Von Däniken's Theory Is True
There is intelligent life on other planets.	0.43*
Biorhythm theory is true.	0.38*
Extrasensory perception exists.	0.52*
Miracles actually happened just as the Bible says they did.	0.01
Darwin's theory of evolution is true.	0.01

* Significant beyond the 0.001 level; the others are not significant. (Students were asked to indicate the chance each theory was true on an eleven-point scale, marked from "0%" = no chance that the theory is true, to "100%" = absolute certainty that the theory is true.)

The insignificant correlation ($r = 0.01$) with belief that "miracles actually happened just as the Bible says they did," suggests that the traditional religious culture is unrelated to belief in chariots of the gods. While von Däniken draws on a variety of religious sources in his books, some of the most intense criticism of his work has come from religious writers (Wilson 1970, 1975). Three items in Questionnaire B measured involvement in traditional religious culture: "I definitely believe in God," "God has a very powerful influence on my life," and "Much of the time, suffering comes about because people don't obey God." Added together, these three produced a reliable index of conventional religion. This index showed no relationship with acceptance of von Däniken's theory (tau = -0.01), or with rejection of it (tau = 0.07). The final item in Table 4, the opinion that "Darwin's theory of evolution is true," measures *both* acceptance of modern science and rejection of fundamentalist religion. Persons who believe in biblical miracles tend to reject Darwin's theory ($r = -0.46$). It is clear that neither the culture of modern science nor the culture of traditional religion is related either positively or negatively to acceptance of von Däniken's theory.

Table 5, based on data from Questionnaire B, confirms that the idea of ancient astronauts is part of a generalized occult and pseudoscientific subculture. Very strong associations link it with belief in UFOs, astrology, ESP, and such exotic spiritual practices as Yoga, Zen, and Transcendental Meditation. Only the last item in the table is unrelated. This suggests that believers did not directly learn von Däniken's theory from participation in the subculture. Apparently, the subculture rests on some very basic modes of thought that encourage acceptance of any new belief that fits the pattern.

TABLE 5

Correlations (tau) Between Opinions About
von Däniken and Other Occult and Pseudoscientific Theories

(N = 121 students)

Other Occult Statements	Von Däniken's Theory Is True	Von Däniken's Theory Is False
"UFOs are probably spaceships from other worlds."	0.54	−0.59
"UFOs are probably illusions."	−0.42	0.40
"I myself have had an experience which I thought might be an example of extrasensory perception."	0.30	−0.35
"Extrasensory perception probably exists."	0.39	−0.35
"Some Eastern practices, such as Yoga, Zen, or Transcendental Meditation, are probably of great value."	0.36	−0.23*
"There is much truth in astrology."	0.40	−0.38
"I was not really at all familiar with von Däniken's theory about ancient astronauts until I saw the movie yesterday."	−0.09**	0.11**

* Significant beyond the 0.005 level
** Not significant

(All other coefficients are significant beyond the 0.001 level.)

One explanation frequently given for the contemporary occult revival blames the influence of the youth counterculture that emerged in the mid-1960s (Wuthnow 1976, forthcoming). Table 6 examines the relationship between acceptance of von Däniken's theory and four plausible measures of involvement in this subculture. Contrary to what the hypothesis would lead us to expect, we find that left-wing politics, drug use, severe criticism of the government, and opposition to police surveillance of dissidents show no connection to acceptance of the theory.

TABLE 6

Correlations (tau) Between Opinions About
von Däniken and Youth Counterculture Items

(N = 121 students)

Counterculture Items	Von Däniken's Theory Is True	Von Däniken's Theory Is False
Student is politically left-wing.	0.05	−0.07
"I have experienced being 'high' on drugs at least once, not counting medical uses."	0.02	−0.07
"Our form of government needs a major overhaul."	0.08	−0.02
Disagreement with the idea that "the police should keep their eye on members of revolutionary groups."	−0.10	0.08

Cultural theories are easily integrated with trait theories. A person's behavior may be determined by his characteristics, but those characteristics are often a result of involvement with cultural alternatives. Unfortunately, this means that the task of pulling the two kinds of theory apart is frequently very difficult. Perhaps Tables 4 and 5 indicate nothing about a subculture but reflect the fact that persons with a certain character type are susceptible to deviant ideas.

Table 7 lists six traits and opinions that might indicate underlying characteristics of the person. We see, first of all, that males and females respond equally favorably to the theory of ancient astronauts.

TABLE 7

Correlations (tau) Between Opinions About
von Däniken and Personal Characteristics of Student

(N = 121 students)

Personal Trait or Opinion	Von Däniken's Theory Is True	Von Däniken's Theory Is False
Sex of student is female	−0.01*	0.08*
"It is good to live in a fantasy world every now and then."	0.18	−0.17
Rank of science fiction among five kinds of adventure fiction.	0.19	−0.18
"It is all right for an unmarried couple to have sexual relations."	0.23	−0.20
Student has a low grade-point average (GPA)	0.20	−0.20
Number of "true" responses on true-false astronomy quiz.	0.01*	−0.09*

* Not significant. (All others are significant beyond the 0.01 level.)

There is a weak but statistically significant association between acceptance of the theory and agreement with the statement that "it is good to live in a fantasy world every now and then." One item in Questionnaire B asked students to rank five kinds of adventure fiction, from the one they liked best to the one they liked least. Preference for science fiction correlates weakly with acceptance of the theory. Although we should not rely too heavily on these weak associations, one possible interpretation is that some people accept von Däniken's theory because they characteristically respond favorably to stories of fantasy.

There is a moderate but respectable association between acceptance of the theory and the feeling that "it is all right for an unmarried couple to have sexual relations." To some extent this item reflects involvement in the youth counterculture and estrangement from traditional religious culture. But neither of these cultures relates to acceptance of von Däniken. The item on sexual freedom may therefore reflect an important character trait as well. The trait might be a readiness to follow feelings and desires rather than be guided by facts and daunted by practical limitations. Such a trait could find expression in acceptance of emotionally stimulating theories as well as in approval of sexual freedom.

Another moderate association links acceptance of the theory with low grade-point average. Because I wanted to protect my students' confidentiality, I did not look up their actual grades at the university but relied on their responses to an item on Questionnaire B. Inaccuracy in student reporting of GPAs may have reduced the apparent magnitude of the association. Grade-point average reflects several factors, most notably intellectual ability and motivation to achieve. Earlier we found that acceptance of the theory was not related to a number of school variables, including responses to the statement that "a college education is pretty much a waste of time." Students who feel this way almost certainly lack motivation to achieve in college. Therefore, it is probable that here GPA is just a weak indicator of intellectual ability. Less intelligent students are more likely to believe in ancient astronauts. Is this trait equivalent to gullibility?

The true-false astronomy quiz in Questionnaire B can be used to test the hypothesis that some students simply accept anything they are told. The quiz consisted of eight statements, four of which were in fact true, and four false. The average score of students was 55 percent correct. Students who are "gullible" might be more likely than others to give "true" responses on the quiz, accepting its statements whether true or false. As Table 7 shows, however, the number of "true" responses was not associated with acceptance of von Däniken's theory. Other tests of this hypothesis, based on measurement of students' tendencies to agree or disagree with several other questionnaire items, and too complex to present in detail here, gave the same negative result.

A very different exploration of "gullibility" was carried out with Questionnaire A that focused on the pseudoscientific biorhythm theory. Like a simplified brand of astrology, biorhythms is an occult means for predicting one's future fortunes. Using the sole empirical input of a person's date of birth, it estimates for any day whether the person will be above or below average in three aspects of his condition: physical, emotional, and intellectual. As explained elsewhere (Bainbridge 1978a), each student was given a personalized copy of Questionnaire A, including on its last page what purported to be his own biorhythms for that day. Although these were in fact fake biorhythms, determined by the flip of a coin, students accepted them as correct descriptions far more often than would be expected by chance—74.7 percent of the time, rather than the expected 50 percent.

As Table 8 shows, students who believed the biorhythm theory were especially likely to accept the fake statements of their rhythms for that day. Students who accepted von Däniken's theory and students who believe in ESP were also likely to accept the fake rhythms. We might think that the seemingly authoritative movie and my lecture had persuaded impressionable students to believe the theories presented in them. But no movie or lecture had presented the idea of extrasensory perception, and the highest coefficient in the table (0.30) indicates that ESP believers were likely to accept the fake rhythms. Apparently, some enduring characteristic, perhaps a habitual mode of thought, facilitated acceptance of the fake rhythms. It was not simply a matter of some students being more susceptible to authoritative appeals. Whether this characteristic is a true trait of individual gullibility or an inferior style of intellectual response taught by the occult subculture is a question for future research.

TABLE 8

Correlations (Pearson's *r*) Between Acceptance of
Fake Biorhythms and Acceptance of Six Theories

(N = 114 students)

Acceptance of Theory	Student Accepts the Fake Biorhythms as Accurate Descriptions of His Own Condition
Biorhythm theory is true.	0.26*
Von Däniken's theory is true.	0.26*
Extrasensory perception exists.	0.30*
There is intelligent life on other planets.	0.16
Miracles actually happened just as the Bible says they did.	0.15
Darwin's theory of evolution is true.	−0.02

* Significant beyond 0.005 level. (All others are not significant.)

Conclusions

We have seen that two popular theories of deviance fail to explain acceptance of von Däniken's theory by the students. *Strain theory* fails because dissatisfied students were not especially likely to accept the theory. *Control theory* fails because acceptance is not related to the number or kinds of college classes the student has taken, nor to the student's evaluation of college, nor to the student's attitudes toward science and technology.

Both *cultural-deviance theory* and *trait theory* seem much more promising, although we cannot yet specify the exact sources of von Däniken's support. The strongest associations linked belief in ancient astronauts to other occult and pseudoscientific beliefs. Weaker but significant correlations suggested that personal habits of thought were also involved. The most obvious conclusion would be that the occult subculture was responsible for poor individual judgment. This interpretation rests on a consideration of the nature of subcultures.

The word *subculture* is used in two very different senses, referring to two conceptually distinct kinds of social phenomena: (1) a cohesive *group of people* sharing opinions, values, and habits not shared by members of the larger society; (2) a coherent *set of ideas* that are not necessarily held by any specific group of people but are spread throughout society and appeal to various individuals. An example of the former would be an ethnic subculture in which people are closely linked by ties of language, tradition, and family structure. The occult subculture is probably an example of the latter. Although cohesive groups frequently emerge within the occult, as a whole the subculture is not tightly organized and recruits people from many different segments of society (Lofland 1966; Evans 1973; Bainbridge 1978b). The occult is a *set of ideas* that share common root qualities. This fact explains the high coefficients in Tables 4 and 5. The subculture is not based in any cohesive *group of people*. Individuals may find the occult attractive for a large number of different reasons. We would expect to find relatively weak associations linking these factors with acceptance of occult beliefs, perhaps considerably weaker than the associations connecting the beliefs into a set. Clearly, the findings reported in this article demonstrate the need for further research to overcome our ignorance of the sources of deviant beliefs.

We are also ignorant of the consequences of belief. Certainly belief in quack medicine can lead to tragedy. The consequences of belief in nonmedical occult and pseudoscientific theories are less clear. Although acceptance of von Däniken was associated with low grade-point average in this study, students were apparently quite able to complete college without ever being forced to give up favorable attitudes toward the occult. In a recent sociological book on the development of modern space rocketry (Bainbridge 1976), I speculated that popular movements excited by notions of flying saucers or other extraterrestrial fantasies might provide significant financial and political support for real future space projects. Some of the miscellaneous findings from Questionnaire B appear to confirm this conjecture. Students who accepted the theory of ancient astronauts were likely to favor increased expenditures for the space program (tau = 0.24) and tended to feel that the space program will have a big payoff for the average person (tau = 0.23). These students seem ready to take the next logical steps beyond von Däniken's doctrine. They give strong support (tau = 0.35) to the proposition that "we should attempt to communicate with intelligent beings on other planets, perhaps using radio."

References

Anonymous 1973. "Anatomy of a World Best-Seller: Däniken's Message from the Unknown." *Encounter* 41 (August): 8–17.

Bainbridge, William Sims 1976. *The Spaceflight Revolution.* New York: Wiley-Interscience.

—— 1978a. "Biorhythms: Evaluating a Pseudoscience." Skeptical Inquirer 2 (Spring/Summer): 40–56.

—— 1978b. *Satan's Power.* Berkeley: University of California Press.

Däniken, Erich von 1971. *Chariots of the Gods?* New York: Bantam Books.

—— 1972. *Gods from Outer Space.* New York: Bantam Books.

—— 1974. *The Gold of the Gods.* New York: Bantam Books.

—— 1975a. *Meine Welt in Bildern.* Munich: Knaur.

—— 1975b. *Miracles of the Gods.* New York: Dell.

Evans, Christopher 1973. *Cults of Unreason.* New York: Dell.

Faris, Robert E., and Dunham, H. Warren 1939. *Mental Disorder in Urban Areas.* Chicago: University of Chicago Press.

Gallup Poll 1976. "13 Million Look to Stars for Help in Conducting Daily Affairs." *Gallup Opinion Index* #132 (July): 25–27.

Glock, Charles Y., and Bellah, Robert N., eds. 1976. *The New Religious Consciousness.* Berkeley: University of California Press.

Hirschi, Travis 1969. *Causes of Delinquency.* Berkeley: University of California Press.

Landsburg, Alan 1974. *In Search of Ancient Astronauts*. A motion picture produced by Alan Landsburg, directed by Harold Reiml, narration written by Don Ringe, a Tomorrow Entertainment, Inc. Presentation.

Landsburg, Alan, and Landsburg, Sally 1975. *The Outer Space Connection*. New York: Bantam Books.

Lofland, John 1966. *Doomsday Cult*. Englewood Cliffs, N.J.: Prentice-Hall.

Lofland, John, and Stark, Rodney 1965. "Becoming a World-Saver: A Theory of Conversion to a Deviant Perspective." *American Sociological Review* 30: 862–874.

Lunan, Duncan 1974. *Interstellar Contact*. Chicago: Henry Regnery.

Oberbeck, S. K. 1973. "Deus ex Machina." *Newsweek* (October 8): 104–106.

Omohundro, John T. 1976. "Von Däniken's Chariots: A Primer in the Art of Cooked Science." *The Zetetic* 1 (Fall/Winter): 58–68.

Ostriker, Alicia 1973. "What if We're Still Scared, Bored, and Broke?" *Esquire* (December): 238–330.

Smelser, Neil J. 1963. *Theory of Collective Behavior*. New York: Free Press.

Stark, Rodney 1975. *Social Problems*. New York: Random House.

Story, Ronald 1977. "Von Däniken's Chariots of the Gods." *The Zetetic* 2 (Fall/Winter):22-35.

Tiryakian, Edward A. 1972. "Toward the Sociology of Esoteric Culture." *American Journal of Sociology* 78: 491–512.

Truzzi, Marcello 1972. "The Occult Revival as Popular Culture." *The Sociological Quarterly* 13 (Winter): 16–36.

Wilson, Clifford 1970. *Crash Go the Chariots*. New York: Lancer.

———— 1975. *The Chariots Still Crash*. New York: Signet.

Wuthnow, Robert 1976. *The Consciousness Reformation*. Berkeley: University of California Press.

———— forthcoming. *The Post-Christian Periphery*. Berkeley: University of California Press.   ●

Investigating The Sirius Mystery

Ian Ridpath

Did amphibious beings from the star Sirius visit the earth 5,000 or more years ago and leave advanced astronomical knowledge that is still possessed by a remote African tribe called the Dogon? This astonishing claim was put forward in 1976 by Robert Temple in his "ancient astronaut" book, *The Sirius Mystery*. An astronomer, familiar with the Sirius system, would say no, because astronomical theory virtually precludes the possibility that Sirius is a suitable parent star for life or that it could have habitable planets. But most of Robert Temple's readers would not know enough astronomy to judge the matter for themselves. Neither would they find the relevant astronomical information in Temple's book, most of which consists of brain-numbing excursions into Egyptology. (Isaac Asimov has been quoted by Temple as having said that he found no mistakes in the book; but Temple did not know that the reason for this, according to Asimov, was that he had found the book too impenetrable to read!*) Even the BBC-TV Horizon investigation on ancient astronauts (broadcast as part of the PBS "Nova" series in the United States), which did an otherwise excellent demolition job on the more extreme fantasies of Erich von Däniken, left the Sirius problem unanswered because of its extreme complexity. Yet an answer is needed, because the Dogon legends about a companion to Sirius are claimed to originate before any terrestrial astronomer could have known of the existence of Sirius B, let alone its 50-year orbit or its nature as a tiny, con-

*Editor's note: See Asimov's essay, "The Dark Companion," in his *Quasar, Quasar Burning Bright* (Doubleday, 1978), in which he says he is embarrassed by his stupidity in not specifying that his comment, made only "to get rid of him [Temple] and to be polite," not be quoted. "I assure you I will never be caught that way again."—K.F.

densed white dwarf star, all of which the Dogon allegedly knew. So what is the truth about the Dogon and Sirius? Does astronomical and anthropological information omitted by Temple help us to resolve this most baffling of all ancient astronaut cases?

First, let's recap Temple's story. At the center of the mystery are the Dogon people living near Bandiagara, about 300 kilometers south of Timbuktu, Mali, in western Africa. Knowledge of their customs and beliefs comes from the French anthropologists Marcel Griaule and Germaine Dieterlen, who worked among the Dogon from 1931 to 1952. Between 1946 and 1950 the Dogon head tribesmen unfolded to Griaule and Dieterlen the innermost secrets of their knowledge of astronomy. Much of this secret lore is complex and obscure, as befits ancient legends, but certain specific facts stand out, particularly those concerning the star Sirius, with which their religion and culture is deeply concerned. In the information imparted to the French anthropologists, the Dogon referred to a small and super-dense companion of Sirius, made of matter heavier than anything on Earth, and moving in a 50-year elliptical orbit around its parent star. The white dwarf companion of Sirius which answers to this description was not seen until 1862, when the American optician Alvan Graham Clark spotted it while testing a new telescope; the super-dense nature of white dwarfs was not realized until the 1920s. But the Dogon Sirius traditions are at least centuries old. How can we account for the remarkable accord between ancient Dogon legends and modern astronomical fact?

Temple's answer, since espoused by Erich von Däniken (of course!), was that the Dogon were told by extraterrestrial visitors. A Dogon legend, similar to many other tales by primitive people of visits from the sky, speaks of an "ark" descending to the ground amid a great wind. Robert Temple interprets this as the landing of a rocket-powered spacecraft bringing beings from the star Sirius. According to Dogon legend, the descent of the ark brought to Earth an amphibious being, or group of beings, known as the Nommo. "Nommo is the collective name for the great culture-hero and founder of civilization who came from the Sirius system to set up society on the Earth," Temple explains in his book. The Nommo were amphibious, he presumes, because water would keep them cool and absorb short-wavelength radiation from the hot star Sirius.

Much of Temple's book is devoted to establishing that the Dogon share common roots with Mediterranean peoples. This explains the central place occupied by Sirius in Dogon beliefs, because the ancient Egyptians, in particular, were also preoccupied with Sirius, basing their calendar on its yearly motion. But is there any explanation of the apparent Dogon belief in life in the Sirius system?

First, let's look at what astronomers know about Sirius to see if it is at least theoretically plausible that advanced life might have arisen in its vicinity. Sirius A, the brightest star in the night sky as seen from Earth, has a mass 2.35 times that of the sun. Its white dwarf companion, Sirius B, has a mass of 0.99 suns. Stellar evolutionary theory tells us that the most massive stars burn out the quickest, so that originally Sirius B must have been the more massive of the two, before burning out to become a white dwarf. Probably Sirius B spilled over some of its gas onto Sirius A during its aging process, so that the original masses of the two stars were approximately the reverse of what we see today.

A star with twice the sun's mass, as Sirius B probably had, can live for no more than about 1,000 million years before swelling up into a red giant; this does not seem long enough for advanced life to develop. But had life evolved, it would have disappeared during the red giant stage of Sirius B, when any nearby planet would have been roasted by the star's increased energy output, followed by a stellar gale for at least 100,000 years as hot gas streamed from Sirius B to Sirius A. During this mass transfer the two stars would have moved apart, thereby destabilizing the orbits of any planets in the system. According to observations of Sirius B as analyzed by H. L. Shipman of the University of Delaware, Sirius B has been a cooling-down white dwarf for at least 30 million years. Sirius B is now emitting soft x-rays, so that life in the region of Sirius would not be very pleasant today. But in any case, Robert S. Harrington of the U.S. Naval Observatory has recently shown that planetary orbits in the "habitable" zone around Sirius, defined as the region in which water would be liquid, are unstable. So there are unlikely to be any amphibious beings living on planets in the Sirius system today, if indeed any such beings ever lived there.

Temple offers one prediction which allows a test of his theory. In his book he says: "What if this is proven by our detecting on our radio telescopes actual traces of local radio communications?" To help in my investigation of the Sirius mystery, I asked radio astronomers Paul Feldman at the Algonquin radio observatory, Canada, and Robert S. Dixon at the Ohio State University radio observatory, both of whom are carrying out searches for extraterrestrial signals, to listen to Sirius. They would normally have paid the star no attention, because of the extreme unlikelihood of its supporting life. In April 1977 both radio astronomers listened to Sirius on different wavelengths, without detecting any artificial signals.

With this information in mind, let's look more skeptically at the Dogon legend. Immediately, we encounter a surprise: the Dogon maintain that Sirius has two companions, not one. These companions have male and female attributes, respectively. It seems that they are not to be interpreted literally as stars, but as fertility symbols. Nowhere is this better shown than in a Dogon sand diagram of the complete Sirius system, shown in the illustration redrawn here from a paper by Griaule and Dieterlen. Its description, given in the caption from information by Griaule and Dieterlen, is clearly symbolic; Temple chooses to interpret it literally. On pages 23 and 25 of his book he gives his own modified version of this diagram, retaining the symbol for Sirius, one of the positions of Sirius B, and the surrounding oval; all else is omitted. He then interprets the surrounding oval meant to represent "the egg of the world," as the elliptical orbit of Sirius B around Sirius A, even though the symbol equated with Sirius B is drawn as lying within the oval, not on it. This is Temple's basis for saying that the Dogon "know" Sirius B orbits Sirius A in an ellipse.

The Dogon are also supposed to know that Sirius B orbits every 50 years. But what do they actually say? Griaule and Dieterlen put it as follows: "The period of the orbit is counted double, that is, one hundred years, because the Siguis are convened in pairs of 'twins,' so as to insist on the basic principle of twinness." The Sigui ceremony referred to is a ceremony of the renovation of the world that is celebrated every 60 years (not 50). And the "twinness" referred to here is an important Dogon concept which explains why they believe Sirius must have two companions.

Is there any astronomical evidence that Sirius has more than one companion star? Some astronomers in the 1920s and 1930s thought they had glimpsed a third member of the Sirius system, but new and more accurate observations reported in 1973 by Irving W. Lindenblad of the U.S. Naval Observatory, Washington, D.C., showed no evidence of a close companion to either Sirius A or Sirius B.

The whole Dogon legend of Sirius and its companions is riddled with ambiguities, contradictions, and downright errors, at least if we try to interpret it literally. But what can we make of the Dogon statement

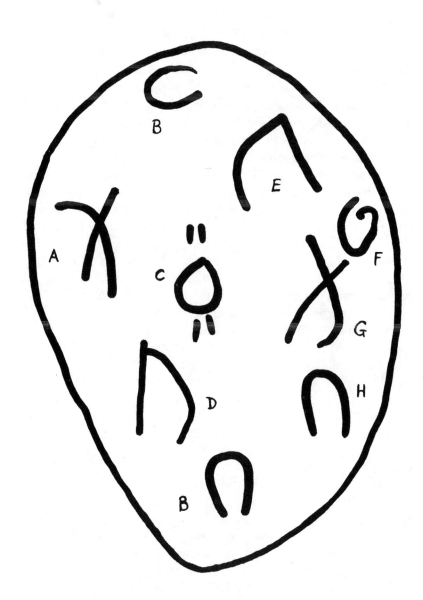

Dogon sand drawing of the complete Sirius system, after Marcel Griaule and Germaine Dieterlen. A, Sirius; B, *po tolo*, the object equated with Sirius B, shown in two positions; C, *emme ya*, the sun of women, equated with Sirius C; D, the Nommo; E, the Yourougou, a mythical male figure destined to pursue his female twin; F, the star of women, a satellite of *emma ya*; G, the sign of women; H, the sex of women, represented by a womb shape. The whole system is enclosed in an oval, representing the egg of the world.

that Sirius B is the smallest and heaviest star, consisting of a heavy metal known as *sagala*? Sirius B was certainly the smallest and heaviest star known in the 1920s, when the super-dense nature of white dwarfs was becoming understood; the material of which white dwarfs are made is indeed compressed more densely than metal. Now, though, hundreds of white dwarfs are known, not to mention neutron stars, which are far smaller and denser. Any visiting spaceman would certainly have known about these, as well as black holes.

Perhaps one would forgive Robert Temple for believing that the Dogon had been visited by men from Sirius if their legend specifically stated so. But it does not! Nowhere in his 290–page book does Temple offer one specific statement from the Dogon to substantiate his ancient astronauts claim. The best he does is on page 217, where he reports that the Dogon say: "*Po tolo* [Sirius B] and Sirius were once where the Sun now is." Of this ambiguous statement, Temple comments: "That seems as good a way as any to describe coming to our solar system from the Sirius system, and leaving those stars for our star, the Sun." But this cannot conceal the fact that the whole Sirius "mystery" is based on Temple's own unwarranted assumption.

The parts of Dogon knowledge that are admittedly both ancient and profound, particularly the story of Nommo and the concept of twinning, are the parts that bear least relation to the true facts about Sirius. The parts that bear at least superficial resemblance to astronomical fact are most likely trimmings added in this century. Indeed, in view of the Dogon fixation with Sirius it would surely be more surprising if they had *not* grafted on to their existing legend some new astronomical information gained from Europeans, picking what fitted their purpose and ignoring the rest.

Carl Sagan has underlined how easily information gained from Westerners can be absorbed into native culture. He recounts the true case of the physician Carleton Gajdusek in New Guinea, who was approached by a scientific colleague who had found that some local natives believed that a certain disease was transmitted in the form of an invisible spirit that entered the skin of a patient. The native informant had sketched with a stick in the sand a circle outside which, he explained, was black, and inside which was light. Within the circle the informant drew a squiggly line to represent the appearance of these invisible malevolent spirits. How did the natives get such an astounding insight into the transmission of disease by microbes? Years earlier, Gajdusek himself had shown the

natives the appearance of a disease-causing germ through his micro-
scope, and the sand drawing was simply the natives' recollection of this
deeply impressive sight.

It is all too easy for Westerners to think of African tribes as isolated,
uneducated, and ignorant. But the Dogon are not isolated. They live near
an overland trade route, as well as close to the banks of the Niger River,
an important channel of trade. Any number of travelers could have come
into their midst, or Dogon tribesmen could have journeyed to the coast,
where they might have met astronomically informed seamen. The Dogon
have been in contact with Europeans since at least the late nineteenth
century.

Nor are they uneducated and ignorant. Peter and Roland Pesch of
the Warner and Swasey Observatory in Ohio have pointed out that
French schools have existed in the Dogon area since 1907. Dogon
tribesmen wishing to pursue their education have been able to do so in
nearby towns. Then there are missionaries, who would naturally be interested in the legends of the natives. Missionaries from the White
Fathers made contact with the Dogon in the 1920s. It is tempting to speculate that certain of the more specific details about Sirius B were grafted
onto the existing Sirius legend at that time, because it was in the 1920s
that astronomers were discovering the true nature of Sirius B as a tiny,
super-dense star, and white dwarfs were being accorded the same kind of
publicity as attends black holes today. Alas, there is no mention in the
missionaries' summary reports of their activities that they discussed
Sirius with the Dogon; if more detailed notes were published, these might
throw more light on the origin and antiquity of Dogon myths.

The point is that there are any number of channels by which the
Dogon could have received Western knowledge long before they were
visited by Griaule and Dieterlen. We may never be able to reconstruct the
exact route by which the Dogon received their current knowledge, but
out of the confusion at least one thing is clear: they were not told by beings from the star Sirius.

References

Harrington, R. S. 1977. *Astronomical Journal*, 82: 753.
Lindenblad, I. W. 1973. *Astronomical Journal*, 78: 205.
Pesch, P. and R. Pesch 1977. *The Observatory*, 97: 26.
Ridpath, Ian 1978. *Message from the Stars*. New York: Harper & Row.
Shipman, H. L. 1976. *Astrophysical Journal*, 206: L67.

Cult Archaeology and Biology

Anthropology Beyond the Fringe

Ancient Inscriptions, Early Man, and Scientific Method

John R. Cole

Anthropologists study countless topics, a disproportionate number of which are alien to the general public and to other scientists. And we also study *ourselves*—belief in UFOs or Bigfoot or Creationism or ESP or witches is grist for the anthropological mill and can itself be studied scientifically. There *is* science in this ecumenicism, however; and however unorthodox our subject matter, most of us distinguish between our discipline and our subject. Without contempt for believers in strange explanations of the world, we fairly often distinguish between the explanations of "insiders" and the empirical evidence we try to explain scientifically. The doctrine of "cultural relativism" is like a scientific ACLU, arguing that everyone has the right to their beliefs without censure from a Western tradition "establishment."

Still, there are weird things afoot to worry even the most cautious anthropologist in the public perception of the things we study. When a folk tradition explains the origin of the world as the result of the actions of a Great Turtle creating heaven and earth, we can respect the tradition without endorsing its scientific reality. But when people invoke the name of science to prove such claims, we need to speak out as scientists skeptical of these claims without disrespect for the people outside the scientific tradition who believe them. We can argue that "sacred cows" are not literally sacred while recognizing the underlying adaptive values of Hindu "cow worship," which is an ideological rationalization of quite reasonable and realistic ecological forces: killing every cow in India might give everyone a hamburger, but it would eliminate that country's source of fuel for fires, traction, fertilizer, construction mortar, milk and cheese, and even methane electrical generators.[1] Traditional Western ideology, then, does not have all the answers to the anthropologists familiar with alternate solutions.

But studying the question of what it means to be human leaves anthropology at the mercy of cranks, crackpots, and the politically inspired cynics—not to mention the perhaps naive sensationalists capitalizing upon irrational, antiscientific cultural currents.

Let us grant that "anything is possible," without rejecting as impossible even the weirdest claims about humanness, while retaining our demand as scientists that revolutionary claims prove themselves rather than be accepted simply because they are asserted.

Major themes of current unverified (to be charitable) anthropological science include several "paranormal" topics: "creationism" of various forms purporting to disprove Darwin; "psychic archaeology" finding sites and interpreting prehistory by séance rather than scientific excavation; acceptance at face value of ESP and magic; resuscitation of hoary Atlantis and Lemuria myths; the "ancient astronauts" syndrome popularized by von Däniken, despite dissections by Ronald Story and others; near-religious belief in pre-Columbian colonization of the Americas by Europeans and Middle Easterners as an explanation of everything noteworthy ever accomplished by Native Americans (and the cyclical "discovery" of American inscriptions and altars that "prove" such pre-Columbian contacts); claims that humans descend from an ancient amphibious female creature, for which there is no known evidence; arguments that contemporary "races" can be ordered hierarchically according to IQ when even the identification of races is an open question among anthropologists and biologists; claims such as Robert Ardrey's that humans are natural killers; and so forth. The list is nearly endless, but the political implications are often rather transparently ominous.

A final category of anthropology potentially awry is the uncritical acceptance of informants' versions of reality without empirical testing. At the 1977 meeting of the American Anthropological Association in Houston, for example, one paper reported on Haitian stories of a coffin traveling the streets under its own power. This paper said the "simplest" explanation of the "evidence" was that either it had happened or there was at least something going on unexplainable by normal science. This paper was the only one I saw reported in the Houston press. I suppose this made better news copy than hundreds of other more orthodox papers read at the meeting, but as "science" it was, to say the least, a bit lacking.

The cultural relativist viewpoint, which refuses to judge *people*, need not rule out scientific evaluations of their *claims* as objective reality. Like most scientists, anthropologists are simply not trained to judge alleged paranormal phenomena. I have this old-fashioned desire to see controlled experiments and empirical evidence rather than simple testimonials assuring me that rules of scientific evidence are outmoded or are a slavish misapplication of Western concepts to phenomena beyond the ken of brainwashed science-worshipers.

Like other sciences, and perhaps even more than some with longer traditions of "sciencing" rather than speculation, anthropology is vulnerable to exploitation by naive or crank claims and by speculations that ignore the criteria of scientific proof while still claiming the mantle of "science." Margaret Mead, for example, is one of my culture heroes, a pioneering and peerless writer and ethnographer, and a woman who has contributed immeasurably to our understanding of the world around us. But when as president of the American Association for the Advancement of Science she convinced that body to accept as an official affiliate the Parapsychological Association, I believe she gave in to the cultural relativist position of uncritical acceptance of *possibilities* rather than insisting on the scientific discipline that she has championed for half a century. Scientific claims simply cannot be voted upon, even if the voters are thoroughly respectable and sincere.

Not every strange claim in anthropology can be dealt with in a short paper, but I can at least stress that anthropology includes a strong scientific, empirical tradition: at the least, sensational claims do not represent "orthodoxy" in anthropology, for better or worse. However, I will discuss briefly two current themes in anthropology and on its fringes: the current fad for "hyperdiffusionism," which derives Native American cultures from Old World sources; and the perhaps more subtle claims for an extreme antiquity for American peoples (which sometimes go so far as to claim that humans evolved in America rather than Africa).

Erich von Däniken's claims of an extraterrestrial origin for human cultural florescence have been disproved devastatingly, but what of diffusionist claims that assert that American cultures did not evolve according to the apparent laws of cultural evolution because they were strongly influenced by European or Asian or African cultures long before Columbus? Without ruling out the possibility, or even probability, of rather small-scale transoceanic diffusion, there is simply no valid empirical evidence of claims for massive pre-Columbian diffusion between the Old and New Worlds, such as those recently revived by Barry Fell in his book *America, B.C.* (New York: Quadrangle, 1977) or Cyrus

Gordon's *Before Columbus* (New York: Crown, 1971). They resuscitate centuries-old claims and arguments repeatedly disproven, to the point that anthropologists no longer bother to argue. "Indians" simply were not too dumb to pile up dirt or stones into mounds and pyramids, despite the prejudices of early European colonialists. Early missionaries tried to prove that Native Americans were descendents of even earlier Old World colonists, sometimes to prove that their charges were worthy of respect rather than annihilation. Today there seems to be a more directly racist tinge to such claims, and in fact claims that Europeans "owned" real estate in the Americas before they were displaced by "savages" has direct political utility in the burgeoning spate of Native American lawsuits challenging European usurpation of tribal lands. Less clear-cut than this perhaps unconscious but significant racism, advocates of "ancient inscriptions" appeal to a romantic idea that a European history for the Americas is somehow more exciting and mysterious than mundane reality based on serious archaeology and physical anthropology.

Fell and his friends have found inscriptions in Celtic, Ogham, Latin, Phoenician, Egyptian, Cretan, Minoan, Greek, Carthaginian, Iberian, Libyan, and many other languages in New England, Ecuador, Canada, Oklahoma, Minnesota, Iowa, and other localities that they claim have been "ignored" by the professionals. At a 1977 "conference" in Castleton, Vermont, celebrating Fell's work, one speaker alluded to "dozens" of major Egyptian cities in Ohio systematically "covered up" by the closed-minded establishment—while offering no evidence beyond the description of some burial mounds! Fell himself said he had evidence (unrevealed) proving that Jesus was known in the Americas "as early as the fourth century A.D." He also cited "evidence" such as Ecuadorian "gold tablets" whose source he could not reveal; when I pointed out that they were from the collection of one Padre Crespi of Cuenca, Ecuador (Erich von Däniken's major source for his *Gold of the Gods* book), discounted as a naive crank by people familiar with his collection (which includes a copper toilet-tank float identified as Inca gold!), Fell refused to answer questions.

He cites the "Iowa Tablets," known where I grew up as the "Davenport Conspiracy" because of actual confessions by the original hoaxers, who used steel tools to carve slate roof slabs traced to a Davenport, Iowa, building. The Minnesota "Kensington Stone" was exposed rather convincingly as a hoax by Erik Wahlgren's book *The Kensington Stone: A Mystery Solved*. "Mystery Hill," New Hampshire, *may* not be fully understood, but its stone buildings have been traced by professional archaeologists to early historic times with an underlay of Native American artifacts dating much earlier but unrelated to European "Megalithic" cultures such as Stonehenge. Its "sacrifice stone" seems to be identical to grooved stones used throughout New England for two centuries in the production of lye and potash from wood ashes. "Celtic inscriptions" prove to be in modern Celtic, not in the form of the language used thousands of years ago at the time of alleged Celtic colonization. American Ogham inscriptions (an Irish shorthand version of Latin, consisting of patterned hash-marks) seem to be plow scratches on rocks, natural geologic fault-fractures, or outright hoaxes, such as an engraved stone

Structure at "Mystery Hill" in New Hampshire. No need to invoke European "Megalithic" cultures such as Stonehenge. From *Exploring the Unknown* (Plenum, 1979) by Charles Cazeau and Stuart Scott.

displayed at the conference with fresh scratches, which, given the type of stone, could not be older than a century or so. Other inscriptions and paintings on pottery are simply Native American pictographs or geometric designs that may resemble some Old World symbols but prove nothing. Something like a "starburst" pattern or asterisk is a worldwide design element depicting the sun or stars, not evidence of diffusion. A mark such as "///" *could* be a tally or just three lines, but certainly is not a definite Roman numeral three! Slashes, triangles, squares, and crosses are so simple to reinvent that their presence is meaningless as proof of diffusion from European languages using similar symbols.

Perhaps buried in all of this hodgepodge of "ancient inscription" claims there *are* some real Old World inscriptions, but the vast majority are either hoaxes or misinterpretations. When Fell and his cohorts accept them all, they discredit possibly *valid* diffusionist arguments, and we should avoid simple guilt by association. Yet every inscription claim that has been investigated by archaeologists, linguists, and geologists seems so far to be spurious. How long must the burden of proof lie with the non-extremists? When Fell "deciphers" an "inscription" such as that on the Canadian "Sherbrooke Stones" in at least three radically different ways in one year—and even accepts the idea that they may be ancient worm-holes in sedimentary mudstone!—how can his decipherments be taken seriously in general?[2] How can one argue with a man who identifies frost-pocking of rocks as "cuppules" with religious significance? When oral tradition, historical memory, and archaeological evidence point to the conclusion that New England's "Megalithic shrines" are simply root cellars and other historic constructions, why should we accept the torturously reasoned claim that they are anything else?

Aside from the dubious nature of these alleged inscriptions and shrines, one must ask: Why did ancient voyagers leave behind nothing but inscriptions and religious structures? Where are the habitation sites and tools and pottery one would expect from the many centuries of claimed Old World colonization? (*Not* just contact, but ongoing settlement and trade networks are being claimed.) Native Americans needed horses, sheep, cows, and pigs much more than they needed inscriptions! Significant colonization should yield such domesticated animal bones in pre-Columbian times for the archaeological record, but they do not appear. Artifacts of subsistence and not just ideology should have been left behind, given what we know about the workings of culture. An early Libyan (or whatever) colony would have been in a real position of power if it could have introduced draft animals and domesticated protein sources, as demonstrated by the Spanish Conquest. And why is there no evidence of Old World diseases before the "official" conquest—or of clear genetic admixture with Native American populations, which would be identifiable from human bone remains? According to what we know about cultural and biological processes, any significant pre-Columbian contact (that is, not just a possible stray boat or very temporary settlement such as the Vikings') would simply have had to leave evidence other than religious artifacts.

The "Early Man in the New World" argument suffers some of the same problems. People clearly were in the Americas about 15,000 years ago and perhaps much earlier; fragmentary evidence much in dispute suggests the *possibility* of human occupation even 50,000 years ago. Even earlier dates are possible, but it is *not* possible that humans evolved in the Americas rather than in the Old World, because there are *no* humanlike or ape fossils in the Americas—only those of very primitive monkeys. No *human* remains such as those of Neanderthals or "Peking Man" types have ever been identified in the New World, while they are plentiful in the Old World. Claims of human antiquity of hundreds of thousands of years in America show up in tabloid news stories but not in scientific journals or museum collections that can be examined by skeptics.

There is a definite popular appeal for "Guinness records," such as earliest, best, biggest, and so on, but American claims for them tend to be reasoned about as well as Bermuda Triangle arguments and documented by claims from people who say "Trust me," rather than, "Here's the evidence and its theoretical explanation." *If* people lived in America during or before the latest Ice Age, what did they do for a living? How did they get here? The claim of an "Atlantic land bridge"[3] has no support from the facts of geology. Sources in Atlantis or Lemuria are, if anything, even more ludicrous.[4] When a supposed scientist says that mammals may have migrated to America as many as a million years ago to prove that humans could have done the same, perhaps he sounds convincingly aware of multidisciplinary data.[5] But some of the oldest mammals known are from the Americas, and such a comment suggests ignorance of the data rather than proof of a good argument (and 60- and 120-million-year-old mammals have nothing to do with humans who have only been around for a few million years at most!).

Anyone claiming extreme antiquity for New World humans must

answer several questions: What is the dating technique? In what controlled context were the artifacts found? Could they be "naturefacts" or natural objects? What were tools used for, and how do they fit into a cultural pattern and evolutionary sequence? What effect would human habitation have had upon other animals and the rest of the environment, and what evidence is there for such effects (slaughtered game, burned clearings, postholes in the ground, gathered plant remains, and so on)? And where did the people come from—and how—if there is no evidence for contemporary human occupation of Siberia, which would have been the only reasonable source for early Americans in the time before boats?

At first glance "early man" claims may not seem as "paranormal" as psychic archaeology or the semireligious inscription mania. But the subject tends to become a single-minded obsession, promoted as an argument of faith rather than scientific method and data. Such statements as, "We're working on a frontier; all we ask is that you accept our subject and evidence" (and then maybe we'll show you our data and methods) are romantic appeals in the tradition of ESP advocates. Another statement often heard is, "Maybe there's more out there than traditionally believed—that's really revolutionary." Well, who could argue with that? But when someone advocating a revolutionary approach says, "The burden of proof lies with our critics," adding that nothing is being *claimed*, but just suggested as a *possibility*, there is a classic paranormal-claim syndrome at work. It *is* tempting to give in to the argument "You won't let us get to first base" until one realizes that such an a priori decision is an unwarranted decision to accept a previously ineligible base-runner! Just as in baseball, a "hit" or walk or error is a prerequisite to getting to first base, and errors do not reflect well upon either side![6]

What is most basic is how unorthodox claimants use scientific method and logic rather than simple claims upon belief. If claims are rejected by most scholars, why is this so? If the only answer is "dogmatism" and accusations of jealousy and persecution of pioneer thinkers, one can at least be a little wary. Denouncing "experts" in praise of noble "amateurs" is not the same as scientific proof. Conservatism *may* be in error regarding new claims, but it serves a real purpose in forcing the burden of proof upon new ideas, making them more rigorous—otherwise we would have to regard every idea that came along, however idiotic, as equal to all others, and science would spend all of its time disproving rather than proving.

Methodology and logic can be evaluated to some extent by nonspecialists if they read carefully enough. Arguments such as those quoted above are giveaways that something may be wrong. Too often we all regard a claim as a valid hypothesis simply because it is in print, but hypotheses, unlike some "claims," are testable and suggest their own tests of verification or falsification. Simply requesting a priori belief in basic premises that contradict established premises is not scientific. While we cannot simply take a vote among experts to decide the "truth," the experts' criticisms need to be taken into account by would-be revolutionaries in their very proposals.

One can endorse cultural relativism without forsaking scientific logic, method, and skepticism. People may be equal, but their contradictory explanations of the empirical world are not. Anthropology deals with sensitive issues of "human nature" vulnerable to political, religious and ideological controversy in a way physics and astronomy no longer are. If humans are called "naturally aggressive," there are clear implications about the inevitability of war and authoritarianism. If civilization can only develop in the Americas as a result of Old World influences, serious racist overtones are at least possible in the argument. Treating culture and its diffusion simply in terms of inscribed symbols implies that ideology and belief are much more important than factors of subsistence and technology—a theoretical position with which I would argue and at least want made explicit.

Anthropology, used or misused, has practical implications that make the treatment of it as simply a collection of "mysteries" or details a mistake, even if it were *not* a science already. But far too often that seems to be its fate in the popular press, tantalized more by exotic things "out there" than by the scientific explanation of them. Explanations are far from complete, but anthropologists simply contribute to further misunderstanding when they sometimes remain aloof from their ultimate clients *and* subject matter, the general public.

Notes

1. See, e.g., M. Harris, *Cows, Pigs, Wars and Witches* (New York: Random House, 1972).

2. T. Lee, "If at First You Don't Succeed . . . ," *Canadian Anthropological Journal*, 15:3, 1977. Translation 1: (a) "Expedition that crossed (the sea) in the service of Lord Hiram to conquer territory." (b) "Record by Hata who attained this limit on the river, moored his ship and engraved this rock." Translation 2: (a) "Thus far our expedition travelled in the service of Lord Hiram, to conquer land." (b) "This is the record of Hanta, who attained the great river. And these words cut on stone." Translation 3: (b) "An oracle that was concealed within. This the message we broke open."

3. W. McDonald, "How Old IS American Man?" *National Observer*, May 31, 1975, quoting Dr. Bruce Raemsch.

4. Jeffrey Goodman revives this idea based on Edgar Cayce and other seers in *Psychic Archaeology* (New York: G. P. Putnam, 1977) and other forums. For a history of such bad ideas, see R. Wauchope, *Lost Tribes and Sunken Continents* (Chicago: Univ. of Chicago Press, 1962).

5. W. McDonald (op. cit.) quoting Dr. Bruce Raemsch.

6. The quotes and paraphrases (and baseball analogy) are from the transcribed discussion of two papers making claims about a supposed pre-Wisconsin (i.e., 70,000 years ago or earlier) archaeological site in New York. These papers and discussions (along with other, more "orthodox" papers) are printed in J. Cole and L. Godfrey, eds., *Archaeology and Geochronology of the Susquehanna and Schoharie Regions* (Oneonta, N.Y.: Yager Museum, 1977). Among other things, their publication is an indication of the *relative* openness of the "establishment" to unorthodox archaeology's right to be heard—but not to be accepted automatically.   •

Psychic Archaeology: The Anatomy Of Irrationalist Prehistoric Studies

Kenneth L. Feder

As a scientific enterprise, archaeology is relatively young. The use of objective, consistent, and meticulous field-techniques is barely 100 years old. As recently as 25 years ago archaeological theory—that is, a body of general theory explaining the nature of the relationship between human societies and the material remains of these societies—was so deficient that those who discussed it could state that it lacked even a name (Willey and Phillips 1958, p. 5). In the past twenty years archaeology has experienced a methodological and theoretical revolution characterized by the explicit application of statistical analysis, use of computers, a heavy emphasis on ecological relationships, and the growth of cultural evolutionary theory.

Though scientific or anthropological archaeology is new, archaeology has never lacked fringe-area, pseudoscientific, and, at times, decidedly *anti*scientific approaches and theory. Archaeology and astronomy are probably the two sciences that have attracted the greatest number of serious and dedicated amateurs who go on to make valuable contributions. These two fields have also attracted the greatest number of individuals whose time would be better spent selling incense. It is truly mind-boggling to consider the number of frauds (Piltdown Man, the Cardiff Giant, the Davenport Stones), racist ideologies (Nazism), religions (Mormonism), and just plain crazy theories (ancient astronauts) that have utilized archaeology to "prove" their often preposterous hypotheses.

It should come as no surprise that a "new" field of endeavor involving archaeology and the paranormal has been introduced: "psychic archaeology." After all, what could be more obvious? Psychics can find missing things; why shouldn't they be able to find archaeological sites?

They can, for instance, psychometrize an object—that is, they can perceive detailed personal information about the owner of an object simply by holding it in their hands. Why shouldn't they be able to do the same for a 10,000-year-old spear-point?

It is not surprising that people who have embraced the paranormal would also wish to embrace psychic archaeology. If this was the extent of this phenomenon, it might best be ignored. However, psychic archaeology has gained an unfortunate professional tolerance, if not acceptance, in some quarters. Ivor-Noel Hume, one of the most influential (and, I might add, theoretically conservative) historical-site archaeologists in North America, details in his introductory text the technique of dowsing for the location of buried metal artifacts (Hume 1974, pp. 37-39). In a popular new introductory text, David H. Thomas, who is highly respected, and deservedly so, devotes a two-page inset to psychic archaeology and straddles the fence on the issue (Thomas 1979, pp. 286-287). However, far and away the most disturbing work is that of David E. Jones (1979), a professional anthropologist at the University of Central Florida. His book *Visions of Time,* published by the Theosophical Society, purports to be a "test" of psychic archaeology. Jones is honest enough in describing his interest in parapsychology as stemming from his desire to "escape from the tyranny of a Western 'logic' of the universe." Yet, through procedures that, at least superficially, feign Western scientific methodology, Jones attempts to test the validity of the psychic-archaeology hypothesis and states, "I feel a positive conclusion is unavoidable." Let us see just how "unavoidable" this "positive conclusion" is in the testing of psychic archaeology. Because Jones's work claims to be an objective test, and since Jones is a professional scientist, his work demands professional attention.

In essence, Jones tests some basic claims made by "psychic archaeologists":

1. the ability to locate unknown archaeological sites;
2. the ability to ascertain the location of buried artifacts and features at archaeological sites; and
3. the ability to perceive detailed information of both a general and specific nature about the prehistoric environment, culture, and even the emotional and physical states of individuals who have been dead for thousands of years, by simply handling artifacts. (As Jones states, "The psychic's claim, if it could be demonstrated *rigorously enough* [emphasis mine], was precisely what archaeology needed to make it an historical tool capable of generating cultural descriptions comparable to primary or first-hand observation" (Jones 1979, p. 4).

Let us see exactly how these phenomena are supposed to work.

Site Location

Whether lecturing to a class of archaeological neophytes or speaking to a local civic group, archaeologists are always asked how they find sites. The question is asked, at least initially, much in the same way one might inquire how pirate treasure is found. The notion is that sites are terribly rare, randomly strewn across the countryside, and almost impossible to find without detailed technical knowledge. With this in mind, certainly if a psychic could precisely locate a site it would seem to be impressive evidence that such abilities exist.

I am certain that it would not be giving away any trade secrets to admit that, although our methodologies may be quite sophisticated, involving rigorous sampling techniques, infrared photography, and so on, the theory behind locating archaeological sites is, in fact, quite simple. Human beings do not distribute themselves randomly across the landscape. Especially in technologically unsophisticated societies, people settle areas differentially, as a function of certain rather obvious environmental parameters. Among these are:
1. distance to potable water
2. distance to a navigable waterway
3. gentleness of topographic relief
4. defensibility
5. natural protection from the elements
6. soil type
7. presence of natural resources (stone quarries, clay, etc.)

Thus, while our methods for assessing the probability of site location in a given area are becoming increasingly sophisticated—Thomas (1972, 1973) utilized a computer simulation model—the gross variables considered are often commonsense characteristics. In order to survive, people need fresh water, a flat area large enough to accommodate the community, and so on. How much detailed technical information must someone have before being able to duplicate the accuracy of an archaeologist in finding a site? In many cases, not much.

In my introductory course in archaeology each semester, I distribute copies of United States Geological Survey quadrangle maps (1:24,000 scale) of an area of Connecticut that has been partially tested in the Farmington River Archaeological Project (FRAP). The goal of FRAP is to explicate, as well as explain, prehistoric human settlement in the Farmington River valley, a small to medium-sized tributary valley of the Connecticut River located in the central portion of the state. In this survey, test pits were excavated at 30-meter intervals along randomly selected transects running perpendicular to the river. Each transect cut across the three distinct topographical/depositional zones in the valley floodplain, glacial terrace, and bedrock upland. A large number of prehistoric settlements have been located in this fashion. The vast majority of sites have been situated in the kinds of areas in which one would expect to find human settlements.

After only an hour or so of discussion of human settlement locational choice, where the students suggest the key parameters of this question and with no other information provided, I ask them to suggest areas where sites might be expected. Invariably, in a class of 40 students, a number point out *precise* locations where, unknown to them, sites have already been found by FRAP. Are these students psychic? Of course not. They simply paid attention and applied the reasoning that was discussed. To predict that an area where a swift brook enters a navigable river would have been attractive to human settlement is like predicting there will be cold nights in Alaska this January. You can't miss.

A perfect example of this "can't miss" test of psychic archaeology occurred in a recent program in the "In Search Of" series on one of the major TV networks. Here we are told of a fabulous experiment in underwater psychic archaeology where a group of "sensitives" were asked to identify the location of possible historic shipwrecks from sea charts. The areas suggested by the psychics were examined in a research mini-submarine, and, sure enough, a wreck was found, although not in any of the precise locations predicted by the psychics. The fact that a wreck was spotted (after many hours of searching) was presented as proof positive of the existence and utility of psychic underwater archaeology. The fact that the area chosen for psychic examination was historically one of the most highly traveled regions of the world's oceans, and where a number of wrecks were known to have occurred, was played down in the show.

According to Jones (p. 15), a number of key archaeological discoveries made in the past were based on the psychic abilities of the researchers. He very clearly implies that it was by using their "intuitive" (read *psychic*) powers that Winckelmann discovered artifacts at Pompei; Stephens and Catherwood, the Mayan civilization; Boucher de Perthes, paleolithic tools; Layard, the palaces at Nimrud in Assyria; and Schliemann, the ruins of ancient Troy. With the exception of Layard, who apparently did claim to possess clairvoyancy, these claims are extremely misleading. When a scientist, after years of analysis and meticulous, painstaking research, puts the data together in a new way, makes a discovery, and gains a new insight into a perplexing question, it is called rational thought, or deductive reasoning. There is no reason to call it paranormal.

Does Jones go on to adequately test the hypothesis that psychics can "locate" sites? Not only does he not *adequately* test the hypothesis, he doesn't test this at all! At no point is a map presented to a psychic to predict the location of an unknown site whose existence can then be tested. Instead, Jones presents psychics with artifacts and photographs from *known sites,* whereupon they are asked to determine the location of these sites.

In one such experiment Albert Bowes (a professional psychic) was left *alone* in a room with his wife and a series of artifacts (p. 55). Bowes correctly identified the site as a coastal occupation, a result which is seen as impressive; after all, the site could have been in the mountains or in a desert. However, the fact that among the artifacts provided were shells and some columella beads may have given Bowes a tiny hint.

In another test Bowes was mailed four photographs of a Florida site, which he kept for three days, and was asked to provide locational data (pp. 106-114). Jones claims that the photographs are not diagnostic of Florida. This seems like a particularly poor way to test a hypothesis. Though Bowes correctly identified this and other Florida sites as being near "water," we must remember that (1) almost all prehistoric, as well as historic, habitation sites everywhere in the world are near water; (2) Florida has, in a state covering 58,560 square miles, approximately 1,800 miles of coastline (more than 8,000 miles of detailed shoreline); and (3) when a psychic correctly guesses that a Florida site is near a lake and even suggests the direction of the lake from the site, it is not terribly surprising, since there are approximately 30,000 lakes in the state. It would be difficult for a site in Florida not to be near a lake. Furthermore you can point in almost any direction from an archaeological site in Florida and find a lake nearby.

What evidence is given that psychics can locate unknown archaeological sites? In a word, *none.*

Site Excavation

In the excavation of most archaeological sites, researchers are faced with the problem of deciding precisely where to dig. All but the smallest of archaeological sites are too large to excavate in their entirety. Archaeological science deals with this problem through the application of rigorous sampling strategies designed to ensure a representative sample of the subsurface materials. Archaeological pseudoscience deals with this problem through clairvoyance.

Jones conducts four experiments purported to test psychic abilities applied to archaeological excavation. The bulk of the test results, however, have nothing whatsoever to do with the proposed psychic ability to perceive unseen artifacts. Rather, most of the information provided by the psychics deals with reconstructing past cultures, that is, "sensing" what went on at a site when it was occupied. I will assess Jones's tests of the psychics' abilities at site reconstruction in the next section of this paper.

However, there were some attempts made by the psychics at specifying the material that was found, or would be found, at a particular site on the St. John's River in central Florida. James Randi has introduced the concept of "negative success" as a common rationalization used in psychic research—"when you win, you win; and when you lose, you win" (Randi 1975, p. 8). Jones exploits his own version of this when he states (p. 101): "If the psychic subject stated that a particular artifact was present in the site, the lack of its discovery during excavation would not disprove the presence of the artifact." This statement, while absolutely true, renders this set of experiments entirely meaningless, since a psychic *cannot* be proved wrong. Anything guessed at correctly is a hit. Anything guessed at which is not found does not *prove* a miss; it may merely mean that it has not yet been found. Philosophers of science often judge the utility of a hypothesis by its falsifiability. This particular hypothesis does not meet this criterion in Jones's test.

In terms of the actual testing, a specific list of artifacts found at the site is provided (p. 105). Of all the materials listed, Bowes was correct only twice in his psychic assessment of the artifacts that were present: a single iron nail and some metal pieces (all of non-Indian manufacture and intrusive to the site). The 11 projectile points, 42 flint pieces, fire hearths, shell tools, beads, worked bone, flat (milling?) stone, European ceramics, Indian pottery, and the human burial were not mentioned by Bowes in the first experiment.

Jones attempts to convince the reader that Bowes came close to perceiving the burials; Bowes did not. Instead, he talks about broken necks and heads on poles, but this is meaningless in testing the hypothesis. In fact, he initially states, "I really feel as though this site or area may not show that many bones or people" (p. 113). In the second experiment, where Bowes actually visits the site, he goes so far as to say "if you find skulls here, it would be surprising" (p. 115). Well, they did find a skull. Jones, of course, sees this as verifying Bowes's psychic abilities because, after all, *only* one skull was found.

The most abundant artifact found at the site was pottery, with over 5,000 shards recovered. Bowes, however, never mentions pottery at all until he is brought to the site and sees or is given a potsherd (in the test it is unclear where it came from). This is not a very impressive degree of accuracy.

Site Reconstruction

The claim is made that psychics can provide detailed information about prehistoric groups. In the lexicon of pseudoscience, "vibrations" of individual people and their natural and cultural environment are imprinted on their possessions, bones, and so on, and these "vibrations" can be read thousands and even millions of years later by psychically inclined individuals. According to Jones, what these people can accomplish is "like nothing that any archaeologist could do by simply observing the artifact" (p. 4). I shall ignore for the moment that archaeologists do a bit more than "simply observe" artifacts.)

The bulk of Jones's book is devoted to testing experimentally the psychics' abilities to accomplish this feat: "psychometrizing" an artifact, that is, psychically producing as detailed a description of prehistoric people as an ethnographer could who lives with a contemporary group. All archaeologists admit a deficiency in detail of our cultural reconstructions compared with those of the cultural anthropologist, who observes and interacts with a people in their daily lives. Can psychics recapture this lost detail?

A slight digression must be made here to discuss the methodology employed by Jones in his testing of this hypothesis. The essence of any good experiment, archaeological or otherwise, rests in consistency and objectivity. It is incumbent upon the rational scientist to detail, as explicitly as possible, the problem being addressed, the methodology and operationalization of the experiment to be performed, and, quite important, the precise nature of the results required to accept or reject the hypothesis. Jones goes as far as the first two steps but never discusses the third. We are not told beforehand what exactly will constitute proof or disproof of the hypothesis that psychics can provide detailed archaeological information. In any acceptable test of a hypothesis, it is necessary to present a set of predictions, to spell out precisely those things that *must be true* if the hypothesis is to be considered valid. It makes no sense, in an objective test of a hypothesis, to propose a phenomenon, design an experiment to test the phenomenon's existence, conduct the experiment, and then seek to explain all experimental results within the context of the assumption that the hypothesis is, in fact, valid and true. This constitutes post-hoc rationalization. This constitutes precisely Jones's "reasoning" in *Psychic Archaeology*. Post-hoc rationalization is *not* science. All of the results of the experiment, including those that quite clearly contradict the hypothesis, are interpreted, squeezed, fudged, and rationalized away as actually supporting the hypothesis. Scientists set out to *test*, not *prove* hypotheses. The trouble with trying to *prove* something rests in the self-fulfilling prophecy: if you want to prove something badly enough, you always will.

Further, Jones states that he does not use quantitative or statistical techniques in the analysis of his data; that his discussion in the body of the test constitutes the "hits," rather than the misses, of the psychics. His reasoning seems to be that, while the psychics were *so* wrong part of the time that not even a post-hoc rationalizer could account for it, when they were correct it was so impressive as to alone prove the existence of psychic powers. Jones seems to believe that explicit, admitted avoidance of quantitative techniques obviates all criticism of his not being statistical. This is akin to a surgeon trying to avoid responsibility for the consequences of not sterilizing his or her instruments on the grounds that he/she specifically said it would not be done. That simply is not a good enough argument. The abandonment of statistical analysis by a person with an advanced degree in anthropology is appalling. This critical problem can be explained through the following example.

Suppose you designed an experiment to test whether or not a particular coin was "fixed." In the experiment you performed 100 sets of 10 flips each. Obviously, if the coin was legitimate you would expect (predict) that in the majority of sets of 10 flips you would obtain 5 heads and 5 tails. Statistically you would expect to achieve 10 straight heads or tails $1/2^{10}$ or $1/1024$ times. Suppose, when presenting the results, the coin tester stated: "In the 100 sets of 10 flips each, there were many nonsignificant (5 heads/5 tails) results, but let's forget these. On one occasion I obtained 10 consecutive heads—a highly unlikely result. I therefore conclude that the coin is fixed." This would not be accepted as being valid by any rational individual. This reasoning displays a complete lack of understanding of the meaning of "proof" in a statistical sense. This is precisely the approach used by Jones.

A key element in the statistical reasoning that *must* be used when testing probabilistic phenomena (which all psychics claim psychic phenomena to be) is the recognition that unlikely events *do* occur. The point of a statistical test rests in providing a mathematical measure of the likelihood of a given result in order for us to determine whether or not the result is so unlikely as to reject or accept the hypothesis.

Specifically, it can be certainly stated that none of Jones's quasi-experimental tests renders a conclusion in favor of psychic archaeology "unavoidable."

Jones initially ran a two-stage preliminary experiment. The first stage involved a double-blind reading of wrapped and boxed artifacts. Remember, Jones is giving a detailed report only of the *hits*. The claim is made that Bowes scored positive hits on 50 percent of the 10 items he was presented. The case for the antique baseball in box number 4 is typical. Indeed, on page 20 a section of the transcript of the experiment is quoted in which Bowes mentions that the object in box number 4 pertains to a game and that the thing is a ball. Sounds impressive. Well, when we turn to the full transcript we are able to isolate 15 different specific references to the item in box number 4: (1) teeth, (2) pottery, (3) game, (4) ball, (5) Incas, (6) South American, (7) man with scar or injured jaw, (8) Egypt (9) mummy, (10) building, (11) cave, (12) pyramid, (13) supreme ruler, (14) bones, and (15) jaws. Hardly an extraordinary outcome.

The second part of the preliminary experiment involved a second reading, with the objects now open for view. Interestingly, in almost every case Bowes provides a far more detailed reading of the objects when seen than when they were boxed. Why would *seeing* an object that was being analyzed by another "sense" help in its description? Does seeing a stereo speaker help you hear it? Can you feel heat or cold better with your eyes open than with them closed? I suggest that it's just easier to make up a story about something if you can see it. The length and detail of the second set of psychic observations are far more congruent with this conclusion than any other I can suggest.

In one of the more purely archaeological experiments Bowes was given a set of artifacts and asked to analyze them. These items were from Lindenmeier, a Paleo-Indian site in Colorado. Jones could not have picked a worse example on which to run a test. The Paleo-Indian period, with its easily recognizable "fluted" points, is known to anyone who has seen even a child's book on Indians or American archaeology. During a recent lecture to fourth-graders I was asked about these "fluted" points, which are found across the country and date from 12,000 to 9,000 years ago. Certainly if I were Bowes, and even if I believed I was a psychic archaeologist, I would at least pick up a book or two on archaeology to get acquainted with the field. Anyone who did so would have to be unconscious not to come across at lease a brief discussion of these late Pleistocene inhabitants of the New World.

Even given all this, how well did Bowes do? Not especially well. He locates the site in California, New York, Nevada, and even suggests Georgia and Tennessee, though he correctly prefers the general area of the American west. He correctly identifies the period generally as being before Christ and before Egypt, but he never gets beyond this in specificity. Much of the other information provided by Bowes is problematical, vague, and, in general, meaningless. However, there is one point at which Bowes clearly gives away the source of his meager knowledge of Paleo-Indians. It is decidedly not psychic.

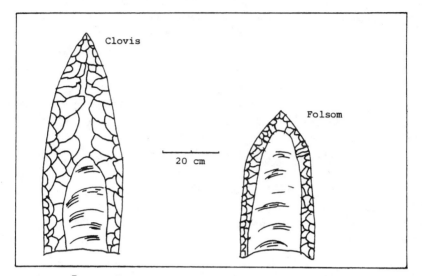

FIGURE 1. The two types of fluted points found in North America

When discussing diet and hunting techniques Bowes states that the fluted points at Lindenmeier were used "to shoot bison, or an elephant or mastodon or something" (p. 82). He later denies the presence of bison (which is rationalized by Jones to be confusion over the fact that an *extinct* form of bison was present at Lindenmeier). While most popular books on North American archaeology will tell you that Paleo-Indians hunted bison and mastodon with their fluted points, most neglect a small, technical piece of information. There were actually two forms of fluted points: Clovis points—long, leaf-shaped points with a channel on both faces rarely reaching beyond the midpoint of the projectile's length; and Folsom points—somewhat shorter leaf-shaped points with channels reaching nearly the entire length of the points (see Figure 1). The Clovis points, which tend to be older, are almost always associated with the remains of extinct proboscidians. The Folsom points are *never* associated with these elephants, which apparently became extinct before the large, Pleistocene bison. Folsom points are almost always associated with bison in the west. The only way Bowes could get the impression of elephants here would be through some faulty psychic connection, since Lindenmeier is a Folsom-period site. The list of faunal species recovered at Lindenmeier includes pronghorn antelope, wolf, coyote, swift fox, red fox, jackrabbit, and bison (*Bison antiquus*) (Wilmsen 1979:23). The list is notably devoid of elephants of any genus or species. Bowes's mention of elephants or mastodons could have come from no psychic vibration from the Lindenmeier artifacts. I would suggest that either he forgot the subtle distinction, or the book he got his general information from did not provide that little detail, which, in any case, seriously calls into question the entire experiment.

The Lindenmeier experiment is typical of the rest of Jones's book. Any time Bowes or the others say anything that might possibly be construed as correct, the only possible explanation is psychic power. When Bowes identifies the source of some Mayan artifacts as the Vikings, it is clearly correct because both Vikings and Mayans were warlike, seagoing, and traders (Jones rationalizes this on page 181). When Mayan astronomical observatories are described as storage areas, it is suggested that they *might* have stored material there. When some very amorphous stones from the Yucatan Peninsula are located by Bowes in North Carolina, Florida, Texas, or South America, it's a hit (each successive guess comes closer to the actual source). When another of the psychics lists Lake Erie, Louisiana, Missouri, and Mexico for the same stones, it's a direct hit (pp. 231-232).

Even when they are clearly wrong, it proves psychic power, since the reading may not relate to the manufacturer of the artifact but, perhaps, to someone else who owned or touched the artifact at any point in time (p. 39), or it may have been contaminated by contact with some other artifact.

Perhaps the best tactic of all in the arsenal of psychic archaeologists is vagueness in the face of a friendly researcher. The vast majority of the data provided by the psychics is untestable personal information about the makers of the artifacts they analyzed. Even on specifics they are vague, though Jones is misleading on this point. When he states, "It would take a highly trained and experienced expert to differentiate a piece of primitive basketry of Bontoc or Mono origin" (39), the implication is that the psychics can do so. They don't.

When he describes Bowes's description of wrapped and boxed toy combs from Boston in the 1840s, he claims that it is "full of references to a mid-1800s Boston" (p. 26). While there are references to a body of water and lots of different kinds of people and buildings, there are *no* specific references to Boston or the 1800s. The only specific reference is to Eskimos. Finally, when detailing the Lindenmeier experiment he states: "No one, not even the most expert New World archaeologist, could look at a Folsom point and tell it came from Lindenmeier" (p. 77). I read into this the implication that the psychic can tell and did state that the material was from *Lindenmeier*. He, of course, never does.

The Jones experiments are not experiments at all—they are rationalized "proofs" of the existence of a highly questionable phenomenon. I do not believe that Jones is dishonest. If he were, he would not have supplied the full transcripts of the experiments in the appendices. Instead, Jones is clearly a researcher who has abandoned science, reasoning, and rationality.

I, for one, am not going to replace my trowels with little Jeane Dixons. Instead, my research in FRAP will plod slowly along, underfunded and understaffed. We will advance one step at a time in our understanding of the prehistoric inhabitants of the Farmington Valley of New England, and, in essence, of all humanity. The human past is a fascinating enough field without wallowing in the mire of the miracles of lost continents, ancient astronauts, or psychic archaeology.

References

Feder, Kenneth L. 1980. "Foolsgold of the Gods." *Humanist* (Jan./Feb.):29-23.

Goodman, Jeffrey 1977. *Psychic Archaeology: Time Machine to the Past.* New York: Putnam.

Hume, Ivor-Noel 1974. *Historical Archaeology.* New York: Knopf.

Jones, David E. 1979. *Visions in Time: Experiments in Psychic Archaeology.* Wheaton, Ill.: Theosophical Publishing House.

Randi, James 1975. *The Magic of Uri Geller.* New York: Ballantine Books.

Thomas, David H. 1979. *Archaeology.* New York: Holt, Rinehart and Winston.

Willey, Gordon R. and Philip Phillips 1958. *Method and Theory in American Archaeology.* Chicago: University of Chicago Press.

Wilmsen, Edwin 1974. *Lindenmeier: A Pleistocene Hunting Society.* New York: Harper & Row.   ●

Science and Evolution in the Public Eye

Laurie R. Godfrey

Many educators have expressed surprise at the extent to which students believe sensationalistic and catastrophic explanations of the origins of cultural and biological traits. Their inclination is to ignore sensationalism as "unworthy" of serious discussion, but they are being hampered by political pressures from the sensationalists, who tend to view themselves as bearers of "true science" and as opponents of outdated scientific beliefs or orthodoxies. Thus these catastrophic and often cryptoscientific views of racial and cultural trait origins are being given increasing exposure in popular literature, on TV, in movies, and in public school and college classrooms.

Among the most notorious examples of this alarming trend are von Däniken's *Chariots of the Gods?* (1970), Barry Fell's *America B.C.* (1976), Jeffrey Goodman's *Psychic Archaeology* (1977), the "In Search of" TV series, and the current UFO mania. Organizations with blatantly racist motives, such as the Nazis and the Ku Klux Klan, who proclaim separate "origins" (or creations) for different "races," are once again growing in visibility. The "orthodoxies" of the anthropological "establishment" are being challenged by students who proclaim separate-origins explanations (a series of invasions from outer space, or "experiments" by a creator) and by some of those proclaiming a single creation.

These sensationalist views are financially supported by evangelistic grass-roots organizations. These organizations are politically active in the sense that each is "spreading the word." The various Bible research groups that hold weekly or biweekly meetings on college campuses engage in peculiar mixtures of odd-fact collecting and religious ceremony. Similarly, followers of Barry Fell and other popular heroes hold public meetings and "conferences" that are in some respects much like religious incantations. At a conference on so-called pre-Columbian colonizations of the New World (cf., Cook, 1978; Cole, 1978), which was held in Castleton, Vermont, in 1977, the organizers gave religious significance to every rock and mark on display in the front room. A woman clutching a copy of Fell's *America B.C.* advised a skeptical bystander that it was ridiculous to think that some of the stone structures found in the area might be colonial root cellars, as archaeologists maintain: "People would not cover root cellars with heavy slabs of rock," she said, "only shrines." "Someday," she admonished, "you, too, will believe."

Many proponents of catastrophic explanations of natural phenomena claim that these theories are well-founded in scientific fact and repeatedly express pained willingness to bear witness to their truth despite extreme antagonism from the scientific elite. Thus these movements combine proselytizing with an odd concept of "sciencing"—a "sciencing" that begins with a premise and denies any means of testing or refuting it.

In the past few years, fundamentalism, which incorporates "scientific creationism," has experienced dramatic growth. In 1978 a network news program carried a three-part report on the phenomenal growth of "born-again" Christianity in America, especially among the educated middle class. Shortly afterward *Newsweek* featured an article on "born-again" wives of national politicians. Simultaneously, courses on scientific creationism appeared in college curricula, along with courses on astrology, Atlantis, the teachings of von Däniken, and so on. Indeed, many academicians have jumped on the paranormal bandwagon.

It is a popular view that education should offer alternative paradigms as "equal but different" explanations of the same phenomena. This has been seen most clearly in recent years in the debate in California, and elsewhere, over teaching scientific creationism on an equal basis with the theory of evolution as an explanation of the similarities and differences among organisms. Unfortunately, "liberal" educators and politicians, in an effort to be "open minded," can be unwitting collaborators in spreading ultraconservative doctrines among our youth. While it is commendable to study unorthodox or unpopular issues without prejudice, the presentation of alternative explanations as "equal but different" implies that there is no way to choose between them.

More alarming, perhaps, is the political opposition to the liberal and open-minded educational programs that grew out of the 1960s. "Man: A Course of Study" (MACOS), an interdisciplinary behavioral-science project for elementary schools, came under severe attack in the United States a few years ago. Consequently, it has been banned in many localities, and its national funding has been crippled. John Conlan, then a Republican congressman from Arizona, claimed that MACOS caused children to reject the values, beliefs, religions, and national loyalties of their parents (Smith and Knight, 1978, p. 4). Conlan singled out three of the authors of the MACOS teachers' guide for criticism. He accused Jerome Bruner of using psychological-warfare techniques, B. F. Skinner of using behaviorism, and Claude Levi-Strauss of using his allegedly dangerous leftist bias to subvert children for "one-world socialism." (Actually, the three scholars' articles were often in opposition to each other, but the guide was nevertheless withdrawn.)

"The consequence of Conlan's attacks on MACOS and Senator Proxmire's criticism of 'those damn fool projects in the behavioral sciences,' which helped prepare the ground for attacking MACOS, was that National Science Foundation funding for MACOS, and all other federally funded curriculum projects, was stopped pending a review by various congressional committees" (Smith and Knight, 1978, p. 5). Smith and Knight document the banning of MACOS in Queensland, Australia, by crusaders using the American experience as a guide. The authors demonstrate by content analysis of the issues raised and key words in the literature that the anti-MACOS and fundamentalist movements are characterized by dogmatism, acceptance of authoritarianism, and totalitarian values that stress state and community control over the individual. On the other hand, according to their analysis, pro-MACOS and pro-evolution literature stresses relativism, freedom of thought, and critical analysis. Anti-MACOS propaganda was found to be high in "coercion" (as opposed to choice) and characterized by high degrees of censorship, ethnocentrism, aggression, and violence (p. 10).

In an article entitled "Public Appreciation of Science," Amitai Etzioni and Clyde Nunn (1974) cite evidence that educated people tend to be less authoritarian than uneducated people and that people who "distrust science" are likely to be more authoritarian. "There is also evidence linking authoritarianism with unscientific beliefs, even though *all* authoritarians are not anti-science" (p. 199). They cite another study showing a correlation between authoritarianism and superstition, pseudoscientific attitudes, and racial intolerance.

What creationists and other sensationalists have in common is the division of the world into true believers heading for salvation and all others heading for damnation. The "saved" group may exclude members of certain racial minorities, social classes, or political factions (socialists or communists), or homosexuals, or evolutionists.

The recent partial disaffection from science of an apparently significant portion of the educated segment of society can be seen as a potentially dangerous outcome of the conservative political climate of the 1970s. Increasing numbers of educated people are accepting, even demanding, simple explanations of complex phenomena. Thus fixed-species explanations have become respectable alternative paradigms whose inclusion in the educational system is receiving increasing legal support. In considering a challenge to the teaching of Darwinism, an article in the *Yale Law Journal* (Bird, 1978) makes a political-legal statement regarding what might be taught as respectable alternatives to Darwinian evolution. Western society may soon witness court rulings on "proper science" similar to legal maneuvers regarding "proper literature" as opposed to pornography. Creationists are advertising new programs to teach scientific creationism at such universities as Michigan State, Wichita State, and the University of West Virginia.[1] Legal arguments advocating laws, court orders, and school policies requiring the teaching of creationism as coequal with evolutionism (such as those summarized by Bird, 1978) are blatant attempts to define "proper science" according to political guidelines and without reference to either predictive advantage or rational explanation. Unfortunately, while scientific creationism is very poor science, most people are ill equipped to evaluate it as such. Many, like Wendell Bird, see the issue as a political struggle between proponents of "equal but opposite" dogmas. The question of what is a good or a poor explanation of similarities and differences in form among organisms in space and time is rarely, if ever, raised.

It is even difficult for evolutionary biologists, who are most cognizant of the data that evolutionary biology attempts to explain, to debate scientific creationism effectively. These difficulties are based on a number of factors, not all of which are easily remedied:

1. Creationist challenges to evolutionary biology bear the earmark of irrationality—they are not simply the presentation of "facts" that are not facts, but they are illogical leaps to conclusions that do not follow from the premises—a kind of "Aha!" complex. For example:

   —There is an error term in carbon-14 dating. Conclusion: Aha! (This is supposed to prove that there are no old fossils, or that there can be no supposition of great antiquity.)

   —Scientists have *not been able to trace the origin* of many suids (pigs) through fossils. Conclusion: Aha! (This is supposed to suggest that extant suids share no common ancestry.)

   What is the use of scientists introducing probability theory or information about kinds of systematic errors in the context of a public debate? The issues are far more complex than the creationists would have the public believe.

2. Creationists tend to appeal to authority—to neat, easy solutions to

complex questions. Such arguments tend to be attractive to the frustrated, the needful, and the alienated.

3. Many people in Western society are ambivalent toward science. Polls show wide, though declining, respect for "science" (Etzioni and Nunn, 1974; Bainbridge, 1978), but they do not define science or test their subjects' understanding of what they say they respect. Respect for authority is *not* respect for science. Creationism, like other simplistic cults, affords people a way of rejecting scientific elitism without seeming to reject science. Indeed, its proponents believe themselves to be the true bearers of scientific facts. They are often more convinced of their righteousness than the "elitist" scientists they accuse of being close-minded. They are convinced that they alone possess the "truth" and they *define* science as that truth. Clearly, there is a basic difference between what scientists understand as science and what creationists understand as science.

4. Scientific creationism is based in large part upon fallacious premises—a misunderstanding of what evolutionary biology is about.

The latter two problems are the most frustrating because they speak more directly to the failure of the educational system to teach rational problem-solving. Not everyone need be well versed in evolutionary biology or anthropological analysis of human biological and cultural variation, but people should be able to recognize sloppy arguments and to choose tentatively between alternative explanations. The fact is that the educated American public is surprisingly unable to cope with even the simplest incorrect premises or illogical, particularistic arguments emanating from scientific creationists and others.

Popular perceptions of evolution, when surveyed directly and indirectly through an analysis of contemporary popular and educational literature, reveal a startling misunderstanding of the basic concepts of Darwinian evolution. Creationists describe evolution as a kind of accidental creationism; despite their allusion to its slow pace, they perceive evolution to be clearly catastrophic. They are degenerationists debating an anthropocentric Doctrine of Progress, which modern evolutionism is *not*. Yet their perception of what modern evolutionism *is*, is not far from that of the general educated public. It is not surprising that educated people are poorly equipped to handle these challenges and succumb easily to political pressures from proselytizing lobbyists.

Numerous popular writers, philosophers, and educators are proclaiming the death of natural selection: "Nobody takes natural selection seriously anymore," "Natural selection is tautological" (cf., Baum, 1975; Bethell, 1976, 1978; Flew, 1967; Himmelfarb, 1968; King, 1972; Koestler,

1.　Natural Selection was Darwin's main idea as to how evolution happened. The fittest survive and the unfit perish. Fine! But where is evolution in this? Because a certain rabbit can run faster or hop higher and therefore may live longer and reproduce more rabbits in no way implies that the rabbit (or its offspring) would be more fit for survival if it were evolving into some other animal. In fact, the opposite is true. Any alteration in a rabbit's physical or mental characteristics would make it less fit, not more fit for survival. Natural selection cannot explain evolution. Nothing can explain how evolution happened, because it never happened! That is a fact and anybody who says it isn't is being an unscientific fanatic.

2.　Students, how many times have you seen and heard the idea that scientists have proof for evolution in the fossil record, in the bones? Here is the truth; let any scientist come forward and deny it if he can: There is not one bone in the entire world that shows one animal evolving into another. All the pictures in your books which are there to convince you that man evolved from apelike creatures (Java Man, Neanderthal Man, Nebraska Man) have all been proven not to be missing links. They have all been proven to be either apes or men, not apes changing into men. (So-called Nebraska Man was built up on the evidence of one tooth which turned out to be the tooth of an extinct pig!) There are no ape-men because evolution is a great deception. The supposed evolution of the horse, eohippus, which is in your science book, no doubt, has been proven impossible. Check your science book for other misleading so-called evidences for evolution. Is the peppered moth there? It's still a moth, isn't it? Is Haeckel's embryological chart there? Your book doesn't tell you Haeckel was kicked out of the university for misrepresenting the truth about embryos, does it? Your book doubtless mentions the fruitfly Drosophila Melanogaster. But does it tell you that exhaustive experiments prove beyond any question that mutations can never explain evolution? Ask your science teacher to tell you about these great deceptions which cannot be carried on under the name of science. Fossils all show complete specimens. They were not changing into anything. The true scientist must look at the vast fossil record and admit that it proves one thing beyond question, namely that no fossil has ever been uncovered which shows one animal changing into another.

3.　Two basic laws of Physics (the Law of Conservation of Matter / Energy, and the Law of Increasing Entropy) flatly contradict the theory of evolution. Everything science knows tells it that these laws are unbendable. They are scientific laws. They are not hypotheses. They are not theories. Yet, evolution theory demands that these iron laws be broken and laid aside. Energy itself does not produce complex, functioning systems. That takes planning and a Planner! Great periods of time do not cause things to improve (get better, become more complex) as evolution theory demands. On the contrary great periods of time cause decay and degeneration, the very opposite of evolution! The First and Second Laws of Thermodynamics prove evolution to be impossible. There is no getting around this truth.

4.　Mathematics proves evolution impossible. Computers prove evolution impossible. Charles Eugene Guye, a Swiss mathematician, has calculated the chances of a single molecule of a protein-like substance being formed by accident at 10^{320} to 1. That means that the odds against even one molecule evolving would be one in 100,000,000, 000,000,000,000,000,000,000,000,000,000,000,000,000, 000,000,000,000,000,000,000,000,000,000,000,000,000, 000,000,000,000,000,000,000,000,000,000,000,000,000, 000,000,000,000,000,000,000,000,000,000,000,000,000, 000,000,000,000,000,000,000,000,000,000,000,000,000, 000,000,000,000,000,000,000,000,000,000,000,000,000, 000,000,000,000,000,000,000,000,000,000,000,000,000, 000,000,000,000,000,000,000,000,000,000,000,000,000. Another mathematician, Dr. Meuller, estimated the probability of the evolution of one horse at one chance in one followed by a million zeroes! (This figure would require 1500 pages just to print!) Then Mrs. Horse had to evolve at just the right time and right place, didn't she? What are the odds on that? Then what about the rest of the whole plant and animal kingdoms? Do you begin to see how incredible (without credibility) evolution theory is? Evolution is just plain impossible. No matter how deeply ingrained the theory is in the world, a person has to deny all science and all logic to believe that evolution theory can explain life on earth.

FIGURE 1. These four anti-evolutionist arguments, taken from a Fair Education Foundation flyer, were used in the questionnaire.

1978; Macbeth, 1971). Their arguments are based on misunderstandings of concepts of neo-Darwinian evolutionary theory—most notably the concept of "fitness" (cf., Dobzhansky, Ayala, Stebbins and Valentine, 1977; Gould, 1977, for discussions of some of the problems). Such widespread misunderstanding of the *basis* of Darwinian evolutionary theory demonstrates most forcefully that there is no general public understanding of Darwinian evolutionary theory as a predictive science (but, cf., Bock and von Wahlert, 1965; Gould, 1970; Godfrey, in press). Clearly, also, the evolutionary theory of many introductory classrooms is little more explanatory than the creationist doctrine it proposes to replace. Students are exposed to a smattering of evolutionary theory and a smattering of "examples" of "well-adapted" organisms. The idea that evolution somehow *produces* well-adapted organisms fosters an image of a static present in which only "well-adapted" organisms have survived. The concept of adaptation is often poorly defined or wrongly equated with "that which has survived" in the present time slice. Students are rarely taught the *data* being explained by evolutionary theory, and they emerge with little or no predictive ability.

For example, when college students with some background in biology, anthropology, and earth science were asked to analyze four specific antievolutionary arguments published by the fundamentalist Fair Education Foundation, of Clermont, Florida (see Figure 1), many were unable to do so with any degree of sophistication. The students were given an open-ended questionnaire that instructed them to state their agreement, disagreement, or uncertainty regarding each of the four arguments presented, and then to discuss them on logical, theoretical and/or substantive grounds.

Here are some comments from one class ($n = 40$) in the survey. Some general reactions:

> I don't understand, so can't argue. This is new to me.

> I agree with these arguments. Evolution has too many impossibilities and I don't believe in it.

> Look, I can't answer these questions. Sorry I can't help you out, but this is beyond me.

> This is a farce. I believe we evolved, but I don't exactly know how.

Replies to specific statements included:

> Number 1 is convincing because of the number of species.

> Number 2 seems very reasonable, but who do I believe—you or them? It all sounds good. But ape and man seem linked in some way—I don't know how. Number 3 sounds good—so why do people still talk about evolution? Isn't the public getting the evidence from the scientists? It's their job to keep us informed.

> Number 1 is a bad argument. Number 2 is a good argument if all the statements are true. Number 3—another good argument if the guy isn't lying. Number 4—the last three arguments taken together form an excellent argument against evolution.

> Number 2 sounds reasonable. It sounds O.K. Actually, I have no idea if it's right or wrong.

> Number 4: If these calculations are valid, perhaps "evolution" as a theory should be reconsidered.

> Number 2: The logic is good—uncovering deceptions and farces. But there must be some way to explain one animal evolving into another. Somebody must have offered an explanation sometime. Number 3: What is "decay" to one person (or one period) might be "improvement" to another.

Number 2 is convincing. An individual could not dispute such an argument. Number 4 would raise doubts in anyone's head.

Number 1: There are many examples. They make you think about each idea, as they seem reasonable. Convincing and logical, but too forceful.

Number 4: Probability is right—it says it could happen and it *did*. A male horse arrived at the right time.

The problem extends beyond a general misunderstanding of "fitness" and "natural selection." (Fitness is *not* survival; natural selection is *not* the result of random or accidental, and therefore unpredictable, differential survival or reproduction.) The term *evolution* is itself commonly misunderstood. Biological evolution refers to change in the genetic composition of populations over time. Many people (not merely fundamentalists, but also the popular press and pro-evolution scientists) ignore genetic commonality and continuity when writing about biological evolution. For them, biological evolution means change, and the mechanisms need not be genetic. For many, "evolution" is imbued with an almost mystical directionality—an inevitable progressionism. But while evolutionary biologists recognize that changes in the genetic composition of populations *occasionally* give rise to greater developmental complexity of organisms, or to more complex interactions between organisms and their environments, neither "progress" nor "directionality" are central to the concept of biological evolution. In this context, however, it is interesting to compare definitions of biological evolution culled from fundamentalist tracts, the popular press, and science writers.

A flier from the Fair Education Foundation, quoting the Bible-Science Association of Western Pennsylvania, said:

Evolution is here defined as a real, natural, self-caused continuing uphill process—in energy, structure and information—which goes from disorganized to organized, from random order to ordered, from lower to higher, from simple to complex, from atom to amoeba, from molecules to man.

A writer for the *Houston Post* (Aug. 23, 1964) said:

Evolution, in very simple terms, means that life progressed from one-celled organisms to its highest state, the human being, by means of a series of biological changes taking place over millions of years.

This definition is cited in *Did Man Get Here by Evolution or Creation?* (Anonymous, 1967), a book distributed by Jehovah's Witnesses. *The Penguin Dictionary of Archaeology* (Bray and Trump, 1970) defines evolution as:

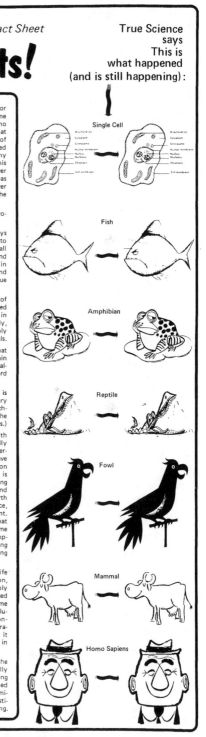

Evolutionary Scientists say This is what happened:

Single Cell

Fish

Amphibian

Reptile

Fowl

Mammal

Homo Sapiens

Evolution Theory Fact Sheet

Students!

There are only two explanations for how you and all other human beings came into existence. That's right: just two and no more. One of these explanations says that mankind started out thousands of millions of years ago when some kind of accident caused non-living matter to change into a very tiny living animal. This explanation says that this tiny animal then changed gradually over these thousands of millions of years and has become every living thing that has ever lived, including you and all the rest of the humans that have ever lived.

You know the theory. It is called evolution.

You know that this theory also says that every form of plant life came into existence by accident. This means that all the foods you know, all the flowers and trees, all the amazing processes that go on in the plant world such as photosynthesis and pollination, came into existence and continue as they are by complete accident.

You probably believe the theory of evolution is scientific. You hear it presented on TV. You hear it in school. You read it in books and magazines. More than likely, you believe that human beings are simply animals which evolved from lower animals.

This paper contains certain proof that evolution is **not true**. It contains certain proof that evolution is **impossible**. We challenge any scientist anywhere to step forward and deny these scientific facts.

Below are **ten reasons** why evolution is impossible. (Ask any science teacher if every reason is not 100% true. Then ask that teacher to help you throw this great lie out of the school systems, off TV, and out of books.)

1) The amazing earth; a globe with water, air, gravity, heat, soil, and literally thousands of other unexplainable characteristics, is assumed by pure evolutionists to have just happened by accident before evolution was even supposed to have started. It is unscientific to base a theory on something that is impossible, and it is impossible beyond any rational question to assume that the earth (and all the universe!) acquired its balance, intricacy, precision, and volume by accident. Yet, pure evolution starts by assuming that this magnificent earth just accidentally came to be like it is. This is an unscientific assumption and anybody who says it isn't is giving you an unscientific opinion and nothing more.

2) It is unscientific to say that life comes from non-living matter. This notion, called spontaneous generation, is simply rejected by scientists. Yet, evolution is based squarely on this rejected idea. At one time there was no life, only dead matter, evolutionists say. Then life came out of this non-living matter. That is spontaneous generation, an unscientific myth. Do you think it is right to teach lies as the truth on TV, in schools, in books and magazines?

3) Evolution theory is also based on the assumption that life evolved very gradually over **thousands of millions** of years during which time conditions on the earth remained virtually the same. This is called uniformitarianism. The dating methods used to estimate such time periods are beyond testing.

True Science says This is what happened (and is still happening):

Single Cell

Fish

Amphibian

Reptile

Fowl

Mammal

Homo Sapiens

FIGURE 2. From an anti-evolutionist group's "fact sheet" poster.

> The gradual change of form of living organisms throughout time, usual-
> ly...towards complexity and functional improvement.

The similarity of these definitions to each other is more striking than
their similarity to modern biological theory. Fundamentalists are not the
only ones who ignore the concepts of genetic commonality and continuity.
Consider the following statement made to *Penthouse* magazine by Robert
Jastrow, director of NASA's Goddard Institute for Space Studies and a
member of the Institute for Advanced Studies:

> We are on the way to being living fossils. But the history of life indicates that
> man is likely to be the rootstock out of which a higher form will evolve. It will
> not be a more intelligent man—man *is Homo sapiens*—but rather a new
> form, something beyond man. The question now is whether this new form
> will be a biological entity having puny limbs and a big head to accommodate
> the progression of intelligence. Will the brain of man continue to be housed
> in some hollow shell of bones, fed by blood vessels, from a model developed
> by the fishes 300 million years ago? Or will it be something different? I say
> that computers, as we call them, are a newly emerging form of life, one made
> of silicon rather than carbon. (Anonymous, 1978)

Freeman Dyson, a leading theoretical physicist, also at the Institute
for Advanced Studies, makes explicit a miracle explanation of human
evolution:

> For apes to come out of the trees, and change in the direction of being able to
> write down Maxwell's equations...I don't think you can explain that by
> natural selection at all. It's just a miracle. (Davis, 1978)

Antievolutionists can find sympathetic scientists to quote. Dr. Louis
Bounoure, Director of Research of the National Center for Science
Research in France, is quoted in one tract (Kutsch, 1978:6) as saying,
"Evolutionism is a fairytale for grown-ups. The theory has helped nothing
in the progress of science. It is useless." Similar quotations abound in
scientific creationist literature.

There is a serious need for innovative approaches to the teaching of
the evolution of complex adaptations. It should not be reserved for graduate courses, and it need not be. To be able to deal with actual data, make
sophisticated predictions, and test hypotheses is not beyond the ability of
the average student, and these are skills useful in everyday life—not just in
theoretical biology. Science should be neither worshipped nor feared.
People practice it daily as they solve problems and cope with reality, and it
seems reasonable to hope that they can be helped to acquire the tools to do
it as well as they can.

Too few people can apply the concept of predictive advantage when dealing with formal sciences, where it should be most explicit, let alone apply it (or other aspects of scientific analysis) to their own worlds. Too often science is simply a fact list in the classroom rather than an example of organized rational thinking.

Educators have a responsibility to resist political pressures urging them to bastardize the educational process by pretending that "all ideas are equal." Indeed, they have a responsibility to *improve* the teaching of sciences such as evolutionary biology at the introductory level in the hopes that educated people will be better prepared to handle the socio-political onslaught of modern life. Evangelists asking for belief in authority should not monopolize the mass media, whether they be creationists or evolutionists. Belief systems affect socio-political decisions, and creationism and other cult movements promising salvation to a select few are urging people to join, hold hands, and wait for miraculous solutions in the face of real economic hardship. Aside from the possible threat of a resurgence of racism and authority-based political oppression, the Western world is faced with the immediate threat of further debasement of the educational system. Can the critical problems facing humanity today be solved by cultists? Much though holding hands may have its psychological rewards, the material-world results may be oppressive.

References

Anonymous 1964. *Houston Post* (August 23).

Anonymous 1967. *Did Man Get Here by Evolution or Creation?* New York: Watchtower Bible and Tract Society.

Anonymous 1978. Interview with Robert Jastrow. *Penthouse* (October): 125-126; 136; 176.

Bainbridge, William S. 1978. "Chariots of the Gullible." *Skeptical Inquirer* 3(2): 33-48.

Baum, R. F. 1975. "Coming to Grips with Darwin." *Intercollegiate Review* (Fall): 13-24.

Bethell, Thomas 1976. "Darwin's Mistake." *Harper's* (February). (Reprinted in *Christianity Today* 21:12-15.)

———. 1978. "Burning Darwin to Save Marx." *Harper's* (December): 31-38; 91-92.

Bird, Wendell R. 1978. "Freedom of Religion and Science Instruction in Public Schools." *Yale Law Journal* 87(3):515-570.

Bock, W. J., and G. von Wahlert 1965. "Adaptation and the Form-Function Complex." *Evolution* 19:269-299.

Bray, Warwick and David Trump (eds.) 1970. *The Penguin Dictionary of Archaeology*. Baltimore: Penguin.

Cole, John R. 1978. "Anthropology Beyond the Fringe," *Skeptical Inquirer* 2(2):62-71.

Cook, Warren (ed.) 1978. *Vermont, B.C.* Castleton, Vt.: Castleton State College.

Davis, Monte 1978. Interview with Freeman Dyson. *Omni* (October):100-106; 173.

Dobzhansky, R., F. J. Ayala, G. L. Stebbins, and J. W. Valentine 1977. *Evolution.* San Francisco: W. H. Freeman.

Etzioni, Amitai and Clyde Nunn 1974. "The Public Appreciation of Science in Contemporary America." *Daedalus* 103(3):191-205.

Fair Education Foundation n.d."Evolution Theory Fact Sheet" (flier). Clermont, Fla.

Fair Education Foundation n.d. "$5,000 Reward and a Challenge to Evolution" (flier). Clermont, Fla.

Fell, Barry 1976. *America, B.C.* New York: Quadrangle.

Flew, Antony 1967. *Evolutionary Ethics.* New York: St. Martin's Press.

Godfrey, Laurie (in press). *Evolutionary Change: A Problem Solving Approach.* Compress, Wentworth, N.H.

Goodman, Jeffrey 1977. *Psychic Archaeology: Time Machine to the Past.* New York: G. P. Putnam's Sons.

Gould, Stephen J. 1970. "Evolutionary Paleontology and the Science of Form." *Earth Science Reviews* 6:77-119.

———. 1977. *Ever Since Darwin.* New York: Norton.

Himmelfarb, Gertrude 1968. *Darwin and the Darwinian Revolution.* New York: Norton.

King, Rachel 1972. "The Inadequacy of Naturalistic Selection" (review of Macbeth 1971). *Christianity Today* (June): 19-25.

Koestler, Arthur 1978. *Janus: A Summing Up.* New York: Random House.

Kutsch, Richard E. 1978. "An Alternative Viewpoint." *Students for Origins Research* (Goleta, Calif.) 1(1):5-6.

Macbeth, Norman 1971. *Darwin Retried.* Boston: Gambit.

Smith, R. and J. Knight 1978. "The Politics of Educational Knowledge: A Case Study." Paper presented to the meeting of the Sociological Association of Australia and New Zealand, May 20. (Cited with permission of the authors.)

Von Däniken, Erich 1970. *Chariots of the Gods?* New York: G. P. Putnam's Sons.

Note

1. Wichita State University professor Paul D. Ackerman, president of the Creation Social Science and Humanities Society, informs me that there is no official creationist program at WSU; his proposal for a creation-science undergraduate or graduate major under the new independent field major program has had no takers so far. But in the future, independent field major programs may be *the* important vehicle for advancing creationist views at universities; so the serious issue remains, regardless of immediate political contingencies.   •

Planetary Pinballs

Scientists and Velikovsky

Scientists Confront Velikovsky. Papers from an AAAS Symposium, San Francisco, Feburary 25, 1974. Donald Goldsmith, Editor. Cornell University Press, Ithaca, N.Y. 1977. 183 pp., illustrated, $8.95.

Reviewed by George O. Abell

The thirty-ninth edition of *Who's Who in America* lists Immanuel Velikovsky as "author, scientist." According to the *World Almanac of the Strange,* "It will probably take years before a definitive verdict is passed on Velikovsky's theories."

Who is Velikovsky, the author-scientist? And what are those theories that have raised such controversy for nearly thirty years? Velikovsky was born in Russia in 1895 and received the degree of Doctor of Medicine in Moscow in 1921. He later studied psychoanalysis in Zurich and he practiced medicine and psychiatry from 1929 to 1934. By the 1940s he had become interested in what he regarded to be interesting correspondences between myths of various early peoples and biblical stories. He invented an astronomical theory with which he hoped to account for these many ancient legends in terms of catastrophic encounters between the earth and other celestial bodies. His main theory was first publicly announced in his book *Worlds in Collision,* originally published by Macmillan and Company in April 1950 but transferred to Doubleday in June 1950.

Briefly, Velikovsky's astronomical hypothesis is that a comet was ejected from the planet Jupiter about the middle of the second millennium

B.C. Over the next several hundred years, the comet passed twice near the earth, causing great upheaval, and at various times rained insects and manna upon the earth. The comet also passed near Mars, diverting that planet from its orbit, so that Mars also passed near the earth, causing additional tidal disruptions. Eventually, about the seventh century B.C., the comet transformed itself into the planet Venus, which has occupied its present orbit ever since.

The astonishing thing to me is why Velikovsky's theory was ever controversial in the first place; it has certainly never been controversial among scientists. Indeed, anyone with even modest training in astronomy or physics would recognize the theory as patently absurd. It is not absurd simply because it is an unproven scientific hypothesis; in fact it is not a physical theory at all. It ignores what is known about the most basic laws of mechanics (such as the conservation of momentum, energy, and mass); it ignores gravitational theory; it violates the most fundamental principles of electricity and magnetism and thermodynamics; it even ignores the very nature of planets and comets. Yet Velikovsky describes his theory as though he were invoking these very physical principles (although in the preface he implies that his ideas may be incompatible with Newtonian theory). Most of us receive, from time to time, letters and tracts from people with highly imaginative "scientific" hypotheses, but which are clearly incompatible with what we know about physical law. Indeed, I have a very sizable "crank file" filled with such documents; yet many of these documents are not more outlandish than Velikovsky's theory.

Why, then, is Velikovsky more notable that the myriads of other cranks? Why should there be any controversy at all over ideas that are agreed to be utter nonsense by virtually all physical scientists? Evidently, it is because Velikovsky managed to have his ideas brought before the public by a major publisher. His first book, *Worlds in Collision,* received highly favorable reviews by such respected journals as *Harper's* magazine. In fact more than 50,000 copies of the book were sold in the three months before its publication was turned over to Doubleday. It is true that *Worlds in Collision* (like Velikovsky's later works) does not read like a crank book; on the contrary, it appears to be a very scholarly work, convincingly presented and well annotated with footnotes. In fact, when I read *Worlds in Collision* it appeared to me that although Velikovsky's astronomy was complete garbage he had really done his homework on archaeology, which of course is heavily involved in his treatment of varoius myths and biblical stories. I recall once having lunch with an archaeologist on our campus, and expressing my admiration for Velikovsky's archaeological treatment. To my surprise my companion, in considerable shock, announced that he felt Velikovsky was quite well versed in astronomy but that his archaeology was complete bunk. My friend informed me that Velikovsky had badly garbled archaeological data, and had even misplaced events by many centuries in time. I suppose it serves as an object lesson on how even the best educated of us can easily be led astray in fields far from our own areas of competence.

In any event, *Worlds in Collision* became a best seller, and was followed by *Ages in Chaos* (Doubleday, 1952), *Earth in Upheaval* (Doubleday, 1955), and *Oedipus and Akhnaton* (Doubleday, 1960). The 1970s have seen a new surge of interest in Velikovsky

have seen a new surge of interest in Velikovsky; in fact, between 1972 and 1974 the Student Academic Freedom Forum published the slick, quality magazine *Pensee*, devoted exclusively to the theories of Velikovsky and alleged proofs of his theories.

Despite his considerable success among the general public, Velikovsky and his followers could rightly claim that he was not only rejected but ignored by the scientific establishment at large. To answer the growing popular criticism of scientists for their attitudes toward Velikovsky, an eminent group led by Professor Carl Sagan urged the American Association for the Advancement of Science to sponsor a special symposium on the theories of Velikovsky, where he and his supporters could meeet face to face with members of the scientific community. The symposium was held at the San Francisco meeting of the AAAS on February 25, 1974. The book *Scientists Confront Velikovsky,* edited by Donald Goldsmith (one of the conference organizers), consists of papers presented at the conference.

In his introduction to the volume, Goldsmith describes the events leading to the conference and the proceedings themselves. Velikovsky was present and presented a contribution. The only person of scienftific training that the organizing committee could find who would support Velikovsky's views was Dr. Irving Michelson, of the Illinois Institute of Technology. (Michelson, according to Goldsmith, claims not to agree with all of Velikovsky's theories but is basically sympathetic.) Unfortunately, both Velikovsky and Michelson declined to submit written versions of their contributions for inclusion in the present volume. There is, in fact, an added invited chapter by David Morrison, Assistant Deputy of the Lunar and Planetary Programs at NASA, and an excellent foreword by Isaac Asimov.

The first formal chapter of the book is by sociologist Norman Storer, treating the subject of the sociological content of the Velikovsky controversy. Storer describes the anti-intellectual atmosphere of the early 1950s (during the height of the McCarthy era) and interprets the scientific community's reaction to Velikovsky as that of an institution feeling itself under attack. He describes the public acceptance of Velikovsky partly as a symptom of a general suspicion of scientists and intellectuals, and partly as a champion of the underdog; Storer suggests that the public's view of Velikovsky is that of a " 'Mr. Smith goes to Washington,' who brings the virtues of unadorned honesty and a simple concern for the truth into the midst of a self-preserving club of arrogant, powerful politicians." He also compares Velikovsky to the "loud, disputatious new kid in the neighborhood who jumps into the middle of an on-going ball game and thoroughly disrupts it for the players."

Here I think Storer misses the point, or at least misses a major sociological phenomenon associated with Velikovsky. I don't believe the scientific community felt itself the least bit threatened by Velikovskyism; by no stretch of the imagination could Velikovsky's ideas pose any challenge to scientific theory. To the scientist the ideas of Velikovsky are as asinine as the notion that an elephant could hatch from an acorn would be to the general public. Rather, I think the scientific reaction was one of frustration—frustration that Velikovsky was ever taken seriously, let alone believed, by the general public. Most scientists, especially astronomers, have long been concerned with public education; indeed, many scientists are also teachers, and a large fraction even give courses for nonscience students to acquaint them with scientific principles. I think the scientific reaction was largely one of bewilderment: "Where have we failed?"

When the average person's television goes on the blink, he calls a TV repairman, not a plumber. When his automobile doesn't start, he goes to an auto mechanic, hardly a shoe repairman. When he feels his appendix is inflamed, he goes to a physician, not a baker. When his house is infested with termites, he calls a pest eradicator, not a jeweler. Yet when it comes to the judgment of a new physical theory, that same person apparently prefers to believe a psychiatrist—one who professes himself ignorant of physical science—rather than the entire community of physical scientists who have devoted their lives to their subject.

Ironically those same people who rally to the support of Velikovsky, and reject the extraordinarily well-documented principles of conventional science, depend on that same science every day of their lives. The television sets they watch, the automobiles they drive to work, and the airplanes that fly them from place to place in the nation, all depend on the working of fundamental principles of science that one would have to reject if Velikovsky's hypothesis were to be correct. Fred Hoyle summed it up nicely by once remarking that if he were aboard a rocket ship going to Jupiter he would certainly hope that the trajectory of the vehicle was calculated according to Newtonian laws and not Velikovsky's.

Of course scientists do make mistakes. Of course there is always debate at the frontier of knowledge. Even unexpected ideas (such as continental drift and meteorites falling from the sky) sometimes turn out to be right. But new advances in science do not negate well-established laws and principles in the regimes in which they are know to apply. Rather they extend our understanding of the physical universe to new realms. The twentieth-century quantum theory and theory of relativity did not make Newtonian mechanics wrong; Newtonian gravitational theory had been extraordinarily successful in landing men on the moon and space probes on Mars, and in sending vehicles to other planets of the solar system. Rather, the quantum theory extended our understanding to the realm of the atom, and relativity extended our understanding of the behavior of objects to those with speeds near that of light, or in the vicinity of extraordinarily intense gravitational fields. Isaac Asimov summed up this matter extremely well in his foreword, where he described the role of the heretic: occasionally the heretic in science turns out to be correct, but almost without exception it is a heretic that comes from within science, and even then only rarely is the heretic right (Asimov estimates one time in fifty). Indeed, we have our heretics in astronomy. Not infrequently the *Astrophysical Journal* contains papers that the majority of astronomers regard as exceedingly unlikely, or even a bit nutty. But if a paper is not demonstrably wrong, it can generally be published, and at least is open for consideration.

I realize it is often, if not usually, difficult for the lay reader to distinguish science from fake science. As the frontier of science is pushed ever forward, it becomes ever more technical and the jargon so specialized that often scientists in closely related fields cannot read each other's technical papers. On the other hand when a scientist goes from one field to another one he generally seeks the advice of an expert in that field, or at least avails himself of review papers written by such experts. He certainly does not rely on a self-proclaimed genius who claims support only from an uninformed public. Although the followers of Velikovsky may think that they have found their Mr. Smith in Washington, they are actually following somebody who may be a bit crazy. For isn't there something psychotic about a person who claims that he alone, in a field with which he is unfamiliar, can fathom the pure truth, while hundreds of thousands of specialists with lifetimes of experience behind them are muddling about in darkness? And doesn't the popular acceptance of such a scientific-religious hero suggest a problem, or at least some kind of an unfilled need, on the part of the follower?

I do not think the reaction of the scientific community was like that toward the new kid from another neighborhood who jumps into the middle of an on-going ball game. A better analogy would be if the new kid jumped into a baseball game, but insisted on going to bat with a *jai alai cesta,* and proclaimed that he had made a home run when in fact he had struck out, and then the referees and fans all agreed with him! Here, I think, is a more significant sociological observation concerned with the Velikovsky phenomenon!

The heart of *Scientists Confront Velikovsky* is the excellent chapter by Carl Sagan. To be sure Sagan's is not complete analysis of all phases of Velikovsky's writings. He refers only briefly to the concordances in myth and legend, and certainly space does not permit a detailed analysis of every aspect of Velikovsky's theory. However, Sagan does address in detail ten particular problems with the Velikovsky model, such as the ejection of Venus by Jupiter, the stopping of the earth's rotation and the subsequent restarting of it with almost exactly the same period, the chemistry and biology of the terrestrial planets, the clouds of Venus, and the circularization of the orbit of Venus and nongravitational forces in the solar system. Sagan's account is brilliant, and although at times witty, he nevertheless takes Velikovsky seriously, and writes in a scholarly and persuasive manner on each of the problems he attacks. Needless to say, Sagan's analysis is devastating to the Velikovskian theory.

Were it not for the singular history of the Velikovsky affair, I might have found Sagan's analysis rather like nit picking. It would have been as though he were attacking the theory that the earth is flat and is supported on the backs of four elephants (I have read that such a view actually was held by at least one ancient civilization) by analyzing the weight distribution of various portions of the earth on each elephant, and considering the structural integrity of the skeletons of the elephants, to see whether they could, in fact, support the required weight. Without background knowledge, one might have been inclined to say it is absurd because the earth is known to be round, and besides, why should there be elephants anyway, what are they standing on, and what do they eat? But if a large fraction of the population takes seriously the view that the earth is supported by elephants, then one must attack the theory on its own grounds. Similarly, Sagan attacks Velikovsky on his own grounds.

Sagan also considers the so-called proofs of the Velikovsky model, such as the claim that the high surface temperature of Venus proves it must be young, or that radio radiation from Jupiter shows that it must indeed have ejected Venus in the recent past. Sagan clearly demonstrates how these proofs are either wrong, actually refute the Velikovsky hypothesis, or are irrelevant. (After all, it is realized that the breathing of the elephants should shake the earth; there are known to be earthquakes, therefore the elephant hypothesis must be correct.)

Sagan's contribution is followed by a chapter by J. Derrall Mulholland, a well-known expert in celestial mechanics in the Department of Astronomy, University of Texas. Mulholland gives a nice review of celestial mechanical consequences of the Velikovsky model which clearly violate observations.

An excellent account of cuneiform evidence for early observations of the planet Venus is given by Peter J. Huber, professor of mathematical statistics, Eigenossische Technische Hochschule, Zurich. Huber's chapter is a very scholarly and carefully prepared paper that clearly lays out the evidence, in the form of cuneiform records, that Venus had been observed far earlier than the time Velikovsky claims it was formed. There are certainly records of observations going back to at least 1900 B.C., and very probably of observations that date back to nearly 3000 B.C. Huber also presents overwhelming evidence that Venus was extraordinarily well observed, and that moreover its orbit must have been essentially the same as it is today, during the period around 1500 B.C. at least seven or eight centuries before it had been formed according to the Velikovsky hypothesis. Huber's arguments alone are sufficient to completely rule out the Velikovsky view.

A final chapter by David Morrison reviews the evidence relevant to Velikovsky's hypothesis that comes from planetary astronomy. He points out clearly that the surfaces of the planets must be old, that the hot surface of Venus cannot have resulted from a recent origin, that the chemical compositions of the atmospheres of Jupiter and Venus are completely incompatible with a Jovian origin for Venus, and that the very nature of the surfaces of the terrestrial planets is completely incompatible with recent (last few millennia) violent encounters between them.

There is a substantial overlap between Morrison's and Sagan's chapters; in fact Morrison's only illustration is a repeat from one presented by Sagan. On the other hand, Morrison's chapter is relatively brief, very clearly written, and is an excellent summary and review; I think, therefore, it is a valuable addition to the book for the lay reader.

It is regrettable that the contributions from Velikovsky himself and from Michelson are not included. However, an article adapted from Velikovsky's address at the AAAS symposium is now available in the November/December 1977 issue of *The Humanist* magazine, along with an abbreviated version of Sagan's chapter in *Scientists Confront Velikovsky*, and also an additional paper, titled *Afterword—1977,* by Velikovsky. The reader will find these additional contributions by Velikovsky interesting, but hardly revealing.

One would like to say that *Scientists Confront Velikovsky* has finally buried this novel notion of the origin of part of our solar system. I'm afraid that it has not. Those who can follow the book's rationality never took

Velikovsky seriously in the first place (although the arguments presented therein will prove useful in debates with believers). On the other hand, those who follow Velikovsky do so not out of reason but out of emotion, as a belief in a sort of neo-religion, and perhaps as a way of rebelling against the authority of science, mistakenly thinking that scientific laws can be changed or overthrown as can those of society, by will or revolution. But for all the Velikovskian protest, Newton's laws will prevail. (As Galileo is apocryphally said to have remarked, "The earth still turns.")

At the time of antiquity, or perhaps even through the dark ages, Velikovsky's hypothesis may have been a viable one. Little was then known about the true natures of the other worlds, nor of the physical laws that govern their behavior. But we've come a long way since then, as our modern technology attests. Velikovsky's ideas concern well-trodden ground with which we are very thoroughly familiar, and are simply incompatible with long-understood and extraordinarily well-documented principles.

Yet I suspect that the new religion of scientific superstition will continue. Our hope is that it will not prevail among the majority of our citizenry. We need the rationality of science to survive the rises of over-population, pollution, depleting energy sources, crime, and man's inhumanity to man, brought about by those very people who reject reason, and turn instead to the murky occult, with its superstition, blind acceptance, and prejudice. As Voltaire said, "Men will cease to commit atrocities only when they cease to believe absurdities."

The Velikovsky Affair

Part I

Ideas in Collision

James Oberg

In the past, revolutionary theories have "turned the world upside down" only metaphorically, but in the case of the writings of the late Immanuel Velikovsky, these words should be taken quite literally. Velikovsky and his followers claimed that they had identified a series of ancient interplanetary cataclysms, during which entire worlds somersaulting in space were involved in disastrous near-collisions. These events involved the planet Earth during the fifteenth and eighth centuries B.C., and memories of them, according to Velikovsky's view, survive in the mythology and folklore of nations all over the world.

Thirty years after the controversy became an issue in the public forum, "Velikovskianism" still exists, but in a form that neither its proponents nor its opponents could have imagined a generation ago. For those who were sure that his theories marked him as a new Galileo, as a genius whose vision would completely overturn the dogmas of the centuries, it would have been unthinkable that he would still be a scientific leper three decades later, his theories shunned and ridiculed by the vast majority of practicing scientists. For those who saw him as a calculating fraud, a crank, and a crackpot, it would have been unthinkable that a vigorous intellectual community would still exist today that not only embraces the world view of Velikovsky but also claims that all scientific progress since 1950 has borne out his vision and brilliance. The Velikovsky "cult" has not shown any signs of withering away as scientists imagined it would.

On the occasion of Velikovsky's death last November in Princeton, New Jersey, at the age of 84, commentators were faced with an unanswerable dilemma. Is Velikovskianism the wave of the future after all, whose day has simply been suppressed by scientific inertia and closed-mindedness? Or has it become just another dead-end of pseudoscience, one of the best case studies in "pathological science" to come along in this century? There are advocates of both views; but, although it's impossible to write the last chapter, most outside observers lean to the latter judgment.

The origin of this fascinating interplanetary-catastrophe theory lies with a man whose life story is fascinating in itself. Born to a Jewish family in Vitebsk, Byelorussia, on June 10, 1895, Velikovsky (whose name in Russian means "the great") was a brilliant young scholar who, after finishing high school in 1913, spent a year traveling in Europe and Palestine. After taking a few preparatory medical courses at the University of Edinburgh, he returned to Russia in the summer of 1914. He went through the horrors of World War I, the Bolshevik Revolution, and the subsequent civil war while a medical student at the University of Moscow. After graduating in 1921, Velikovsky made his way to Berlin, where he met and married a young concert violinist. After working with several psychiatrists in Vienna, he emigrated with his family to Tel Aviv, Palestine, and in the summer of 1939 to New York City, one step ahead of World War II and the anti-Semitic insanity of Nazi Germany.

For the second time in Velikovsky's life the world was engulfed in mass murder, and he turned to ancient myths for intellectual refuge. By the spring of 1940, as Nazi armies smashed into Paris, he had outlined the thesis that would consume his energies for the next 40 years. "I felt," he later recalled, "that I had acquired an understanding of the real nature and extent of that catastrophe." By the fall of 1940, as Nazi warplanes decimated the British Royal Air Force over London and nothing seemed to stand between Hitler and world domination, Velikovsky completed his story of planetary catastrophes, beside which even world war and genocide paled to insignificance. He spent the next nine years perfecting the documentation, during which time (in his own words) he daily "opened and closed the library at Columbia University."

Velikovsky had created an intricate ballet of celestial encounters that repeatedly devastated the earth (he called his work a "reconstruction" rather than a "theory"). Venus was one of the key planets, having sprung from Jupiter during a close encounter with an errant Saturn. The new-born world of Venus was enveloped with gases and appeared as a comet, periodically swinging deep into the inner solar system. After a few hundred, or a few thousand, years, Earth's luck ran out, and it nearly collided with "comet" Venus. The resulting series of disasters and near-collisions caused the parting of the Red Sea, the plagues of Egypt, the fall of the walls of Jericho, and the "sun standing still" over Gideon—such is the "reconstruction" of Velikovsky. Venus, meanwhile, while settling into a more peaceable orbit, knocked Mars off course; the Red Planet, in turn, made a series of equally devastating visitations seven centuries later.

And all this constituted only the final act in a drama that Velikovsky envisaged as going back thousands of years. Earlier, he wrote, Mercury, too, had been involved in planetary encounters; Jupiter and Saturn had been double planets, and Earth had been a satellite of one of them; Saturn had exploded, showering Earth with the waters of the Deluge; the moon had come in from somewhere and had somehow been captured by Earth. All of this was supposed to have occurred "within the memory of man."

In *Worlds in Collision*, Velikovsky laid out this reconstruction. Published in 1950, it was preceded by a publicity campaign involving excerpts and summaries that aroused a great deal of popular excitement before the book appeared. Many scientists assumed it was a hoax designed to prove that the Bible was literally true; and since Macmillan brought out the book through its *textbook* division, the scientific community reacted violently, threatening boycotts and other retaliations. Macmillan passed the rights to Doubleday, even though the book was on its way to becoming a nationwide best-seller. Widely acclaimed by the news media as a new Leonardo da Vinci, Pasteur, Galileo, and Einstein rolled up into one, Velikovsky was denounced and vilified by practicing astronomers—and he returned the feelings.

Fans of Velikovsky made much of his rejection, claiming that it was a symptom of the fossilized nature of scientific neanderthals who would never be able to change their minds about anything. In *Harper's* magazine in June 1951, Velikovsky wrote that the older generation of scholars "are for the most part psychologically incapable of relearning." (This comment was completely negated by the next 20 years of science, in which hard evidence in a dozen fields led these same scientists to change their minds many times—except about Velikovsky!) "How Much of Yesterday's Heresy Is Today's Science?" demanded a headline in a pro-Velikovsky journal in 1972, obviously implying that past heresies automatically became future science. This was a counterfeit debating trick that ignored the fact that most heresies were actually true nonsense, although the converse (today's science is yesterday's heresy) *is* correct. All the crackpot "science heresies" of the past were ignored, and only the success stories were described, leading to a blatantly illogical claim.

In fact, Velikovsky *did* have his antecedents, however much his followers refused to face up to such previous analogues. In three separate cases, lengthy books had chronicled ancient legends as proof of inter-planetary catastrophes, often with uncanny parallels to Velikovsky's reconstructions. Yet, although Velikovsky must have read these books (they were available at the Columbia library) and used them as source materials, no mention of them has ever been made in his writings—perhaps because their authors have been totally discredited as crackpots. But they had not been forgotten by many of Velikovsky's critics, who saw him as only another spiritual successor to previous crank theorists.

The first was William Whiston, a British clergyman and mathematician, who in 1696 published *New Theory of the Earth*. In it, Whiston claimed that before Earth's disastrous encounter with a comet, there were exactly 360 days in a year and 30 days in a month (Velikovsky claimed this as well). The comet came by in 2349 B.C., causing the Noachian flood and the changing of the earth's orbit. Whiston found extensive evidence to support this theory, based on worldwide legends.

In 1882, the American politician-reformer-editor-novelist-crank Ignatius Donnelly published *Ragnorak*, in which he used 200 pages of widely circulated myths to reconstruct a story of a comet causing a worldwide catastrophe, including the sun's standing still over Gideon (Velikovsky also accounted for this by means of the passage of a comet). The critiques of *Ragnorak* sounded strangely like premonitions of the reception of Velikovsky's books 70 years later: according to one commentator, Donnelly's theory was "coherent in all its parts, plausible, not opposed to any of the teaching of modern science, and curiously supported by the traditions of mankind. If the theory is true, it will revolutionize the present science of geology."

The third recycling of this extraterrestrial catastrophe concept was the Hoerbiger-Bellamy "World Ice Theory." First published in Germany in 1913 (Bellamy took over the movement in the 1930s), it was, in the words of Martin Gardner, "one of the great classics in the history of crackpot science." In this case, it was falling moons, not comets, that had caused all of the world's legendary disasters. The theory took an ominous turn in Germany, where it provided ammunition for the Nazis, in the form of dozens of popular books and magazines that made believers out of millions of people whose rejection of "Jewish science" was manipulated by the Hitlerites into political power. Bellamy's books published in English did not have such a taint and were pure crackpottery, telling how a former moon of Earth had crashed into the planet some 13,500 years ago, being soon afterwards replaced by the current moon (which Velikovsky agreed was indeed captured at about that time). Although Velikovsky claimed that the moon was once covered with water that has since disappeared, the Bellamy books claimed that the moon still had layers of ice more than a hundred miles thick.

Velikovsky obviously did not see his work as in any way related to these earlier versions of the space-disaster theory. Indeed, he singled himself out as the one and only person capable of such a reconstruction: "Like the early memory of a single man, so the early memory of the human race belongs in the domain of the student of psychology. Only a philosophically and historically, but also analytically, trained mind can see in the mythological subjects their true content—a mind learned in long years of exercise to understand the dreams and fantasies of his fellowman." It is strange that he was able to claim such a uniqueness, since he must have known that it was only in the details, not in the concepts, that his ideas differed from those of Whiston, Donnelly, Hoerbiger, and Bellamy. But he had good reason, as we have seen, for portraying himself as the *originator* of these theories and for neglecting to draw attention to his predecessors.

As the years passed, Velikovsky claimed a series of important "successful predictions" based on his world view; such predictions vindicated him, or so he and his followers insisted: "Seldom in the history of science have so many diverse anticipations—the natural fallout from a single central idea—been so quickly substantiated by independent investigation," wrote one of his chief disciples, Ralph Juergens, in 1963. "The space age gave my views a record of confirmations," Velikovsky himself claimed in 1977. His detractors called these predictions "guesses," and not particularly good ones at that.

Chief Velikovsky critic Carl Sagan, for example, told a conference in 1974: "To the best of my knowledge, there is not a single astronomical prediction correctly made in *Worlds in Collision* with sufficient precision for it to be more than a vague lucky guess, and there are . . . a host of claims made which are demonstrably false." In 1979, NASA space scientist David Morrison elaborated: "Every important prediction [Velikovsky] made in 1950 concerning conditions on the planets, such as the hydrocarbon clouds on Venus, large amounts of argon in the atmosphere of Mars, recent melting of the lunar surface, large internal heat sources on Venus, large-scale recent cratering of Earth and Moon, and synchronized planetwide volcanism on Earth have been shown decisively to be in error. Every new space mission, such as the recent Pioneer Venus probes, pounds another nail in the coffin . . . The cruel truth is not only that astronomical evidence fails to support Velikovsky, but that a great deal that seemed plausible or at least possible when suggested in 1950 has been shown to be incorrect."

Velikovsky, or course, would have none of this: "The most despicable of all ways of suppression is denying me the originality and correctness of my predictions," he told a conference at Notre Dame in 1974. That same year, speaking in San Francisco, he elaborated: "My work today is no longer heretical. Most of it is incorporated in textbooks, and it does not matter whether credit is properly assigned . . . None of my critics can erase the magnetosphere, nobody can cool off Venus, and nobody can change a single sentence in my books."

Skeptics have suggested that the person most adept at erasing (or changing the meaning of) sentences in Velikovsky's books is Velikovsky himself. That is, many of his most famous predictions were allegedly based on reinterpretations of what he had originally written, which turned out to be false or were ambiguous enough to be reinterpreted to mean practically anything. Specific examples seem to bear out this criticism.

Take Venus, for example: Velikovsky predicted it would be giving off heat because of its short, violent history. Although it turned out to be hotter than scientists had guessed (mainly due to the unpredictably massive atmosphere), it is *not* giving off more heat than it gets from the sun, *nor* is it cooling off (as Velikovsky also predicted). On both counts, Velikovsky's predictions were in error. He also vigorously denounced the "runaway greenhouse theory" for the high surface temperature, asserting repeatedly that such a mechanism was impossible—if it was possible (as the latest space probes indicate), then his major prediction would have been blown away.

Turning to our moon, Velikovsky's followers make much of his prediction of "frequent moonquakes" and "remanent magnetism," but a closer look is less impressive. Moonquakes are *not* "numerous" as Velikovsky predicted—they are weak and rare. He said it was possible that astronauts on the moon for a few hours might "experience" (i.e., feel) a moonquake, but in fact they could remain there for hundreds of millions of years without that ever happening. The traces of magnetism in the lunar lava, which *did* surprise geologists, was in Velikovsky's view the result of the rocks melting and recooling only thousands of years ago. Yet *all* radioactive dating schemes indicate that the moon has not melted for billions of years. The recent melting predicted by Velikovsky supposedly happened in the presence of a magnetic field of a passing planet. (He suggested Mars or Venus, but neither has a magnetic field of anywhere near the needed power; so Velikovsky assumed that they used to have such a field but subsequently lost it.) However, the actual magnetic orientations of surface samples are randomly directed, indicating that they were formed in connection with individual cratering events. They do not show a planetwide consistent orientation as they presumably would in the case of an externally imposed field. Velikovsky predicted that lethal amounts of radioactivity would endanger the lives of the lunar astronauts, but none such was found. He predicted that the surface had been covered with water only thousands of years ago, but not a trace was found. He claimed that the craters had been created by bubbling of the surface and by interplanetary lightning bolts, not by meteors and volcanoes, but no evidence was found. He declared that "within the memory of man" the moon had been captured into orbit around the earth. These and other claims were so much at variance with every possible interpretation of the Apollo moon data that it was no wonder professional space scientists reacted angrily at the mere mention of Velikovsky's name.

Other cases of Velikovsky's false predictions can be found: the bountiful argon of Mars (Velikovsky was wrong, and his followers have been tripping over each other conjuring up mutually contradictory excuses), the canals of Mars (Velikovsky claimed they were fracture cracks, but they were really optical illusions—there are cracks on Mars but not of the kind Velikovsky predicted), the skies of Mars (Velikovsky predicted they would be black, but they are reddish), the temperature of Jupiter (Velikovsky said it would be cold, but he *did* guess that the atmosphere would emit radio waves—a near miss since it does, but the ionosphere emits even more radio waves), the Jovian Red Spot (Velikovsky said it was a scar on the surface of Jupiter, showing where Venus had emerged; but there is no surface of Jupiter, and the "red spot" is just a larger-than-average random hurricane), a Mars "richly populated by microorganisms pathogenic to man" (they avoided Viking samples), and a moon rich in oxygen, chlorine, and sulfur (these are in fact among the most depleted elements on the moon). Velikovsky was wrong on all of these predictions; yet by altering the meaning or by ignoring it entirely, his supporters claimed that he was right.

Philosophers of science have also objected to the often expressed notion that Velikovsky was rejected solely because he disagreed with the dogmas of the age. According to the pro-Velikovsky historian Joseph May, "Apparently [*Worlds in Collision*] challenged too many of those principles commonly believed to be necessary, though lacking in direct proof . . . It provoked the kind of reaction to be expected when the perceived needs of a community take precedence over the purposes of the community. One gathers that an important reason for the resistance to new hypotheses, like Velikovsky's, that require major revisions of theoretical structures lies in the vested interests and ego involvement of those who have devoted years of study under the guidance of the accepted assumptions or who have committed themselves in print . . . How ironic that so many humane people, proud of their liberalism in political or social matters, should vehemently defend the belief that science cannot function unless it intolerantly rejects departures from past belief."

Leroy Ellenberger, another pro-Velikovsky theorist and a professional industrial engineer, wrote in 1979: "Thus it is clear that under the influence of strong ego forces scientists who felt threatened by Velikovsky's thesis reacted with the violence they did. The impact of the original reaction persists to this day as many original participants are still alive and younger scientists appear to follow blindly the judgments made so hastily in 1950 and before."

That is certainly not a flattering view of how science functions today. Frankly, it is downright insulting, as historian Michael Jones wrote last year: "This kind of statement is exactly that which can be calculated to antagonize scholars most effectively, and provide a justifiable reason for thinking that Velikovsky's writing and followers are trivial and uninformed. The appearance of new evidence constantly causes scholars to modify their views," as indeed they have since 1950 in practically every field. Nor should it be ignored that there is considerable ego capital in embracing the pro-Velikovsky movement as well, since it is in effect a "shortcut" around all the details of contemporary science (which has been made obsolete by the new dogma) and a posturing of superiority over all believers in the "old" theories (and who would not, for example, like to be compared with Galileo and Newton and Pasteur?).

So what of the future? With the death of the originator of these latest reconstructions of Earth's entire underpinning, the movement may fade or may descend into internecine ideological struggles, since "holy writ" is interpreted differently by different students and there is no authority to appeal to. Or, perhaps, with the departure of the chief protagonist in what often became a personality conflict (with evidence often glossed over by both sides), any valuable contributions that might be gleaned from the Velikovskian reconstruction can be melded into the mainstream of ongoing (and ever changing) science.   ●

Part II
Passions and Purposes: A Perspective

Henry H. Bauer

For more than three decades the Velikovsky controversy has been unproductive. In all that time, no common ground has been established on which the two sides could carry the discussion further; indeed, the nature and flavor of the arguments remain essentially unaltered.

Immanuel Velikovsky, one can safely assume, wished to have his views accepted by historians and by scientists. He did not succeed: scientists have not performed experiments that he suggested and have not taken his ideas seriously enough to discuss them in scholarly journals.

Velikovsky's critics reveal a variety of motives. To combat pseudoscience is perhaps the chief one, but one gathers that they would also like reviewers to be knowledgeable about science, editors and publishers to promulgate only scientifically sound material, and the public to be critical rather than gullible.

Velikovsky's supporters want his ideas to be taken seriously by the specialists. Beyond that, they want scientists always to behave impeccably, in accord with the highest ideals regarding scientific activity, and to reform their ways when they do not so behave; they also want "catastrophism" to be accepted as a scientifically plausible world-view.

Neither critics nor supporters have been markedly successful. Pseudoscience flourishes; reviewers are much as they were 30 years ago; and much unsound so-called science is published, and the public laps it up. Velikovsky's ideas are taken seriously by only a tiny number of specialists; scientists behave much as they did 30 years ago; "catastrophism" is generally regarded as a cult, not as a hypothesis in the mainstream of scientific thought. It is not easy to find anyone involved in the Velikovsky debate whose objectives were reached.

Much frustration and anger, and much self-defeating behavior, result when actions are aimed at accomplishing the impossible; or when the goals of one's activities have not been clearly formulated; or, even if the goals are clear, if the means used in attempting to achieve them are unrealistic—not capable of leading to those goals. One sees all these factors at work in the Velikovsky affair.

Velikovsky sought to be accepted as a scientist, but he broke the accepted rules by which scientists work. Velikovsky's critics wanted everyone to share their own attitude toward science and scholarship— hardly a realistic goal. Velikovsky's supporters could not bring themselves to accept the realities of what science is and how scientists work.

Further, almost everyone in the affair failed to make a crucial distinction in seeking their objectives. They failed to distinguish between those intellectual activities by which knowledge eventually becomes an accepted part of the human heritage and those political or social activities that are appropriate in attempts to reach a purely contemporary acceptance of some particular point of view. As a result, the discussions were a confused mixture of attempts at logic and weighing the evidence at one extreme, and of propaganda and attempts to exert authority at the other.

Debates are hardly likely to be productive when so many points of confusion exist. Furthermore, in situations where opinion becomes polarized and emotions are aroused, those who seek even-handedly to clarify the issues are welcomed by none of those who are arguing. It would be quite futile, of course, to suggest that participants in such controversies should act in some other manner—if they could, they would. But it might be useful to recognize how an individual who partakes in controversy might best serve his own purpose.

First, one needs to define clearly what one seeks to accomplish. Second, one needs to judge whether the goal is attainable. Third, if it is, one needs to find the best means toward reaching that goal. I am hard put to find anyone in the Velikovsky affair who behaved as though he had followed those simple steps. The seeking of unrealistic goals was characteristic of all concerned, and the means chosen were not well suited to attaining those objectives even if they had been attainable.

Both sides wrote and spoke as though they wanted to persuade the other. How might one best attempt such a task? One way might be to look for evidence or viewpoints that would appeal to the other side. Another might be to use modes of expression least likely to offend those persons who are to be persuaded. Was there anyone in the affair who behaved in that fashion?

From the beginning, Velikovsky let the scientific community know, in no uncertain terms, that he stood in dead opposition, that he considered some major accepted "truths" to be wrong, and that the community was hidebound and psychologically incapable of recognizing the real truths. His supporters followed that lead.

The critics did not take the time and trouble to explain in detail what their objections were; they asserted them and said that it would be an unwarranted effort to go into the details. So the goal was perhaps clear, but the method adopted was one that could not lead to that goal.

So the controversy was rarely waged on an intellectual plane—it was chiefly a battle of propaganda. And in that the critics were much less adept than were Velikovsky and his supporters. For scientists engaged in these types of ventures, it would be very useful to be clear about what they are seeking: an intellectual resolution of the substantive issues, or acceptance by a wide public that the experts are right in their judgment. These are two quite distinct things; but scholars, used to arguing among themselves—where intellectual resolution and acceptance by the peer community tend to go hand in hand—do not often recognize what they need to do if they wish to achieve acceptance of their views outside their own sphere of professional activity.

The Velikovsky affair, then, has been passionate and unproductive—it has been both for the same reasons: interpretations were mistaken for facts, personal wishes were translated into absolute "shoulds." Other controversies frequently evidence the same processes; but it is not inevitable that this be so. Each of us can become more dispassionate and more likely to attain an objective, by pausing before acting or speaking out of emotion and by asking:

Is this occurrence *really* as I picture it? (Can I prove this to be pseudoscience? Am I absolutely sure that this is a conspiracy by the scientific establishment? Is this objectively an insult?)

Which of my beliefs are involved here? Are those beliefs rational? (Can pseudoscience be eliminated? Will scientists ever behave other than in the way they behave?) Am I translating a personal wish into a "should"? (If so, don't.)

What do I wish to accomplish? What are the best available means?

Those who have never systematically practiced such a procedure will be surprised at the extent to which calm replaces anger as one works through these questions. Emotions are to that extent under one's own control: each of us chooses among pleasure, calm, anger, and all the rest—by choice of *interpretation* of an event and by choice of *belief.* Anger can be lessened—even avoided altogether—to the degree that one's beliefs and interpretations are in tune with external reality. If one understands what types of things actually happen, and accepts—since one has no control over it—that they are bound to happen, then one will not say to oneself that they "should" not happen. One may not *like* the occurrences, and one is certainly free to do whatever is possible to make things more to one's liking—but one will not be consumed by helpless rage and will thereby be freer to work effectively for the desired result.

The Velikovsky affair has been overwhelmingly a battle of polemic and propaganda. The commonly held view that substantive points of science or fact were at stake is only partly correct. In any event, the manner in which the arguments were carried on made resolution of substantive points impossible. Passions were generated and released, but never purposefully harnessed.   ●

Part III
The Distortions Continue

Kendrick Frazier

Immanuel Velikovsky may have died last November, but the promotion of his eccentric brand of planetary-origin theory continues undiminished. An advertisement in the April 20, 1980, *New York Times Book Review* symbolizes the start of the posthumous stage of Velikovskianism and serves as an interesting mini case-study of the subject.

The ad, placed by Doubleday, the publisher of all of Velikovsky's books in hardcover, is titled, "Velikovsky: The Controversy Continues." The ad reminds us that all of Velikovsky's books are still in print. It also announces that he left behind manuscripts for several new books "which will doubtless . . . interest intelligent and open-minded readers in search of an understanding of the forces that have shaped our world."

But it also briefly reviews the Velikovsky controversy, labeling it "one of the most heated controversies in the history of science," and goes on to make some specific assertions about recent discoveries of planetary science and Velikovsky's role in supposedly anticipating them.

What particularly caught my eye was the statement that findings of the recent Pioneer probe of Venus's atmosphere supported Velikovsky's theories about that planet's origins. Those theories, you will recall, postulate that Venus was born about 3,500 years ago as a comet ejected out of Jupiter and that this comet grazed Earth around 1500 B.C., causing, among other things, the earth to temporarily stop rotating and accounting for various biblical legends. Then, in the seventh and eighth centuries B.C., the comet knocked Mars into several near-collisions with Earth and the atmospheres of all three intertwined. The comet then settled into a circular orbit and became the planet we now call Venus.

"Velikovsky's assertion that Venus is a young planet, expelled from Jupiter only thousands of years ago," the Doubleday ad states, "has received strong support from the evidence of the Pioneer probe which revealed so little surface erosion and such an abundance of primordial Argon 36 that scientists now speculate that Venus was formed more recently than the other planets and by a different process."

The ad also states that Velikovsky claimed that space is not a void "but is filled with electromagnetic fields and radiations. This claim completely contradicted the accepted view. But the first satellites encountered magnetic fields, solar winds, and radiations, and 'heresy' became fact." It also mentions Velikovsky's claims that Venus should be hot (due to its recent violent birth) and that "Venus should be surrounded by a blanket of petroleum hydrocarbons." Even this last prediction, the ad claims, has now been verified. "Methane (hydrocarbons) has been found on Venus."

I have always been fascinated by planetary astronomy, and as a professional science writer and editor I have closely followed the past decade's striking discoveries about the planets. This is a remarkable age of discovery. In just the past four years our spacecraft have landed on Mars, sent probes down through the atmosphere of Venus, photographed Jupiter and its moons at close range, and flown inside the rings of Saturn. We have discovered rings around Uranus and Jupiter, a fourteenth and fifteenth satellite of Jupiter, another set of rings and several moons around Saturn, and volcanoes in eruption on the Jovian moon Io. We've detected lightning on Venus, directly measured the composition of its dense atmosphere, and found a surprisingly ordered wind system around the planet. Soviet spacecraft have landed on Venus and photographed its surface. We've found the solar system to be extraordinarily interesting.

But I could not seem to recall any report that we've discovered Venus to be a very young planet, born in historic times and scooting about in the neighborhood of Earth as recently as 2,700 years ago. Certainly I couldn't remember any such announcement coming out of the Pioneer Venus mission to that planet in late 1978. But perhaps I missed something.

I decided to send copies of this ad to a number of noted planetary scientists and ask them about its assertions. Most of the scientists queried had participated in the remarkable observations and studies of the planets— especially Jupiter and Venus—in the past few years. They, more than anyone, would know whether, as the ad claims, the recent findings have verified Velikovsky's theories. With only a few exceptions they had not previously spoken out on Velikovsky's theories. I received responses from about half those queried and followed up by phone calls to about half of the nonrespondents.

There was overwhelming agreement from the planetary scientists that the substantive claims in the ad were wrong. Some put it far stronger than that. "The ad is thoroughly dishonest," said A. G. W. Cameron of the Harvard-Smithsonian Center for Astrophysics. "The ad, like the Velikovsky books it is promoting, contains more falsehoods in a paragraph than one can refute in a chapter," said Edward Anders, a University of Chicago cosmochemist who has specialized in chemical studies investigating the early history of the solar system. "No reputable scientist 'now speculate[s] that . . . Venus was formed more recently than the other planets,' " said Anders.

"The statements [in the ad] are not accurate," said geophysicist William M. Kaula of the University of California at Los Angeles. "As a Pioneer Venus co-investigator, I am not aware of any support in the scientific community for Venus being a young planet. The Pioneer probe did not reveal anything conclusive about surface erosion; its relevance to origin is very slight, in any case. The excess of argon-36 suggests, if anything, that Venus is older, since the solar wind swept volatiles out of the nebula with time."

"Several of the statements in the ad are outright lies," said University of Hawaii astronomer David Morrison, particularly the assertion that scientists now speculate that Venus is a young planet. "While the large amount of argon-36 discovered by Pioneer has indeed upset current ideas about the conditions (particularly pressure) in the solar nebula when the planets formed, a recent birth of Venus is not in any way indicated."

"Anyway," said UCLA astronomer George O. Abell, "Velikovsky never mentioned argon-36." Abell and Morrison, a former deputy director of lunar and planetary programs for NASA, are two respondents who have in the past tried to clarify issues in the Velikovsky debate.

According to chemist Vance I. Oyama of NASA's Ames Research Center, the argon-36 abundances for Mars, Earth, and Venus are con-

The publisher's ad and its key claims

sistent with the model that volatiles are incorporated into the grains that formed these planets, provided the large pressure differences, but small temperature differences, characterized the region of the nebula that gave birth to the solar system some 4.5 billion years ago. "The argon-36 amounts in the atmosphere then are suggestive that Venus's position in the solar system is approximately correct," Oyama said.

Michael B. McElroy, a Harvard University atmospheric chemist and a specialist in studies of planetary atmospheres, put the argon/age question into perspective. The noble gases on Venus are indeed abundant and puzzling, he said, and in his opinion Venus may be *slightly* younger than Earth. "But, by slightly, I still mean 4.5 billion years old," he said. McElroy thinks the likely age difference is only about 1 million years. As for the ad's assertion that scientific support has been found for Venus being a young planet, "Obviously, I don't agree," he said.

Even if there had been any serious questions about Venus's age, they would have been settled this past spring. On May 28, scientists announced that the Pioneer Venus satellite's radar indicated the presence of an apparently ancient supercontinent on the planet, with a crust that could be 4 billion years old.

McElroy, who said he had read Velikovsky's *Worlds in Collision*, noted the difficulty scientists have in responding to claims that the author's predictions have been verified. Velikovsky did say Venus should be hot, for example, and it is; but he predicted that it would be hot because he thought it had been born as a comet out of Jupiter 3,500 years ago. "That," says McElroy, "is not true." A pronounced "greenhouse effect" is the leading explanation for Venus's high temperatures. That view has been strongly supported by the Pioneer Venus data.

What about the ad's claim that Velikovsky's prediction of "a blanket of petroleum hydrocarbons" surrounding Venus has been verified by a recent discovery of methane on Venus? That assertion is false, both in fact and in implication.

Oyama was the principal investigator for the gas chromatograph experiment on our Pioneer probes of Venus. "The Pioneer Venus gas chromatograph was capable of measuring methane, ethane, and propane, but none of these molecules could be detected," he said. "If they exist in the Venus lower atmosphere, their level is less than a part per million. It may be concluded that the surface of the planet is hardly likely to be 'blanketed' with a layer of petroleum."

"Hydrocarbons have been found on Venus in only trace amounts," said Kaula. Ethane (C_2H_6) of less than one part in a million is the only one reported from the mass spectrometer, he said, while the gas chromatograph found none and set upper limits of less than 5 parts per million of methane, ethylene, ethane, and propane. "Such drastically low levels are evidence against Venus's having come from Jupiter, since it would be difficult to imagine a process of stripping these gases, abundant on Jupiter, off a planet so completely if it emerged therefrom."

The petroleum hydrocarbon assertion "is an example of the meaninglessness of Velikovsky's 'predictions,' because of their lack of quantitativeness," Kaula said. It should be remembered that Velikovsky's references to Venus's petroleum were quite literal. He wasn't referring to trace gases. He believed that from the comet Venus "oil rained on the desert of Arabia and on the land of Egypt" and that this "rain of fire-water contributed to the earth's supply of petroleum." Was Venus a regular cosmic gusher? Alas, the facts don't support this grand conception.

The respondents similarly labeled as absurd the ad's claim that Velikovsky first predicted the fact that interplanetary space is filled with electromagnetic fields and radiations. "Before Velikovsky, every astronomer knew that space was filled with electromagnetic radiations and fields," said one. "How else could light get to Earth from distant galaxies."

"Everything we see in the sky we see by means of electromagnetic waves from those objects," pointed out Abell. "Velikovsky, of course, didn't know what 'electromagnetic radiation' means."

"That interplanetary space is not a void, and in particular that the earth has a magnetosphere, was predicted, and serious calculations were carried out by scientists long before Velikovsky," noted Kaula. He referred to Stormer's pre-1910 calculations on charged particles in the magnetic field to explain the aurora, Chapman and Ferraro's solar-wind models in the 1920s and 1930s to explain magnetic storms, and work before 1950 by Bierman, Alfven, Forbush, and others on solar-emitted ions and their interactions with comets, the earth's magnetic field, and so forth.

Several respondents also disputed the ad's implication that the Velikovsky controversy was among scientists. "The Velikovsky hypothesis was *never* controversial among scientists," said Abell. "It is, and was recognized at once as, a crank idea." Said another, "There is no controversy about Velikovsky in the science community. It is only between scientists and admirers of Velikovsky who are not experts in astronomy, physics, chemistry, or geology." Those comments were borne out by the other respondents, none of whom considered Velikovsky's views serious scientific hypotheses.

A number of respondents addressed broader scientific and philosophical isues in the Velikovsky debate.

"Science requires quantitative statements presented with likely errors," said Edward L. Fireman of the Harvard-Smithsonian Center for Astrophysics. "Science does not accept predictions (whether right or wrong) unless accompanied by a statement of the required assumptions and the proper deductive formalities. Any soothsayer can make vague predictions about Venus or any other subject and later boast that most of the predictions turned out to be right when the measurements are done. Scientists, in contrast, usually discuss the discrepancies (and errors) between the predictions and the measured results and try to find possibilities for improvements." That point was reiterated by Abell. "Velikovsky's predictions were vague and general, and the so-called predictions said to come true were all formulated post hoc."

Several of the planetary scientists contacted felt that scientific response to the claims of Velikovsky promoters only served to give the mistaken impression that scientists took his views seriously. "I personally don't think [the ad] deserves comment," said one of the principal investigators in our Voyager mission to Jupiter. "I'd be happy to comment on things for which there is some more supportive information."

A NASA research scientist prominently involved in the Pioneer Venus mission said, "Quite frankly, I really don't want to get involved in anything to do with the subject of Velikovsky. I think a mistake was made by some people in the early years in taking him too seriously. It only led to more sales and publicity." As for the Doubleday ad, this scientist said simply, "I don't know where they got some of their claims. There were a number of misstatements." He was one of several who felt the overwrought scientific reaction of the early 1950s was a mistake. McElroy, while not agreeing with Velikovsky's views, also regretted that Velikovsky had been "treated harshly" in the fifties.

One of the most thoughtful replies came from Johns Hopkins University geophysicist Walter M. Elsasser, originator of the dynamo theory of the origin of the earth's magnetic field. "I not only have read some of Mr. Velikovsky's writings but even knew him personally at one period of my life," said Elsasser. "My only drawback is that I am a reasonably successful academic scientist, which in the eyes of some segments of the public makes me a representative of 'vested interests.' I do not, however, consider myself as biased in the matter.

"I have often observed Velikovsky's method of making 'scientific' predictions. He makes a whole string of predictions. Then he waits until some experiments show up, and from the results he picks those that happen to verify some things that he predicted previously, forgetting simply all the other predictions of his that were not verified. This, in my opinion, and in that of anybody familiar with scientific method, is not a scientific procedure at all; it is a form of fraud.

"I do not think that Velikovsky as a person had any intent to defraud . . . He was undoubtedly a man of phenomenal intelligence; and it is a tragedy to see such talent being used in an utterly asocial manner as he did by trying to give his personal fantasies a 'scientific' validation."

To return once again to the ad and its assertions: every scientist contacted for this article felt the ad was at best misleading, at worst dishonest.

"I am used to distortions by the Velikovsky supporters, but this ad seems to be particularly reprehensible," said David Morrison. "I would think Doubleday could make a perfectly good case for selling Velikovsky's books, which have long enjoyed a wide and enthusiastic readership, without sinking to the level of this ad."

"If Doubleday were trying to sell appliances rather than books with such doubletalk," said Edward Anders, "it would be in trouble with the FTC, the Better Business Bureau, and perhaps even the U.S. Chamber of Commerce."

No one disputes the right of a publisher to promote its books in any way it sees fit. We've long learned to accept exaggerations and distortions in advertising of all kinds. And Doubleday, a respected publisher, could probably even make a case that the phenomenal economic success of Velikovsky's books over the decades has helped enable it to market less lucrative but legitimate works of science. There is no reason to believe that as a publisher it is motivated by a desire to promulgate pseudosciences; it has published many excellent science books, including works by Isaac Asimov and astronomer E. C. Krupp that have included strong critiques of Velikovskianism.

But having said all that, scientists and others also have the right to criticize and evaluate the scientific claims a publisher makes on behalf of a work of fringe science it is promoting. Assertions like those made in the Velikovsky ad have long helped reinforce the idea among the well-meaning public, including some portions of the nonscience academic community, that Velikovsky did not get a fair hearing. I have attempted in this article to determine whether several very specific assertions have any validity. They do not.

There is near-zero hope, nevertheless, that the promoters and consumers of Velikovskian confusion will suddenly change their ways. The psychological appeal of Velikovskianism to certain segments of society is too powerful. It's best to keep one's humor and sense of proportion. As William Kaula said: "My normal second thought after being irritated by nonsense in the name of science is like the submariner's response to the urge to exercise: lie down until the feeling goes away, since the likelihood of satisfying resolution is low." But sometimes it is enjoyable, and maybe even marginally useful, to try.　　　　　　　　　　　●

Tunguska Echoes

James Oberg

A mighty midair explosion over a remote Siberian swamp is still sending echoes around the world, seven decades after it happened. Near the Tunguska River, in the summer of 1908, an object from outer space was annihilated in a detonation as powerful as a modern hydrogen bomb. If the object was a natural one, it serves as a warning of a repeated disaster over a populated region today; if the object was not natural, it serves as an interplanetary calling card, announcing to those who can recognize and decipher it that an attempt had been made at interplanetary contact.

The "Tunguska Event" remains a puzzle for science, even as it has become a fertile subject for science fiction and UFO speculation. Recently, two TV drama critics, Thomas Atkins and John Baxter, threw together a book called *The Fire Came By* (Doubleday, 1976), condensed in the February 1978 *Reader's Digest*, which insists that the only possible explanation for the event is that it was an exploding nuclear-powered interstellar spaceship. Soviet scientist Aleksey Zolotov makes the world news wires about once a year with a new version of his claim to have discovered radioactivity at the impact site. The flying-saucer subculture has firmly canonized the Tunguska Event as physical proof of UFOs.

Not surprisingly, traditional astronomers reject that interpretation as fanciful and unscientific, as "not required," and as insufficient. The leading standard theory for the 20-megaton (equivalent to 20 million *tons* of TNT) explosion is that it was a comet nucleus that exploded upon hitting the earth's atmosphere.

Explosion of comet nucleus over the Siberian swamps on the Tunguska River, as conceptualized by astronomers in the standard explanation for the 1908 blast.

It is a first-rate scientific puzzle. Its uncertainties and mysteries provide plenty of dark corners in which all manner of far-out theories can safely lurk. And it is an excellent case study in how different schools of thought use and abuse facts.

So let's look at the Tunguska Event the way it is being reported (and exploited) today and the way it is being scientifically researched. Let's examine the media standards that have been used to inform or misinform the public about this puzzle.

At about 7:15 AM local time on June 22, 1908, hundreds of Russian settlers and Tungus natives in the forested hills northwest of Lake Baykal looked up in amazement. A brilliant white light was racing across the sky, casting shadows on the ground, and dazzling the eyes of many who tried to stare at it. Minutes after it passed, a distant rolling thunder came to the ears of the witnesses.

Witnesses 20 to 40 miles from the impact point experienced a sudden thermal blast that could be felt through several layers of clothing. For several seconds, half of the sky lit up like a hundred suns. A few moments after the flash, the shock wave arrived, a crashing boom that broke windows and knocked people off their feet.

The blast was recorded as an earthquake at several weather stations in Siberia, and the atmospheric shock wave bounced barograph needles in weather stations in western Europe. A forest fire was ignited that burned for days over several square miles of pine forest.

At sunset that day, the inhabitants of northern Europe were treated to a celestial spectacle that puzzled them for many years. It did not get dark that night. The night sky glowed with an eerie light, and at midnight it was possible to read a newspaper or to photograph the landscape. American observatories later noted that the atmospheric transparency was degraded for several months.

Russia was soon to be plunged into war and revolution, and it took 20 years for the first scientific teams to reach the point of impact of what was thought to have been a giant meteorite fall (the expedition was financed by the promise that tons of meteoric iron would be available for Soviet industry). But no crater could be found. Instead, it was discovered that the trees for miles around had been blasted—from above. Those nearer "ground zero" were still standing but were stripped of their branches and bark. Those further away were smashed down in a direction away from the center.

Years later, a Russian science-fiction writer named Kazantsev was struck by the similarity of the blast effects at Tunguska to those at Hiroshima, which he visited in 1945. A year later, he wrote a story in which the blast at Tunguska was represented as the result of an exploding nuclear power plant of a spaceship from Mars, which was seeking fresh water from Lake Baykal. Many fictitious elements of Kazantsev's imaginative story have since become confused with the real features of the Tunguska story.

Two other Russians adopted—and adapted—the story. A Moscow junior-college lecturer in astronomy named Feliks Zigel, taking time out from studying flying saucers and the Abominable Snowman, became a spokesman for the "spaceship theory" of Tunguska. Physics professor

Three more speculative proposals: matter-antimatter collision, mini-black-hole, interstellar space-vehicle explosion.

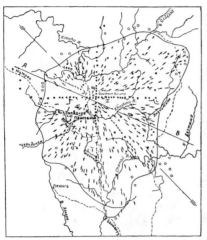

Actual smashed-tree pattern of the Tunguska Event (map) looks very similar to the "butterfly" pattern simulated in physical tests in Moscow. No extraordinary physical or artificial events need be conjured up to account for the explosion.

Aleksey Zolotov organized several college expeditions to the Tunguska site and made a series of announcements of "abnormal radioactivity," followed by embarrassed retractions.

These reports of radioactivity at the Tunguska site still persist, but they seem to be groundless. Only Zolotov can find them. Other expeditions, initially as dedicated as Zolotov to the spaceship theory, concluded in disappointment that there was no abnormal radiation beyond traces of fallout from secret Soviet H-bomb tests. Nor did the anomalous fast growth of trees, often a sign of radioactive exposure, impress a special team from the Soviet Forestry Ministry. They discovered that the faster growth followed the distribution of forest fires set off by the blast, to the exclusion of regions closer to ground zero (and hypothetical radioactivity) that did not happen to catch fire. Accelerated growth is an established effect of forest fires, and the Soviet scientists concluded that this had been the case at Tunguska.

In the West, carbon-14 expert and Nobel laureate Willard Libby reported on results of measurements of tree rings in Arizona. There was a small rise in the carbon-14 level following 1908, to be sure, but it was matched by other erratic ups and downs over the years and could have been a random jump. Libby calculated the force of the blast at Tunguska and the radiation that would have been released had the blast involved thermonuclear reactions. He concluded that the low levels of carbon-14 he found, even if completely attributable to the blast, could still only account for no more than 15 percent of the whole force of the detonation. Other carbon-14 tests in Norway, meanwhile, showed absolutely no rise

The old story that won't die

Every year the "Tunguska UFO" is reborn when some unsuspecting Western newsman in Moscow, searching for an interesting story, falls into the company of Feliks Zigel or Aleksey Zolotov. Within hours, the world's news wires are humming again with an old story that has more lives than the fabled cat: "Top Russian scientists have proved that an atomic UFO crashed in Siberia in 1908."

As more and more evidence piles up indicating that the Tunguska blast was actually caused by a small comet's impact, the handful of "spaceship buffs" seem to have grown more desperate, but no less effective, in corralling the public's attention. Zolotov, for example, has vowed to let his Tolstoyan beard grow until the world adopts his theories. (Needless to say, no professional Soviet geologists or astronomers think much of such tactics or theories.) Zigel, a longtime devotee of UFOs and the Abominable Snowman (Zolotov's major non-Tunguska specialties are faith healing and body auras), poses as "a respected scientist" and a "cosmonaut instructor," but is, in fact, a lecturer in astronomy at a Moscow junior college.

Zigel's latest publicity tactics are exemplified in an Associated Press story out of Moscow on October 22, 1978, based on an interview Zigel gave Tass. He boasts that the book *The Fire Came By* independently corroborates his *own* theories, when in fact the authors are actually only quoting Zigel's assertions. Nor are the authors "University of Texas scientists," as the story claims: they are TV drama critics currently enjoying the commercial success of their latest dramatization (some would say fictionalization), while refusing to debate the subject with informed specialists.

Once again, there is no evidence that AP made even the slightest effort to verify the sensational claims with reputable scientists in the Soviet Union or America. It's an old familiar pattern, and it will take more power than that of the multimegaton Tunguska blast to sweep it away.

—*James Oberg*

Soviet Scientist Gives Theory

Siberian Catastrophe Linked to UFO

MOSCOW (AP) — A cataclysmic explosion that scarred a vast area of Siberia 70 years ago and has baffled scientists ever since may have been casued by a flying saucer, says a respected Soviet astronomer.

The explosion, known as the "Tunguska Mystery" devastated a remote forest region in central Siberia June 30, 1908. It leveled trees over a 1,250-square-mile area and was followed by intense radiation and a great fire.

A score of scientific expeditions to the area and dozens of scientific research papers have put forth theories ranging from the crash of a meteorite or comet to the collision of a "black hole" with the Earth. "B___k holes," whose existence have not been pr_____ ___dly _____llapsed stars.

fied flying object which exploded over the forest in 1908 was an extraterrestrial probe," Zigel said Saturday in an interview with the official Soviet news agency Tass.

Zigel based his theory in part on the fact that scientists have determined that the object flew in an "enormous loop," first northward then westward, before crashing — a trajectory that apparently cannot be explained by any natural phenomenon.

He also supports his theory with data showing that the explosion deposited a high quantity of zinc, bromine, sodium, iron, lead and "other elements that are completely atypical of come t nuclei but are quite possibl____ ar a___ cial structu___

Wire-service dispatch distributed Oct. 22, 1978, unquestioningly relates, once again, Zigel's claims of a Tunguska "UFO."

at all following 1908, so these tree-ring experiments are strong evidence that the Tunguska Event did not involve significant nuclear reactions, if any.

But wait a minute! That's not what the modern Tunguska myths tell. Why does the book *UFOs Behind the Iron Curtain*, by Ion Hobana, state explicitly that "Libby is one of the leading proponents of the nuclear theory"? Why does Zolotov still get worldwide attention when he periodically reissues his "discovery" of Tunguska radiation? Why does *The Fire Came By* declare that the thermonuclear nature of the Tunguska blast is inescapable and irrefutable? Why indeed?

In *The Fire Came By*, the spaceship theory has reached its zenith. I had the dubious honor of debating one of the authors on radio about a year ago when the book was being serialized nationwide by the *New York Times*. In the considered opinion of many listeners I "made mincemeat of the poor guy." He hadn't done his homework. He hadn't bothered with contrary evidence and didn't know how to begin to answer it (I had to let him off the hook several times myself, lest the "debate" grind to a halt). He was a TV drama critic who had seen a chance for a successful

Approach of comet was from sunward side of the earth, which explains why it was not seen beforehand.

The comet's tail extended "down sun" (westward), which meant impact in Siberia would have resulted in the tail's dust entering the atmosphere over western Europe.

book, given the right slant and a minimum of responsibility. In terms of its commercial success, he guessed right. Despite friendly words exchanged at the end of the radio debate, he never responded to my letters requesting clarification and support of many of the major claims in the book. His point of view is often expressed in the news media. But I remain willing and eager to reopen the Tunguska debate with the authors or with any other proponent of the spaceship theory in order to get all the facts on record.

So, as a theory, the spaceship hypothesis loses out. Now, follow the arguments for the "comet hypothesis," perhaps not as exciting, but probably a lot more important. Tunguska was not an isolated event; it could happen again, and we better get ready for it.

The "glowing night sky" over Europe was the comet's dust in the upper atmosphere, lit by the midnight sun of the famous midsummer "white nights" (which I have seen from Sweden and Leningrad). An astronomer took a spectrogram and showed that the "eerie glow" was pure reflected sunlight. The comet-earth encounter geometry was such that the object came "out of the sun" and was masked in the daytime sky. Besides, it wasn't a very big comet. When it hit, the tail was extending "down sun," toward the west, precisely across Europe.

And cometary or meteoric material often does detonate violently on contact with the atmosphere. This is not a chemical or antimatter reaction—it's just the kinetic energy of the speeding object being converted suddenly to tremendous internal heat. Often, only a few hunks of ice and some fine sooty dust will be all that reaches the ground.

It has happened before and since, at places other than Tunguska. Over western Canada in 1965, a 10-kiloton midair explosion sprinkled meteoric dust over freshly fallen snow, prompting astronomer Ronald Oriti to suggest that the Tunguska Event could have been simply a larger object of the same type. The iconoclastic books of Charles Fort chronicle numerous cases of midair explosions, falling ice, and "black rains." Secret United States barographic nuclear blast monitors have been picking up strange random explosions in the upper atmosphere several times a year, often of kiloton size or larger. Even a moderate-sized comet air blast today might not be recognized, resulting only in a spate of angry phone calls to the nearest Air Force base. It is possible that many of the "earthquakes" of recorded history, which killed tens of thousands and which leveled cities, may actually have been Tunguska-sized blasts over inhabited areas. The old records must be reread with this new perspective.

This underscores the real danger: there are few uninhabited wastelands on earth today, and even the open sea is covered with human traffic. The records of nature suggest that such giant blasts, or larger ones, can be expected again.

The idea of a Tunguska explosion occurring again, as a human tragedy instead of a scientific (and pseudoscientific) curiosity, is not so far-fetched. But to defend against such future cosmic bombs it is first necessary to recognize them for what they are. Here the spaceship theory and the distortions, omissions, and fabrications of its proponents (both well-meaning and otherwise) remain a major obstacle, and a major danger.

Ultimately, both radar and optical sensors on earth and in space will watch for incoming objects. Missiles based deep in space, armed with multimegaton warheads, will stand guard. Science fiction has long prepared for this scenario, and fact science is now preparing for that eventuality to come to pass. The fictional science and pseudoscience of the "Tunguska spaceship" should be discredited and dismissed as quickly as possible so that we can get on with defending the earth against future "Tunguska comets."   ●

UFOs By Any Other Name

NASA, the White House, and UFOs
Philip J. Klass

When any government agency is asked if it would like to expand the scope of its activities, the answer would seem to be a foregone conclusion, according to the well-known Parkinson's Law. Yet last year when the White House asked the National Aeronautics and Space Administration whether it believed that still another government investigation of Unidentified Flying Objects (UFOs) should be conducted, under NASA's auspices, the agency's negative response seemed to deny the findings of Professor C. Northcote Parkinson as well as the strident claims of the UFO buffs. A decade earlier, the U.S. Air Force seemed to fly in the face of Parkinson's Law when it eagerly jumped at the opportunity to get out of the UFO business after twenty years, after a University of Colorado UFO study report confirmed USAF findings that there was no evidence of extraterrestrial visitations or any other extraordinary phenomenon.

Parkinson's Law has not been repealed, nor is it fundamentally invalid. But there is a more basic law of self-preservation that says that no government agency is anxious to take on a new task when it is the political equivalent of walking barefoot through a heavily seeded mine-field. And the UFO mine-field was more heavily seeded in 1977 than ever before. One reason is that President Jimmy Carter himself had a UFO-sighting back in 1969 when he was governor of Georgia.

If NASA had agreed to launch a new UFO study, certainly Carter's own sighting would have deserved a high priority on the agency's list of cases to be investigated, since the President could hardly be dismissed as a "UFO kook" or someone whose veracity could be questioned. And if NASA were to do a rigorous investigation, its findings would be embarrassing because the Carter UFO almost certainly would turn out to have been the planet Venus. This was the conclusion of Robert Sheaffer, a member of the the UFO Subcommittee of the Committee for the Scientific Investigation of Claims of the Paranormal, after his own lengthy investigation and talks with persons who had been with Carter on the night of January 6, 1969. When these witnesses said that the UFO had appeared to them to resemble a bright star, Sheaffer, who studied astronomy, turned to his astronomical records. He discovered that a very bright Venus had been at the same azimuth and elevation angles that night at the time of Carter's sighting (*The Humanist*, July/August 1977).

It ought not embarrass the president to learn this, because Venus and other bright celestial bodies, especially when viewed through layers of haze, probably generate more UFO reports than any other single source. But because of the cloak of infallibility that always envelopes any occupant of the White House, and especially a Naval Academy graduate trained in celestial navigation, a NASA finding that the president's UFO was really Venus could hardly enhance the agency's political standing in White House circles, especially at budget-review times.

It was this Carter UFO-sighting, and an interview given during his presidential campaign to the tabloid *National Enquirer*, that subsequently embroiled NASA in the UFO issue. A long-standing cornerstone of the dogma of UFO buffs is that the U.S. Government "really knows the truth about UFOs" but that administration after administration has conspired to keep this truth under deep security wraps for more than thirty years. (This conveniently ignores the inability of the Nixon Administration to keep the Watergate scandal under wraps and the number of Central Intelligence Agency indiscretions that have emerged under Congressional scrutiny.)

The explanation for this alleged cover-up, according to UFO dogma, is that the "government is afraid that the public might panic" if faced with the prospect of extraterrestrial visitations. In support of this contention, the UFO buffs cite the aftermath of the famous Orson Welles radio dramatization of "The Invasion from Mars," broadcast on Halloween night in 1938. Yet the new Steven Spielberg-produced UFO movie about an extraterrestrial visitation, *Close Encounters of the Third Kind*, is playing to packed houses without producing panic in the theaters.

The *National Enquirer*, which gives a big play in its pages to UFO reports, put a reporter on Carter's campaign trail after learning of Carter's UFO-sighting. When the reporter asked Carter if he would release all of the government's classified UFO information if elected, it got a useful quote which was featured on the front page of its June 8,

1976 edition under the headlines "Jimmy Carter: The Night I Saw a UFO," with the subhead ". . . If elected I'll make all the Govt.'s UFO Information public." The precise quotation, contained in the accompanying article, was: "If I become President, I'll make every piece of information this country has about UFO sightings available to the public, and the scientists. I'm convinced that UFOs exist because I have seen one." Not a surprising response from a candidate who was then crusading for greater candor in government operations.

This Carter statement was widely hailed by the UFO buffs in their publications. At long last, after thirty years of secrecy, if Carter were elected, the public would finally learn the truth about UFOs. The new president had barely learned to find his way around the White House before the avalanche of letters and telegrams began to arrive. One typical letter, from a man in California who claimed he had been "zapped" and injured by a UFO, began as follows (and is reproduced exactly as written):

DEAR MR. PRESIDENT,

DURING YOUR PRESIDENTIAL CAMPAIGN; YOU HAD MADE THE COMMITT-MENT, THAT YOU, SIR, WOULD RELEASE, TO THE AMERICAN PEOPLE, ALL OF THE *U.F.O.* SECRECY NOW HELD IN THE ARCHIVES OF THE FEDERAL GOVERN-MENT, CONSISTING OF VARIOUS AGENCIES; YOUR ARTICLE WITH THIS COM-MITTMENT APPEARED IN THE *JUNE 8, 1976* ISSUE OF *THE NATIONAL EN-QUIRER* NEWSPAPER. I HAVE VOTED FOR YOU, MAINLY FOR THIS REASON. I DO FAITHFULLY HOPE YOU WILL NOT DISAPPOINT ME, AS ONE OF MILLIONS OF CITIZENS WHO HAS VOTED FOR YOU . . .

When Dr. Frank Press, a noted geophysicist, was named the Presidential Science Advisor, his office was assigned the task of responding to the letters from the UFO buffs, many of them charging that the Defense Department, the USAF, and/or the CIA were withholding significant information on UFOs. Acting as the president's agent, Press's office wrote to the Defense Department and to the CIA to inquire about such alleged secrets and was officially informed that there were none. The Pentagon pointed out that all of the USAF's UFO files were now open to the public, in microfilm form, at the National Archives, and interested citizens could even purchase microfilm copies of the entire Air Force files on the subject.

Dr. Press sent a memo to the president reporting the results of his queries but its contents seemingly were not carefully read by some of the president's top aides. One, believed to be press secretary Jody Powell, in a background briefing with a reporter for *U.S. News & World Report*, dropped a juicy tidbit that prompted the magazine to publish the following item in the "Washington Whispers" column of its April 18, 1977, issue:

> Before the year is out, the Government—perhaps the President—is expected to make what are described at "unsettling disclosures" about UFOs—unidentified flying objects. Such revelations, based on information from the CIA, would be a reversal of official policy that in the past has downgraded UFO incidents.

This was good news to the UFO buffs. Clearly the president had not forgotten his campaign promise! At least one UFO buff, from Ft. Smith, Arkansas, during a trip to Washington, visited Dr. Press's office to volunteer his services to assist in any way in the big event. (My own response was to write a letter-to-the-editor, published in the May 9, 1977 issue of *U.S. News & World Report*, offering 100:1 odds that no such "unsettling disclosures" on UFOs would occur by December 31, 1977. I had expected that such generous odds might induce a number of "takers," but I received no response—not even from a reporter on the magazine!)

Meanwhile, after receiving official denials that the Defense Department or CIA was withholding anything of significance on UFOs, Dr. Press's office was responding to the increased flow of mail from the UFO buffs with a form letter indicating that the government was not withholding vital information on the subject. But this did not prompt the UFO buffs to question their own dogma. Rather it brought vitriolic responses that President Carter, like his many predecessors, was trying to "keep the truth from the public."

In some instances, the White House asked the Defense Department to help it respond to the barrage of letters. But when the man from California, cited earlier, received a form-letter response from the office of the Secretary of the Air Force denying that UFO information was being withheld, the indignant UFO buff responded with a letter that included the following paragraph (unedited):

> YOUR STOCK-LETTER REPLY TO ME IS ONE THE REPITITIOUS [*SIC*] STATEMENTS ABOUT "PROJECT BLUE BOOK," ETC., WHICH I AM TOTALLY FAMILIAR WITH SINCE 1969 . . . WHAT THE HELL HAS ANYTHING YOU SAY, IN

YOUR LETTER, OF MAY 25, 1977, HAVE TO DO WITH A REPLY TO MY LETTER TO: *PRESIDENT CARTER*? NOTHING! . . . A COPY OF THIS LETTER-REPLY IS BEING SENT TO PRESIDENT JIMMY CARTER* ALSO: COPIES OF THIS COMMUNICATION AND YOUR STOCK-LETTER REPLY WILL BE SENT TO MY CONGRESSMAN, VARIOUS CONGRESSMEN AND SENATORS FOR THEIR COMMENTS AND PERUSAL.

Clearly, White House efforts to respond to letters from the UFO buffs were not winning any potential second-term votes for the president. And it would not help matters if such letters were simply ignored and left unanswered. And so, on September 14, 1977, Dr. Press wrote to the NASA administrator, Dr. Robert Frosch, asking his agency to take over the task of responding to letters from the public on the UFO issue. This was not an entirely new assignment, inasmuch as the space agency, understandably, had been the recipient of such queries prior to the new administration. NASA sent out a standard information sheet (76-6), dated July 1976, saying that "NASA is not involved in research concerning unidentified flying objects. Reports of unidentified objects entering U.S. air space are of interest to the U.S. military as a regular part of defense surveillance, but no government agency is conducting an ongoing investigation of UFOs at this time."

This statement flatly contradicted a part of the current UFO-buff dogma—that the U.S. Government had not really gotten out of the UFO business in 1969 when the Air Force closed down its Project Blue Book UFO office. At least some UFO buffs were sure this was simply a ruse and that government UFO investigations still were going on, secretly, in another agency.

The NASA information sheet also quoted the conclusions of a National Academy of Sciences panel, created to review the results of the University of Colorado investigation: "On the basis of present knowledge the least likely explanation of UFOs is the hypothesis of extraterrestrial visitations by intelligent beings." And the NASA statement concluded by providing the names and addresses of two private UFO groups engaged in the investigation of UFOs.

Within a few weeks the word was out that NASA had been asked for its views on whether it should launch a new government-funded UFO study. The timing could not have been worse for NASA, because Columbia Pictures had opened its multi-million-dollar publicity campaign to promote the new Steven Spielberg UFO-thriller, which previewed in New York and Los Angeles during the third week in November. The story published by the *Christian Science Monitor*, November 17, 1977, reported: "A White House request to the National Aeronautics and Space Administration asks that the space agency consider becoming the government's focal point for a 'national revival' of interest in reports of UFO sightings." The article quoted an unidentified NASA project officer as expressing some reluctance to become involved in a new UFO investigation.

The widespread news-media coverage included an article by Deborah Shapley in the December 16 issue of the respected magazine *Science*, published by the American Association for the Advancement of Science. The article concluded: "Truth is as strange as fiction. The Air Force, officials say, indeed classifies some results of its inquiries made after UFO 'sightings'—many of which are made near military bases, and by men trained to observe the skies, and a few of which are investigated by Air Force men going up in planes. Press's office says that these facts, together with the conflicting responses the government hands out to UFO buffs who write in, keep alive this belief in a cover-up. Policies like these, officials say, need review and perhaps changing."

(When I called Ms. Shapley to ask whether she had checked out the claim that the Air Force "classifies some results of its inquiries made after UFO 'sightings'—many of which are made near military bases . . . ," she told me that she had not, and had accepted the statements given to her by persons in the office of the Presidential Science Advisor. I told her that I believed she had been badly misinformed.)

In late December, NASA's Dr. Frosch wrote the following letter to Dr. Press informing the White House of its conclusions:

Dear Frank:

In response to your letter of Sept. 14, 1977, regarding NASA's possible role in UFO matters, we are fully prepared at this time to continue responding to public inquiries along the same line as we have in the past. If some new element of hard evidence is brought to our attention in the future, it would be entirely appropriate for some NASA laboratory to analyze and report upon an otherwise unexplained organic or inorganic sample. We stand ready to respond to any *bona fide* physical evidence from credible sources. We intend to leave the door clearly open to such possibility.

We've given considerable thought to the question of what else the United States might and should do in the area of UFO research. There is an absence of tangible or physical evidence for thorough laboratory analysis.

And because of the absence of such evidence we have not been able to devise a sound scientific procedure for investigating these phenomena. To proceed on a research task without a disciplinary framework and an exploratory technique in mind would be wasteful and probably unproductive.

I do not feel that we should mount a research effort without a better starting point than we have been able to identify thus far. I would therefore propose that NASA take no steps to establish a research activity in this area or to convene a symposium on this subject.

I wish in no way to indicate that NASA has come to any conclusion about these phenomena as such. Institutionally we retain an open mind, a keen sense of scientific curiosity and a willingness to analyze technical problems within our competence.

When those who had a hand in composing the NASA letter reviewed their final product, they probably saw it as the best response under the circumstances. It could not possibly give offense to any of the citizens who had reported seeing a UFO, including the president. Nor did NASA dismiss completely the possibility, however remote, of extraterrestrial visitations.

Instead, Frosch's letter sought to place the burden of proof where it rightfully belongs, on those who promote the extraterrestrial hypothesis, to come up with "tangible or physical evidence for thorough laboratory analysis." Knowing that thirty years of UFO reports had yet to produce a single piece of *credible* physical evidence of extraterrestrial visitations, NASA officials seemed to believe that there was in fact little if any *claimed* physical evidence. In this they were grossly in error. Having made this offer without having a thorough knowledge of UFOlogy, or consulting with those who have, NASA may soon regret it.

During the coming years, I predict, NASA will receive hundreds of pieces of tree branches that allegedly were broken by a UFO and burned grass, charred twigs, and soil samples allegedly taken from spots where UFOs reportedly landed. It will receive soil samples, some carefully prepared by hoaxers, to challenge the skills of the U.S. Department of Agriculture's Research Center, not far from NASA's own Goddard Space Flight Center.

If rigorous laboratory analysis of five hundred such pieces of "tangible evidence" shows nothing extraordinary, perhaps the five-hundred-and-first will, the UFO buffs will insist. If NASA's patience runs thin, the determination of the UFO buffs is far more long-lived. One such piece of physical evidence submitted earlier to the USAF was the broken head of a hunting arrow, allegedly fired by the submitter at robots seen near a landed UFO!

But still greater pitfalls await NASA, for it soon will discover that it must take far greater security precautions with UFO samples than it needed in handling lunar samples. For the latter, it was only necessary to ensure that none were stolen or diverted. For UFO samples, NASA must protect itself against later charges that the soil sample or tree branch it returned after analysis was not the same one submitted to it. There will be charges of substitution and claims that the original artifact now resides deep in underground security vaults at NASA or some other governmental agency.

This undoubtedly would seem far-fetched and paranoid to NASA officials today, because they probably are unaware that a NASA scientist already has been charged with such "hanky-panky" in a recent book by UFO-buff Ray Stanford. On April 24, 1964, a lone policeman reported that he saw an egg-shaped UFO land, in broad daylight, on the outskirts of the small town of Socorro, New Mexico. When Stanford visited the site shortly afterward as a UFO investigator, he picked up a rock that reportedly contained "metallic scrapings," seemingly left by the UFO as it brushed the rock.

On July 31, 1964, Stanford came to Washington and together with two other UFO buffs, including Richard Hall, then deputy director of a large UFO group with headquarters in Washington, drove to the Goddard Space Flight Center to give the rock to a NASA scientist there who agreed to analyze its "metallic particles" on an unofficial basis. Stanford claims that he asked the NASA scientist to "leave one-half of the particles on the stone's surface, so that I retain half the evidence" and the scientist agreed.

Stanford also alleges that the NASA scientist later told him: "I am virtually certain that the alloy involved here is not manufactured anywhere on Earth . . . I would make a statement to that effect, if you need it." But subsequently, Stanford charges, the scientist denied having made any such statement and said the "scrapings" on the rock were simply silica, a natural constituent. Stanford also charges in his book that when the Socorro rock sample finally was returned to him all particles had been removed, thereby depriving Stanford of any opportunity to have an independent analysis conducted. Stanford also accuses fellow UFO-buff Hall of having joined forces with NASA to suppress the Socorro evidence, an allegation that Hall flatly disavows along with Stanford's claim of hanky-panky by a top NASA scientist. But the Stanford book has gained wide acceptance in UFO circles.

Thus, unless NASA handles every broken tree branch and soil sample as it would the Hope diamond, it can expect that Stanford's earlier, if ill-based, charges of hanky-panky will be raised again to "substantiate" the more recent allegations.

(At present, Stanford is director of Project Starlight International, which operates a million-dollar UFO research facility near Austin, Texas. The facility includes elaborate flashing lights to attract UFOs, radar and telescopes to spot UFOs, cameras, and a laser that could be used for communication with the UFO. Stanford has never disclosed the source of the funds for what certainly is the best-instrumented UFO facility in the United States, if not in the entire world.)

Although NASA's decision not to initiate a new UFO investigation will be criticized by many UFO buffs, they can take solace in Frosch's reference, on two occasions, to "these phenomena." Seemingly this implies NASA recognition that UFOs exist as a phenomenon. In reality, the only thing known to exist with absolute certainty are UFO *reports*, suggesting that Frosch might better have used the term "these reported phenomena." For, as President Carter has demonstrated, a "reported UFO" can turn out to be a well-known phenomenon that was identified, and better named, long ago.   ●

Astronaut "UFO" Sightings
James Oberg

The glamour and drama of manned space-flights has been transferred to the UFO field via a highly publicized group of "UFO sightings" and photographs allegedly made by American and Russian space-pilots. Hardly a UFO book or movie fails to mention that "astronauts have seen UFOs too."

Careful examination of each and every one of these stories (and they total more than 20 or 30) can produce quite reasonable explanations, in terms of visual phenomena associated with space flights. On a visit to the NASA Johnson Space Center in Houston in July 1976, Dr. J. Allen Hynek, of the Center for UFO Studies, concluded that none of the authentic cases (as opposed to the majority of reports, which are fictitious) really had anything to do with the "real UFO phenomenon."

Skeptical investigators, while pleased that Hynek had dismissed all "astronaut UFO reports" as unreliable, have insisted that this body of stories has quite a lot to do with the major problems besetting the UFO community. How, they ask, can a body of stories so patently false and unreliable obtain such seeming authenticity simply by being passed back and forth among researchers without ever being seriously investigated? Is this a characteristic of UFO stories in general; and if so, the skeptics ask, can a study of how the "astronaut UFO" myth began and flourished help us to understand better the UFO phenomenon in general?

Hynek's disavowal of the stories came after his book *Edge of Reality* (coauthored with Jacques Vallee) carried a long list of astronaut sighting reports. Hynek told colleagues that the inclusion of the list (compiled by George Fawcett) in the book was Vallee's idea, not his, but that even so he just wanted to generate interest and discussion. He insisted that inclusion of the list was not a judgment on his belief in its credibility and that readers had no right assuming that the data had actually been verified just because they were included in his book. Fawcett, on the other hand, claims that he just assembled the list from all available sources and assumed that somebody *else* would check the accounts before publication. "Maybe 1 percent of the stories are *true* UFOs," Fawcett suggested in 1978.

This is the complete "Fawcett List" as printed in *Edge of Reality.* Following each incident, we supply, in italics, the most likely explanation of the report.

February 20, 1962—John Glenn, piloting his Mercury capsule, saw three objects follow him and then overtake him at varying speeds. *Glenn also said that these "snowflakes" were small, and seemed to be coming from the rear end of his capsule. Later flights also observed them and were able to create "snowstorms" by having astronauts bang on the walls of their capsules. Verdict: Significant data was withheld, totally altering the nature of the incident.*

May 24, 1962—Mercury 7: Scott Carpenter reported photographing fireflylike objects with a hand camera and that he had what looked like a good shot of a saucer. *Carpenter did see "fireflies," as well as a balloon ejected from his capsule. The claim that he reported photographing a "saucer" is counterfeit. His photo, taking into account the glare of sunlight, smeared window, and gross enlargement of the small image, has been widely published as a "saucer" but was in fact the tracking balloon.*

May 30, 1962—X15 pilot Joe Walton photographed five disclike objects. *This story appears to be a complete fabrication. The real pilot's name was Joe Walker.*

July 17, 1962—X15 Pilot Robert White photographed objects about 30 feet away from his craft while about 58 miles up. *Right—and as he reported, the objects were small, "about the size of a piece of paper," and were probably flakes of ice off the super-cold fuel tanks. Verdict: Important information withheld by authors.*

May 16, 1963—Mercury 9: Gordon Cooper reported a greenish UFO with a red tail during his fifteenth orbit. He also reported other mysterious sightings over South America and Australia. The object he sighted over Perth, Australia, was caught on screens by ground tracking stations. *Cooper has recently denounced all stories of UFOs on his space flights as fabrications. The multicolor UFO is based on a deliberate mis-quotation by an author of Cooper's postflight report on a sighting of the Aurora Australis. Verdict: Fraud.*

October 3, 1963—Mercury 8: Walter Schirra reported large glowing masses over the Indian Ocean. *Indeed he did, referring to lightning-lit cloud masses over the ocean a hundred miles below. The author of this story deliberately quoted out of context. Verdict: Fraud. And note: Wrong date (really 1962), and Mercury 8 follows Mercury 9 in this "reliable" chronology.*

March 8, 1964—Voskhod 2: Russian cosmonauts reported an un-identified object just as they entered the earth's atmosphere. *Several hours before returning to earth the cosmonauts spotted a cylinder-shaped object they assumed (probably correctly) was just another man-made satellite. Such sightings were becoming more and more frequent as the number of manned flights and unmanned satellites rose.*

June 3, 1964—Gemini 4: Jim McDivitt reported he photographed several strange objects, including a cylindrical object with arms sticking out and an egg-shaped UFO with some sort of exhaust. *This is the most famous "astronaut UFO" case, and it has been embellished and distorted in dozens of publications. McDivitt saw a "beer-can shaped" object which he took to be another man-made satellite (some observers believe it was his own booster rocket), and tried to take a few photos which did not turn out. A still from the movie camera (which McDivitt insists he never touched during the sighting) was mistakenly released without the astronaut's review, showing what turned out to be a light reflection off his co-pilot's window, according to McDivitt. UFO buffs took this photo and acclaimed it as one of the best UFO photos ever taken, showing (they claim) a glowing object with a plasma tail. McDivitt never saw anything like that in space. Verdict: Gross exaggeration and distortion on the part of UFO writers. Also, the year is wrong—it should be 1965.*

October 12, 1964—Voskhod 1: Three Russian cosmonauts reported they were surrounded by a formation of swiftly moving disc-shaped objects. *This story appears to be a complete fabrication, but UFO buffs cling to it while challenging skeptics to "prove it did not happen."*

December 4, 1965—Gemini 7: Frank Borman and Jim Lovell photographed twin oval-shaped UFOs with glowing undersides. *This famous photograph is a blatant forgery, in which light reflections off the nose of the spacecraft are made to look like UFOs by airbrushing away the vehicle structure around them. Verdict: Fraud.*

Famous McDivitt "UFO" photo (actually a still movie film), which has been reprinted in dozens of UFO books and magazines, is only a reflection of the sun on co-pilot's window, according to McDivitt. The astronaut *did* see a nearby satellite, but did not succeed in getting a photo; he never even touched the movie camera. Yet UFO groups have selected this as one of the "four best UFO photos ever taken" and theorize that the "tail" is actually a "plasma jet" for interplanetary propulsion.

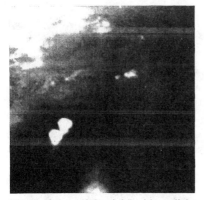

NASA Hiding UFOs From You

Photo of nose of Gemini-7 with sunlight gleaming off two rocket thrusters. Dark nose almost blends in with dark earth background.

With the help of airbrushing, the Gemini-7 photo became a view of two glowing UFOs. Forgery was first published in Japan and subsequently in an American tabloid newspaper. Hoax eventually appeared in *Edge of Reality*, by Dr. J. Allen Hynek.

July 18, 1966—Gemini 10: John Young and Mike Collins saw a large, cylindrical object accompanied by two smaller, bright objects, which Young photographed. NASA failed to pick them up on screens. *The astronauts reported two bright fragments near their spacecraft soon after launch, presumably pieces of the booster or of some other satellite. No photos were taken. They were out of range of NASA radar at this point anyway. Dramatization of ordinary space event.*

September 12, 1966—Gemini 11: Richard Gordon and Charles Conrad reported a yellow-orange UFO about six miles from them. It dropped down in front of them and then disappeared when they tried to photograph it. *The astronauts described the close passage of another space satellite, identified by NORAD as the Russian Proton-3 satellite but later shown to have been some other object. The men got three fuzzy photos, which, much blown up, have been widely published. But their eyesight accounts describe a solid satellite-looking object on a ballistic nonmaneuvering path.*

November 11, 1966—Gemini 12: Jim Lovell and Edwin Aldrin saw four UFOs linked in a row. Both spacemen said the objects were not stars. *Indeed they were not, since the astronauts were talking about four bags of trash they had thrown overboard an hour earlier! Deliberate misquotation by a UFO-book author.*

December 21, 1968—Apollo 8: Frank Borman and Jim Lovell reported a "bogie"—an unidentified object—ten miles up. *Actually, Borman referred to a "bogie" on his first space-flight three years before, describing some pieces of debris associated with his spacecraft's separa-*

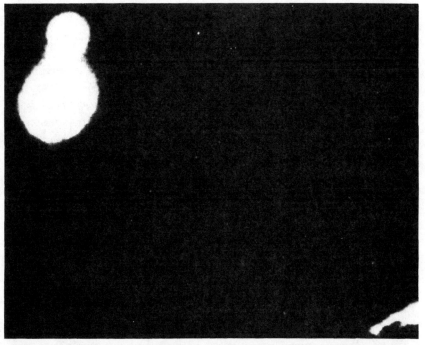

Photo of "Snowman UFO" reportedly made by "Buzz" Aldrin in lunar orbit (also described as "a mass of intelligent energy" in *Science Digest*). A series of pictures of this event appeared in a Japanese magazine, were widely publicized in America by Bob Barry (20th Century UFO Bureau), and have now been entrenched in UFO folklore.

The forgeries have been widely printed in the UFO media, along with fabrications about "NASA cover-ups."

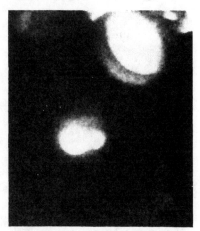

The "Snowman UFO" photo is a forgery. Scenes such as this one from Apollo-11 film magazine "F" were heavily retouched to eliminate other reflections and accentuate sharpness of primary reflection of interior lights.

tion from the booster rocket. The reference to Apollo 8 is careless, possibly even fictitious.

July 16, 1969—Apollo 11: This was a mission on which a UFO reportedly chased the spacecraft. *"Reportedly," indeed, but not very accurate. Actually, several UFO stories have attached themselves barnacle-like to man's first moon landing. A photo of an insulation fragment taken soon after third-stage separation has been widely published as a "UFO." The astronauts watched their booster through a telescope on the way to the moon. A series of "UFO photos" allegedly taken by astronaut Aldrin in lunar orbit are actually forgeries by a Japanese UFO magazine. An alleged "astronaut radio conversation" describing a UFO ambush is a hoax.*

November 14, 1969—Apollo 12: Astronauts Pete Conrad, Alan Bean, and Dick Gordon said a UFO accompanied them to within 132,000 miles of the moon, preceding them all the way. *No, they never said that. They were joking with the ground control room about a tumbling piece of their booster rocket that was flashing in the sky. UFO buffs completely misunderstood the meaning of the conversation and conjured up a UFO. On the way back to earth, the astronauts were puzzled by a light between them and the earth, which turned out to be the reflection of the moon behind them on the night-time Indian Ocean below them.*

Many other "astronaut reports" have been added to this list, including photographs from Skylab (of a passing satellite, distorted by some

Strange squiggle photo from Skylab is probably a film or camera fault, since crew testimony and other photos show this was a bright point source—clearly a nearby artificial satellite.

camera artifact), from Apollo lunar flights (movies showing debris floating around inside the cabin), and from other Mercury and Gemini flights. None, when investigated with an appreciation of the actual space-flight environment, appears to be "extraordinary" or "unusual," although many sightings of passing satellites remain technically "unidentified," since the actual satellite has never been named (nobody has taken the trouble to spend several hours of computer time searching memory banks).

The entire phenomenon of the "astronaut UFO sightings," however, does explicitly demonstrate the carelessness and lack of verification among UFO circles eager to exchange the latest hot stories without any regard for authenticity or accuracy. Skeptics have claimed that this characteristic is not limited to the "astronaut UFO sightings." The topic is not one to which UFO specialists can point with pride in their own behavior and standards of reliability.

Closing note: A common claim is that there is some sort of NASA "cover-up" of secret photographs and/or voice transcripts from space. In fact, every photograph taken by NASA in space is available for publication and can be inspected by accredited news media representatives (there are tens of thousands of photos—there is no way to arrange public viewing). And volumes and volumes of voice transcripts are readily available at Houston.

Often astronauts are quoted about UFOs. Sometimes they are referring to experiences they have had before or after their roles in the space program. In other cases they are making general statements based on reading the news media. Many quotations are fictitious. Only one astronaut claims to have seen a UFO in space, and that is Jim McDivitt, who stipulates that *his* definition of a UFO covers the probability that his object was some other man-made satellite which has not been identified. He does not think it was an alien space vehicle or any such similar "real UFO" manifestation. •

UFOs, the CIA, and the New York Times

Philip J. Klass

For many years the *New York Times* has been sharply criticized by those who seek to promote public belief in unidentified flying objects, because of the paucity of its coverage of the subject. Now the *Times* is being hailed by these former critics, and with good reason.

UFO promoters are praising the *New York Times Magazine* for its October 14, 1979, feature by free-lance writer Patrick Huyghe, which was headlined "UFO Files: The Untold Story," with a subhead that read "Though officials have long denied that they take 'flying saucers' seriously, declassified documents now reveal extensive Government concern over the phenomenon."

If news officials of the *Times* believed the thrust and contents of the Huyghe article, clearly the situation must be a "Cosmic Watergate" coverup, as UFO-lecturer Stanton Friedman has so often characterized it. For example, the article began by highlighting a series of UFO reports from U.S. Air Force SAC air and missile bases that occurred four years earlier, in the fall of 1975, that suggested that these vital installations were being visited by craft of unknown origin. If extraterrestrial craft were reconnoitering our key military facilities, having penetrated national air

448

space undetected by air and defense radars, perhaps a "Cosmic Pearl Harbor" was in prospect and alleged efforts to suppress news of such intrusions (if true) would indeed amount to a "Cosmic Watergate."

One would have expected the *Times* to have promptly formed a team of its best investigative reporters for a journalistic assault on what appeared to be the biggest story of all time. Yet, so far as I can determine, news officials at the *Times* simply ignored the Huyghe article.

Similar "journalistic oversight" occurred many months earlier at the *Washington Post,* whose investigative efforts had exposed the original Watergate scandal. The January 19, 1979, issue of the *Post* carried a front-page feature, by Ward Sinclair and Art Harris, recounting the same SAC-base UFO incidents under the headline "What Were Those Mysterious Craft?" The syndicated article, which was carried by major newspapers around the country, began: "During two weeks in 1975, a string of the nation's supersensitive nuclear missile launch sites and bomber bases were visited by unidentified, low-flying and elusive objects, according to Defense Department reports." Yet in the many months since this ominous-sounding article appeared, the *Post's* news officials also seem to have been derelict about digging into this apparent coverup.

When I later talked with Sinclair about his article, he admitted that when he wrote it he was not aware that a major feature story about the same SAC-base incidents had been published only a month earlier in the December 10, 1978, issue of *Parade* magazine, the popular Sunday supplement distributed with the *Washington Post.* The *Parade* article, written by Michael Satchell, carried the headline "UFOs vs. USAF: Amazing (but true) Encounters." Curiously, the essence of Satchell's article had been published a year earlier by the *National Enquirer,* in its December 13, 1977, issue, under the headline, "UFOs Spotted at Nuclear Bases and Missile Sites."

Considering the readership of the *National Enquirer, Parade,* the *New York Times Magazine,* and the *Washington Post,* as well as other major newspapers that also carried the *Times* and the *Post* articles, it is curious that there were no investigative reporters, eager to win a Pulitzer Prize and achieve world fame, who were smart enough to recognize the "Cosmic Watergate" implications of these articles—*if* they believed what they read. Even if the "UFOs" over SAC bases were not extraterrestrial craft, as the article implied, and were "only" Soviet or Cuban aircraft on reconnaissance missions, it would seem to be a "helluva story."

(Although my more than 13 years of investigating famous, seemingly mysterious UFO incidents have made me a skeptic, as a senior editor of *Aviation Week & Space Technology* magazine, I decided that the SAC-base incidents did warrant further investigation. The results of my investigation, which will be detailed in a book now in progress, indicated that neither extraterrestrial nor foreign aircraft were involved in the incidents.)

Huyghe's article in the *Times Magazine* also covered the contents of the once-classified government files dealing with UFOs, principally from the Central Intelligence Agency, which were made public in December 1978 through the Freedom of Information Act. Huyghe reported: "Official records now available appear to put to rest doubts that the Government knew more about UFOs than it has claimed over the past 32 years." Having personally studied the nearly one thousand sheets of UFO-related material released by the CIA in late 1978 (only a third of which were of import), I can vouch for the accuracy of this statement.

Yet shortly after these files were made public, the *New York Times*, in their January 14, 1979, issue, carried a long news story quoting William Spaulding, the head of a national UFO organization, who claimed that the CIA files revealed that "the Government has been lying to us all these years." The article said that, according to Spaulding, "after reviewing the documents, Ground Saucer Watch believes that UFOs do exist, they are real, the U.S. Government has been totally untruthful and the cover-up is massive."

Huyghe was grossly inaccurate when he wrote that newly released files showed that "the [UFO] phenomenon has aroused much serious behind-the-scenes concern in official circles. Details of the intelligence community's *protracted* obsession with the subject of UFOs have emerged." (Emphasis added.) Huyghe went on to claim: "But it is the CIA that appears to have played the key role in the controversy, and may even be responsible for the Government's conduct in UFO investigations throughout the years."

What the CIA files really reveal is that the agency first became actively—but only briefly—interested in UFOs more than a quarter-century ago, in the summer of 1952, in the wake of several incidents in which unidentified radar blips appeared on the displays of a radar installed at Washington's National Airport, prompting the USAF to launch inter-ceptor aircraft to investigate. (A subsequent formal investigation by the then Civil Aeronautics Administration showed that the spurious radar blips were the result of anomalous propagation due to temperature inver-sions and had been experienced at numerous other such radars without precipitating a UFO incident.) The several incidents over the nation's capital, which had made headlines around the country, had prompted inquiries by the White House to the CIA, which, understandably, had triggered its official interest.

The CIA had secretly convened a panel of distinguished scientists, headed by H.P. Robertson of the California Institute of Technology, which met in mid-January 1953 to consider the most impressive UFO incidents then in the USAF's files. After examining these "best cases," the panel concluded that *none* were "attributable to foreign artifacts capable of hostile acts."

Prior to the meeting of the Robertson panel, once-secret CIA papers revealed that *some* agency officials had been anxious to have the National Security Council authorize the CIA to initiate a major UFO investigation. Such an authorization would be needed because the USAF earlier had been given primary responsibility for investigating UFOs. When a copy of the Robertson panel report was transmitted to the Intelligence Advisory Committee on February 18, 1953, by committee secretary James Q. Reber, he wrote, "The results of the panel's studies have moved the CIA to conclude that no National Security Council Intelligence Directive [authorizing the CIA to launch a UFO investigation] on this subject is warranted."

The once-secret CIA files contain a number of memoranda from second and third-tier officials discussing what should be done with UFO material accumulated during the previous six months. One memo, dated March 31, 1953, expressed the view that "very little material would be worth saving except as samples of indicative or unusual reports. The rest I recommend be destroyed." The memoranda make it clear that none of the officials wanted his own division to be saddled with the task of maintaining UFO files and analyzing new UFO reports. The head of the division that was finally designated for the unwanted task wrote a memo on July 3, 1953, saying he "planned to handle the project with part-time use of an analyst and a file clerk."

The CIA files also contain a memo dated August 8, 1955, from the chief of the Physics & Electronics Division, to his superior, the Acting Assistant Director for Scientific Intelligence, recommending that his division's responsibility for monitoring new UFO reports be terminated. The CIA official noted that during the two years that his division had been following UFOs "no intelligence of concern to national security has been developed from the project." And he *complained* that his division had been spending "between 10 and 25 analyst hours per month" (a small fraction of one full-time employee's efforts) in reviewing UFO reports and "about half that much clerical time."

That provides an accurate indication of the extent of the CIA's official interest in UFOs *nearly a quarter of a century ago*. And the CIA files show a decline of interest in UFOs since that time. The CIA files contain internal memoranda on the subject of UFOs as recent as April 3, 1976, which offer added confirmation that the CIA had no official interest in the subject. The memoranda—names are censored because of the Privacy Act considerations—concern a report submitted to the agency by a U.S. scientist speculating on UFO propulsion systems. The scientist wanted to know if the paper should be classified for national security reasons; he also asked if the U.S. government, possibly the CIA, was secretly investigating UFOs. (In late 1969, the USAF announced that it was officially closing down its UFO investigative office and getting out of the UFO business.)

The internal CIA memorandum of April 3, 1976, says: "It does not seem that the Government has any formal program in progress for the identification/solution of the UFO phenomena. . .At the present time, there are offices and personnel within the agency who are monitoring the UFO phenomena, but again, *this is not currently on an official basis."* (Emphasis added.) The memo continued: "We wish to stress again that there does not now appear to be any special program on UFOs within the intelligence community."

This is what the once-secret CIA files on UFOs really reveal, as I stressed to Patrick Huyghe when he interviewed me in preparation for writing his article for the *New York Times Magazine.* But what did Huyghe write and the *Times Magazine* publish? "Ever since UFOs made their appearance in our skies in the 1940's, the phenomenon has aroused much serious behind-the-scenes concern in official circles. Details of the intelligence community's protracted obsession with the subject of UFOs have emerged over the past few years with the release of long-withheld Government records obtained through the Freedom of Information Act. . .
It is the CIA that appears to have played the key role in the controversy, and may even be responsible for the Government's conduct in UFO investigations throughout the years."

A Controlled UFO Hoax: Some Lessons

David I. Simpson

For many years it has been fashionable to argue a high probability that intelligent life has developed elsewhere in the universe (Shklovskii and Sagan 1966; Cameron 1963). The logic involved, however, suggests an extremely small likelihood that any such life would journey to Earth even once, with the probability of daily visits being negligible. According to the majority of those interested in unidentified flying object phenomena, this conclusion is at variance with the wealth of evidence indicating that our skies are often frequented by extraterrestrial visitors: the pilots of UFOs.

To resolve this apparent paradox, it is important to appreciate the abilities and motives of an enthusiastic group of people who make themselves responsible for investigating and reporting UFO sightings: the UFOlogists. Since Kenneth Arnold's sighting in 1947 (Arnold and Palmer 1952), when UFOs were first described as "flying saucers," the world has witnessed an ever growing number of UFOlogists, UFO "research" groups, and related magazines; there are at present approximately 250 such organizations throughout the world.

Starting in 1967 and examining in excess of 200 UFO reports from Britain, my investigations failed to discover a single case that could reasonably be argued to indicate anything more exotic than misidentified natural or man-made phenomena. A number of these reports contained insufficient data to reach a conclusion, a few were thought to be hoaxes, and a few more the result of mental illusions. The cases chosen ranged from simple lights in the sky, through UFOs that stopped cars, to photographic evidence and claims of alien contact. On examining the same cases, however, other UFO commentators usually published alarmingly different conclusions, often disregarding plausible but mundane explanations. Confronting these authors with alternative solutions provoked many accusations that I was a "nonbelieving skeptic with a closed mind."

It is sometimes difficult to convey to uninformed third parties the highly partisan nature of investigations undertaken by most UFO enthusiasts. I was therefore prompted to illustrate my opinions by perpetrating a series of controlled hoaxes. They were designed to attract the attention of UFOlogists directly, not the general public, with the aim of comparing known details of fabricated "UFO" stimuli with the issued statements of investigators. Since the experiments yielded broadly similar data, just one is detailed here.

* * * * *

Throughout the world there are certain locations famous for attracting the attention of UFOlogists; they have been called UFOcals. Cradle Hill near the Wiltshire town of Warminster is one such place; UFOlogists make pilgrimages there most weekends, and it was the setting for the opening scene of this experiment on Saturday, March 28, 1970.

At 11 P.M. a 12-volt high-intensity purple spotlamp was directed from a neighboring hill toward a group of about 30 sky-watchers on Cradle Hill, three-quarters of a mile away. The lamp was switched on for 5, and then 25, seconds, with a 5-second pause between. During the second "on" period, a bogus magnetic-field sensor, operated among the sky-watchers by a colleague, sounded its alarm buzzer, apparently indicating the presence of a strong magnetic field. (UFO folklore states that strong magnetic fields are a characteristic of UFOs, so this sensor was not an unusual sight.) In practice, the alarm was simply synchronized to sound while the distant spotlamp was on. The "strangeness" of the purple light was thereby enhanced.

Norman Foxwell (another colleague stationed among the sky-watchers) pretended to photograph the purple light with a camera mounted on a tripod. Part of his film had already been exposed, however, and bore two latent images, each showing part of the distinctive night view of the streetlamps observable from Cradle Hill with a spurious UFO superimposed. (See figure.) Neither photograph included the site of the spotlamp.

Frame one showed a cigar-profile UFO with a semicircular blob above and below center. With respect to the sky-watchers, it was approximately 22 degrees horizontally removed from the spotlamp site. Frame two showed the same UFO but farther removed by 8 degrees, slightly lower, fainter, and blurred. Shortly after the "sighting," Foxwell took two genuine time-exposure photographs so that the developed film would show a total of four relevant negatives, two with UFOs and two without, on successive frames. They were designed to present substantial inconsistencies that would allow any moderately critical investigator to cast strong suspicion on their authenticity. Not only did the first pair of negatives show a UFO image quite unlike the observed UFO and on a different part of the horizon, but their magnification was 10 percent greater than the genuine negatives on subsequent frames. Also, the faked negatives were prepared from originals taken the previous year, when two lamps from the distinctive streetlamp pattern were not working. Therefore, two streetlamps that appeared on the genuine pictures were missing from the adjacent faked ones.

Foxwell was briefed to give the film from his camera to any UFOlogist on the hill who would be prepared to have it developed privately. Surprisingly, he managed to do this without raising suspicion. The recipient was John E. Ben, who had connections with *Flying Saucer Review (FSR)*, a glossy international UFO magazine.

For two and a half years, the hoax nature of this "sighting" was kept secret, during which time UFOlogists' letters, published articles, and general comments were collected. To quote the entire file would require more space than is available here. I therefore refer to just a few items that may provide insight into the way UFO enthusiasts investigate and record UFO reports.

Ben was employed by the Wellcome Institute of the History of Medicine, and the film was developed in their photographic department. In early communications he sought permission from Foxwell to take the photographs to a meeting of the *FSR* consultative committee, adding that the top six men in Europe were fortuitously due to attend. After this meeting *FSR* wanted to examine the negatives in their laboratory. On May 26, Ben wrote to Foxwell: "Mr. Charles Bowen of *FSR* [the editor] has contacted me this morning to tell me about your Warminster photographs. I am pleased to inform you that they have now proven the negatives to be genuine beyond all doubt."

The Warminster photographs were first publicized by *FSR* in their July-August 1970 issue, with an artist's impression of the purple light on the front cover. Drawn by Terence Collins, who had been with the sky-watchers on Cradle Hill, the general details were correct, although with

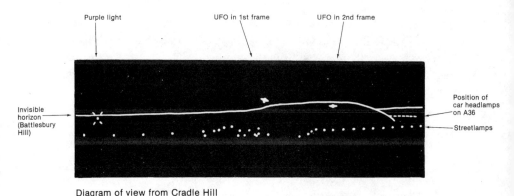

Diagram of view from Cradle Hill

respect to his streetlamps his purple light subtended an angular diameter roughly ten times too large. Inside the magazine, Ben's report, entitled "Photographs from Cradle Hill" (Ben 1970), described the stationary grounded light, which was visible for 30 seconds at an elevation of approximately zero degrees, in the following terms:

> At 11:02 P.M. an object was seen at an elevation of approximately 20 degrees in the eastern sky. The object appeared very suddenly as if it came through the clouds, and appeared to the eye as a very bright ovoid light—purple in colour with a periphery of white. Two members of my group who observed the object through binoculars both remarked they could see a crimson light in the centre; this was also attested to by witnesses with good vision.
>
> The object remained stationary for approximately 30 seconds, during which time Mr. Foxwell was able to take the first of his photographs. The object then moved slowly to the right—towards the town—and lost a little altitude in the process. At one stage in the movement it dimmed considerably as though obscured by low cloud. The object continued moving for approximately 20 to 30 seconds, and then stopped again. The light then increased considerably in intensity, though we could not be sure if the object was moving directly towards the observation point, or if it remained stationary. At this point the alarm of a detector sounded and a witness ran to switch it off. After 10 to 20 seconds the light dimmed and went out as though concealed by cloud. However, we were all certain that the object had not moved once more. The sighting had lasted for approximately one to one and a half minutes.

It would perhaps be unfair to criticize the duration estimate and even the "20 degrees in the eastern sky." Of more interest is the movement described, it being inconsistent with the observed stationary light but consistent with the implied movement of the UFO in the fake photographs. Neither Ben nor any subsequent investigator ever commented, to my knowledge, on the fact that the photographs did not include that part of the horizon on which the purple light was located.

In the same issue of *FSR* Percy Hennell, a photographic consultant to *FSR*, reported: "Let me say at the outset that there is nothing about these photographs which suggests to me that they have been faked in any way" (Hennell 1970). And later, because his enlargements showed the fake UFO to be slightly elongated at one end, he suggested that "some propulsive jet may have been operating to move the object to the right."

Both Ben and Hennell identified car headlamps on the pictures (see figure) but, seemingly unaware of the A36 main road, assumed that they were caused by a single vehicle on a track on Battlesbury Hill.

R. H. B. Winder, a consultant to *FSR*, was presented with the artist's impression of the purple light and observed: "These colours are reminiscent of the colours associated with ionisation in air" (Bowen 1970).

After examining the negatives, Pierre Guerin, director of research at the Astrophysical Institute of the French National Centre for Scientific Research, published a "tentative interpretation" of the Warminster photographs (Guerin 1970).

> In my opinion there is no question of the object photographed being in any possible way the result of faking. The question that arises is why the appearance of this object on the photographs is so different from its appearance to the eye according to the descriptions given by the witnesses [Bowen 1970].
>
> In this connection it should be noted that the eye is not sensitive to the ultra-violet radiations of wave-lengths of less than 0.36 microns, whereas all photographic films are, whether panchromatic or not. On the other hand, the sensitisation of the panchromatic films in commercial use (such as the Ilford HP4 emulsion) drops off very sharply in the red area for wave-lengths of more than 0.63 micron, while the eye remains sensitive to them up to around 0.70 micron and even a bit beyond that.
>
> Consequently the interpretation of this divergence between what the witnesses "saw" could be quite simple: namely, that the object photographed was emitting ultra-violet light, which the eye does not see. Around the object, however, a ruby-red halo, probably of a monochromatic colour and doubtless due to some phenomenon of air ionisation, was visible only to the eye and in actual fact has made no impression on the film.
>
> If this interpretation is correct, the consequences which we can draw from it are important. As will be known, in a recent issue of *Flying Saucer Review* (Vol. 15, No. 4), John Keel disputed the presence of any solid material object inside the variable luminous phenomena which he calls "soft sightings," claiming thereby that the solid phase of the UFO phenomenon is only one of the aspects—and no doubt the least frequent aspect—of the phenomenon in question. The Warminster sightings do indeed appear to furnish us with an example of "soft sighting" linked with the presence, at its centre, of a solid object not visible to the eye but emitting ultra-violet light.
>
> That the UFOs can appear, or disappear, on the spot, when leaving or entering our usual four-dimensional space-time is probably true. But it would be rash to assert that they do not always possess a material, solid body

right from the moment that they have penetrated into this space-time. Despite the claims of John Keel, the "soft sighting" could in fact very well be merely secondary effects of the presence of solid objects, whether or not visible to the eye, in the gaseous medium of our atmosphere. This hypothesis had already been formulated long ago, and the Warminster sightings seem to confirm it.

The March-April 1971 issue of *FSR* published five reports related to the Warminster photographs (Bowen 1971; Scammell 1971; Ben 1971; Collins 1971; Samuels 1971). Charles Bowen's "Progress at Cradle Hill" included a print of the negative strip showing all four photographs. The images are small but a ruler is the only apparatus required to measure the magnification discrepancy outlined earlier (by comparing the distance between ten streetlamps on negative one with the distance between the same ten streetlamps on negative four). Scammell (a land surveyor), Ben, and Collins each attempted to pinpoint the position of the photographic UFO with respect to Battlesbury Hill. Each generated gross errors, largely through assuming that the A36 main road car headlamps were on the side of Battlesbury Hill. Collins calculated that the UFO was 60 feet long and, including the "globes," 30 feet in diameter. Michael Samuels, an "independent consulting photographer," widened the debate with a three-page article discussing erroneously the effects of ultraviolet radiation on photographic emulsions.

Their investigations continued. The "Case of the Warminster Photographs" rapidly became a UFO classic, and it incorporated qualities rarely found together: multiple "independent" witnesses and good photographic data (the film was chaperoned from camera to developing tank), and the prime investigators were linked through *FSR*, a magazine regarded in the field as a forum for dispassionate UFO research.

It was therefore unfortunate that, when presented with a case of such potential importance, the investigators failed to learn the geographical layout of the sighting area and no effort was made to examine the basic data critically. The negatives contained glaring inconsistencies that were never discovered, and in more than two years no attempt was made to interview the prime witness, Foxwell. Yet without his photographs the sighting would have been insignificant.

* * * * *

Regrettably, my experiences in the UFO field have shown that the investigator incompetence demonstrated by this particular experiment, far from being exceptional, is typical. Of course very few UFO reports stem from calculated hoaxes like this one; but when reading or hearing of any "sighting," it is important to be aware of the general caliber of UFO enthusiasts, even if they do not appear to have been directly involved in the case. Their irrational thinking is infectious and has frequently provided the media with entertaining headlines. As a result, certain members of the general public, on seeing something in the sky that is strange to them, describe not what they saw but what they think they ought to have seen.

That unidentified flying objects exist is undeniable, as hundreds of thousands have been reported globally since biblical times. Argument otherwise would suggest, absurdly, that every human being has always been fully conversant with the multifarious causes of visual phenomena. UFOlogists have, therefore, plenty of material to contemplate and often plead to the scientific community for assistance and recognition of their exotic theories. Assistance is sometimes forthcoming—for example, the University of Colorado's *Scientific Study of Unidentified Flying Objects* (Condon 1969)—but acceptance of the alien nature of UFOs is not. This is simply because the evidence, when subjected to detailed critical analysis, fails to provide the degree of integrity required. UFOlogists are reluctant to accept that scientific evaluation requires inconclusive, suspicious, or self-contradictory testimony to be classified as such and that a hypothesis based on disreputable evidence or myths remains weak, unconvincing, and adds nothing useful to the understanding of our world. Instead, they prefer suggesting that government and scientific authorities are party to a world-wide conspiracy to prevent the "truth" from being known, demonstrating remarkable faith in governmental unity and little knowledge of the scientific fraternity. More than once they have suggested that science should be "modified" in order to cope with UFO phenomena and have actively encouraged the growth of UFOlogical pseudoscience. Occasionally individuals with relevant technical backgrounds become involved; it is disturbing to witness the abandoning of their mental disciplines and common sense. Unfortunately, credibility is given to dubious evidence when it is endorsed by people of high professional status—such as Dr. Pierre Guerin in the controlled hoax.

In conclusion it is felt that the wealth of UFO reports available for study represents nothing more significant than the relatively simple events listed in paragraph three. There is no logical reason whatever to decide that a more exotic, perhaps extraterrestrial, solution is justified. If ever there is subtle evidence suggesting extraterrestrial visitation, it is unlikely to be discovered by a typical UFOlogist, and care must be taken to ensure that the signs are not swamped or destroyed by nonsense.

References

Arnold, K. A., and R. Palmer 1952. *The Coming of the Saucers.* Amherst, Wisc.: Amherst Press.

Ben, J. E. 1970. "Photographs from Cradle Hill." *Flying Saucer Review* 16, 4 (July-Aug.).

———. 1971. "Continued Investigations at Warminster." *Flying Saucer Review* 17, 2 (Mar.-Apr.).

Bowen, C. 1970. "What the Eye Sees." *Flying Saucer Review* 16, 4 (July-Aug.).

———. 1971. "Progress at Cradle Hill." *Flying Saucer Review* 17, 2 (Mar.-Apr.).

Cameron, A. G. W. (ed.) 1963. *Inter-Stellar Communication.* New York: Benjamin.

Collins, T. 1971. "A Further Examination of the Warminster Photographs." *Flying Saucer Review* 17, 2 (Mar.-Apr.).

Condon, E. U. 1969. *Scientific Study of Unidentified Flying Objects.* New York: Bantam Books.

Guerin, P. 1970. "The Warminster Photographs: A Tentative Interpretation." *Flying Saucer Review* 16, 6 (Nov.-Dec.).

Hennell, P. 1970. "The Warminster Photographs Examined." *Flying Saucer Review* 16, 4 (July-Aug.).

Samuels, M. 1971. "Unexpected Photographic Effects at Warminster." *Flying Saucer Review* 17, 2 (Mar.-Apr.).

Scammell, S. E. 1971. "A Surveyor's Criticism." Ibid.

Shklovskii, I. S., and C. Sagan 1966. *Intelligent Life in the Universe.* San Francisco: Holden-Day.   •

The Persian Gulf UFO: Fiery End to Drama

James Oberg

Yet another UFO report from the *National Enquirer* has demonstrated the unreliability of UFO eyewitnesses and their unwillingness to consider prosaic explanations in the excitement of having seen a "true UFO." And if further demonstrations were needed, the case also shows the ease in which extremely poor UFOs—i.e., those solved with less than five minutes worth of investigation—can slip into the UFO data base, there to serve as traps for the unwary.

This case occurred on the evening of August 24, 1979, over the oil-rich Persian Gulf. "Flaming 6,000 m.p.h. UFO Explodes in Midair" was the tabloid's headline the following November 6, when the case was finally reported. The lead paragraph was especially provocative: "In one of the most widely reported UFO sightings ever, hundreds of witnesses—including veteran airline pilots, air traffic controllers, and government officials—watched in stunned disbelief as a metallic UFO plunged toward earth in the Middle East."

The first to see the object was a British Airways pilot, who said it was flying at an incredible speed between him and a Japan Airlines craft. Kuwait Airways flight 370 then reported "a bright metallic object." Additional reports came from Scandinavian and Swiss airliners.

Ground observers saw it as well. According to Shoori Ghawas, an air traffic controller in Bahrain, "It was shiny and metallic, with red flames shooting from the rear. Then suddenly the center of it turned a bright, glowing red and there was an explosion and it disappeared from sight." Ghawas was particularly baffled by a feature that was reported from other airports as well: the object did not appear on radar!

George Williams, editor of the English-language *Gulf Daily News,* also was a witness. "We were all just standing there hypnotized as the UFO zig-zagged across the sky," he told an interviewer. "It seemed to stop and hover over a brightly lit mosque for a few seconds . . . I'm convinced I saw a spacecraft from another world." Another witness was a veteran off-duty pilot named Norman Vacher. "I was stunned," he recalled. "I have never seen anything like it. I quite honestly could not believe what I was seeing."

Official government probes were launched in Bahrain and Qatar, and both groups concluded that it had been a genuine UFO. The leader of the Qatar group was quoted as saying, "My final conclusion was that the object was definitely a UFO under its own power which eventually exploded." The Bahrain group calculated that the object was going in excess of 6,000 miles an hour. Further, the group wrote: "This is the most positive sighting by reliable witnesses ever in history . . . This was definitely not refracted or reflected light, nor a meteor or satellite, or any known flying craft made by man."

As the months passed, no new data came out; but the tabloid's story was picked up by pulp UFO magazines and by various UFO club newsletters. Its appearance in the next generation of UFO books is only a matter of time.

Nobody seemed to want to check up on the case, however. To experienced UFO investigators not out for sensationalism or exploitation, the descriptions sounded very much like a re-entering satellite or a meteor—despite the disclaimer of the "official" probes. The odds are that meteors leave no records behind, but falling satellites do.

This one did. A telephone inquiry to the North American Air Defense Command hit paydirt. "We have a match," public affairs officer Sgt. Mike Bergman reported a day after my phone call. "The rocket booster of Kosmos-1123 was coming down in that area at that time along a north-to-south trajectory." Using computer tracking data supplied by NORAD, it was possible to plot the path of the dying satellite; it zipped right across Bahrain and Qatar at an altitude of about eighty miles and a speed of 18,000 miles per hour (ironically, the *National Enquirer's* speed estimate had been *low*!).

The object had helped carry the Kosmos-1123 spy satellite into orbit three days earlier. Now the 20-foot-long cylinder, its mission accomplished, was falling back to a fiery death over the western Indian Ocean.

The incorrect perceptions of the eyewitnesses are not at all uncommon in such cases. Pilots can almost be counted on to estimate fireball meteors and burning satellites as being "very close" when they are scores of miles away. Balls of fire will be described as "metallic" when all they are, are balls of fire. People on the ground will misjudge, misremember, and misreport the motion of lights in the sky, and will attribute motivations to them (such as the reconnaissance of the brightly lit mosque) without any basis in fact. And another factor was typical: all efforts to contact the "official organizations" and the newspapers that carried the UFO story were fruitless. No retractions were ever printed.

In this case, we were lucky enough to be able to determine the actual cause of the "UFO," but it will not help since most of the public won't hear about it. It *will* help, however, in judging the reliability of similar reports from witnesses with similarly impressive credentials, even though we are unlikely to stumble across recorded data on the actual prosaic stimuli that lead to such reports. Such cases as this one go a long way toward explaining why a few percent of UFO reports will always remain "unsolved" even though no alien-extraordinary cause may be involved.

The Semantics of UFOs

Anthony Standen

If I ask someone, "What do you think of UFOs?" I am likely to be told whether the speaker believes, or does not believe, that intelligent beings from some other planet, or solar system, or galaxy, or whatever, are flying around in our atmosphere in their own ingenious flying machines.

But that is not quite an answer to the question. I am asking whether there are really things to be seen, in our atmosphere, that are unidentified, or unexplainable by science, things that could not occur naturally. It doesn't matter whether they are engines made by intelligent beings or not.

It seems to be true that the great majority, at least, of UFO sightings can be explained as things that are perfectly well understood. But it may be that a persistent minority have not yet received a rational explanation depending only on nonintelligent forces. There might be such an explanation, but we don't know it in all cases. One possibility is "electrical phenomena, of a nature not yet understood." Among electrical phenomena, ball lightning is reported to do the most extraordinary things; and that something roughly similar occurs occasionally, up in the air, is a natural possibility that cannot be ruled out.

But I am not discussing whether "unknown electrical phenomena" can or cannot explain UFOs. I am asking why it is that when such a question is asked, it seems to lead immediately to the much wider question, "Do you really believe that extraneous intelligent beings are visiting our planet?"

The reason for this seems to lie in the terms in which the question is asked. This is one of those cases where a clumsy phrasing of a question puts misleading ideas into the mind. It is an example of the power of words to falsify thinking.

"UFO" means Unidentified Flying Object. Now what is an *object?*

The very word seems to suggest something round, hard, and distinct; *this* object and *that* object, like so many billiard balls. It suggests to the mind that the things seen are of that character. But they might be no more "objective" than, say, mirages seen in the desert. There are plenty of clouds up in the air; they have vague boundaries, and they sometimes merge, or fall apart. Do they qualify as objects?

Even more misleading is the term *flying*. Birds are flying around in the air. So are bats, and many insects. But are the clouds flying? No one would say so. The word *flying* then, suggests to our minds an animate (not necessarily intelligent) being. And so the phrase "unidentified flying object" suggests ideas that are irrelevant. The phrase was probably started by members of our Air Force, who like to be able to "identify" any "object" they see as, say, a fighter plane, or a bomber plane, or perhaps a private plane, or a balloon, or even, perhaps, an enemy spy craft.

Now the Air Force personnel have admirable expertise in flying; no one doubts this. But they are not necessarily expert in semantics. They could have used a phrase like "Unexplained Aerial Appearance"—UAA. This would not put into people's minds any preconceived notion that should not be there. But as it is, the phrase Unidentified Flying Object was a psychological disaster, and to ask about it is to ask a question that misleads by the very terms in which it is asked.

UAA *si,* or "see" (perhaps). UFO no.

List of Contributors

GEORGE O. ABELL is a professor of astronomy at the University of California at Los Angeles whose specialty is the large-scale structure of the universe. He is author of several astronomy textbooks and co-editor of *Science and the Paranormal.*

JAMES E. ALCOCK is a social psychologist at York University in Toronto and author of *Parapsychology: Science or Magic?,* a critical book about science and the paranormal.

ISAAC ASIMOV is author of well over 200 books and is a professor of biochemistry at Boston University.

WILLIAM SIMS BAINBRIDGE is an associate professor of sociology at the University of Washington in Seattle and author of *The Spaceflight Revolution* and *Satan's Power: Ethnography of a Deviant Psychotherapy Cult.*

RALPH W. BASTEDO is a professor of political science at the Laboratory for Behavioral Research at the State University of New York at Stony Brook.

HENRY H. BAUER is a professor of chemistry and dean of the College of Arts and Science at Virginia Polytechnic Institute and State University. He has a special interest in controversies on the fringes of science and is writing a case study of the Loch Ness Monster controversy and awaiting publication of his study on the Velikovsky affair.

GARY BAUSLAUGH is dean of instruction at Malaspina College in Nanaimo, British Columbia. He is a chemist and has published articles and essays on a variety of topics.

VICTOR A. BENASSI is an associate professor of psychology at California State University, Long Beach, involved in research on paranormal belief, illusion of control, human judgmental biases, and operant conditioning.

JOHN R. COLE is an anthropologist at the University of Northern Iowa, Cedar Falls.

KENNETH L. FEDER is an instructor of anthropology at Central Connecticut State College in New Britian and is now working on the Farmington River Archaeological Project.

KENDRICK FRAZIER is editor of the *Skeptical Inquirer* and former editor of *Science News* in Washington. He is a science writer with special interests in astronomy, the geophysical sciences, and social and philosophical issues of science.

MARTIN GARDNER is the long-time author of the Mathematical Games section of *Scientific American* and author of many books about science and one on crank science, the classic *Fads and Fallacies in the Name of Science.*

LAURIE R. GODFREY is an assistant professor of anthropology at the University of Massachusetts, Amherst, with special interests in primatology, functional anatomy, processes of evolutionary change, and issues of public understanding of evolution.

JACK K. GREENBERG is a professional magician and a measurement and evaluation consultant in Pittsburgh.

BENNETT GREENSPAN is a resident in the Department of Radiology, Wadsworth Veterans Administration Hospital, Los Angeles, and is doing research in the

Department of Radiology at the University of California at Los Angeles.

TERENCE M. HINES is an experimental psychologist with the Department of Neurology, Cornell University Medical College, New York City.

MICHAEL HUTCHINSON is an amateur magician, an investigator of the paranormal, and secretary of the United Kingdom branch of the Committee for the Scientific Investigation of Claims of the Paranormal.

RAY HYMAN is a professor of psychology at the University of Oregon with special interests in the psychology of belief systems and in the analysis of parapsychology research.

RICHARD KAMMANN is an associate professor of psychology at the University of Otago, Dunedin, New Zealand, and co-author of *The Psychology of the Psychic.*

PHILIP J. KLASS is a senior avionics editor of *Aviation Week & Space Technology* magazine, author of *UFOs Explained,* and chairman of the UFO Subcommittee of the Committee for the Scientific Investigation of Claims of the Paranormal.

PAUL KURTZ is a professor of philosophy at the State University of New York at Buffalo and chairman of the Committee for the Scientific Investigation of Claims of the Paranormal.

LARRY KUSCHE is author of the highly acclaimed book *The Bermuda Triangle Mystery—Solved* and of the recently published *The Disappearance of Flight 19.* He lives in Tempe, Arizona.

DAVID MARKS is a senior lecturer in the Department of Psychology at the University of Otago, Dunedin, New Zealand, and co-author of *The Psychology of the Psychic.*

DONALD H. MCBURNEY is a professor of psychology at the University of Pittsburgh. His special interest is in sensory processes and perception, and he teaches a course in ESP and Pop Psychology.

JOHN D. MCGERVEY is a professor of physics at Case Western Reserve University doing research in positron annihilation. The second edition of his textbook *Introduction to Modern Physics* is in preparation.

ROBERT L. MORRIS is a senior research scientist at the Psi Laboratory of the School of Computer and Information Science at Syracuse University.

SCOT MORRIS, a senior editor of *Omni* magazine, is a science writer and humorist. His background is in clinical psychology.

JAMES OBERG works at Mission Control at NASA's Johnson Space Center in Houston. He is also a prolific writer on space activities and on "outer-space mysteries," including UFOs.

JOHN T. OMOHUNDRO is an associate professor and chairman of the Department of Anthropology, State University of New York, College at Potsdam.

VERNON R. PADGETT is a doctoral candidate in social psychology at Ohio State University. He has written on religious cults and socio-cognitive biases in applied research.

JAMES RANDI is an internationally known conjuror with 35 years of experience in evaluating both demonstrations of a purported paranormal nature and the rationalizations that some scientists offer for their failures and unorthodox protocol in parapsychology. He is author of *The Magic of Uri Geller* and the recently published *Flim-Flam*.

IAN RIDPATH is a science writer and broadcaster and is editor of *The Encyclopedia of Astronomy and Space*. He lives in Brentford, Middlesex, England.

RONALD A. SCHWARTZ is a clinical psychologist who has been engaged in pharmaceutical research for the past ten years. He is director for scientific affairs of a pharmaceutical company in Illinois.

CHRISTOPHER SCOTT is a statistician, psychologist, and former Council member and research supervisor of the Society for Psychical Research in England.

ROBERT SHEAFFER is a systems analyst for an aerospace company in San Jose, California, a frequent writer on scientific subjects, an investigator of UFO reports, and author of the newly published *UFO Verdict*.

DAVID I. SIMPSON is a physicist at the National Physical Laboratory in England engaged in applied research using laser interferometry.

BARRY SINGER is a professor of psychology at California State University, Long Beach, interested in approaching controversial and sensitive subjects, such as sexuality, racism, sexism, and paranormal beliefs, in rational ways. He is co-editor of *Science and the Paranormal*.

ANTHONY STANDEN is author of *Forget Your Sun Sign* and *Science Is a Sacred Cow* and was formerly executive editor of the *Encyclopedia of Chemical Technology*.

RODNEY STARK is a professor of sociology at the University of Washington in Seattle and author of four books on the sociology of religion.

JAMES R. STEWART is associate professor of sociology in the Department of Social Behavior at the University of South Dakota, Vermillion.

RONALD STORY is a free-lance journalist who undertook a detailed critical study of the "ancient astronaut" theory and published his critiques in two books, *The Space-Gods Revealed* and *Guardians of the Universe?* He lives in Tucson.

ERNEST H. TAVES is a psychoanalyst and writer and co-author with the late Donald H. Menzel of *The UFO Enigma*. He lives in Cambridge, Massachusetts.

MARVIN ZELEN is a professor of statistical science at Harvard University.